S0-EDH-265

HOLOGNOSIS

Beyond Relativity and Consciousness

Asha K. Amir-Jahed, M.D.

Library of Congress Control Number:		2008909166
ISBN:	Hardcover	978-1-4363-7771-3
	Softcover	978-1-4363-7770-6

This book was printed in the United States of America.

To order additional copies of this book, contact:
Xlibris Corporation
1-888-795-4274
www.Xlibris.com
Orders@Xlibris.com

51155

ACKNOWLEDGMENT

I owe much to my parents for their legacy of the quest for truth that they left in my mind. After so many years without them, I still feel their superego figures, hear their wise words, and remember and appreciate their deeds. My father ignited the fire of inquisitiveness in my mind at my early age and taught me to aim beyond the visible. I owe to my mother the purest sense of loving, for she taught me to give love and gain happiness.

It is a difficult task to express thankful affection to one's own children, but I have to word my gratitude to my son, Sepehre, and my daughters, Shahla and Shahrezad, for being the mirrors of my hopes, giving me the beautiful reflections that I only wish to live long enough to appreciate fully. Similarly, I express my gratitude to my sister, Kaynoosh Partamian, PhD, for her expert psychologically elaborated remarks and for providing me with some key references beyond my reach. The same gratitude goes to my cordial family advisors, Dr. Krikor Partamian and Mr. Arman Partamian, for their constructive criticisms on some metaphysical issues that incited me to rethink and redefine my views more clearly.

I have greatly benefited from Javad Rahimian, PhD, through his expert views in many discussions on subjects of interest in quantum physics related to this work. Javad's gift of some rare volumes in physics also particularly enriched my modest library. His caring attitude keeps reminding me of his much appreciated friendly concern and encouragement.

I keep fond memories of times spent with Mr. Cyrus and Mrs. Farah Massoudi, my best friends of all times who, throughout the years of my preparation of this work in English and other writings in Persian, showed keen interest as spiritual supporter in the progress of my work and never ceased their unfailing encouragement. Through their family bonds, I came to know many fine personalities and developed lasting friendship with some prominent figures of the highly educated Persian society in Los Angeles, namely the late Colonel Massoudi, whose spiritual insight I felt particularly comforting in my meditating times. Another much-appreciated acquaintance, Mr. Bijan Hosseinian, helped me in the computer preparation of this work as an adamant supporter, and I am indebted to him.

To my former students and friends, I am indebted for their continued spiritual encouragement, especially to my valued close associate and colleague, Dr. K. Bakshandeh, whose expertise in Judaic metaphysical field brought the needed light to some of my obscure religious issues on metaphysical views.

Expression of a just homage is due to all my teachers who shared knowledge with me at the Medical School in Lausanne where the spirit of friendship, truth, and logic reigned above other interests, leaving me with the fondest cherished memories. Names of all eminent professors and friends to whom I owe so much for my happy memories for those memorable days of learning are too numerous, but their memories remain adorned in my mind with gratitude. My dearest remembrances go to my friends Dr. M. Kobreh and Dr. H. Sadeghi with the spiritual constancy that the three of us have kept the old torch of the school camaraderie alight for the last fifty years. I can still enjoy its shining through our shared memories. I have incessantly felt their supportive spirits in writing of this book.

I also feel honored by consideration and respect from my faculty friends, colleagues, and students at Pahlavi University in Shiraz where I spent seventeen years of my own formative spiritual inquisitiveness, along with my teaching surgery to residents. That inquisitiveness initiated the questions whose answers led to the line of thought entertained in this work. The high level of academic standing of that young university in an atmosphere of progress-oriented search for excellence, of which I also profited much and felt being endowed, was exemplary. To Dr. Mohsen Ziai, Dr. Farrokh Saidi, Dr. Karim Vessal, and the late Dr. Ali Farpour, I owe my initial attraction to that academic milieu. I appreciated the shining spirit of friendship and inquisitiveness in teamwork that reigned in securing the pursuit of the highest level of teaching and research focused on the care of the patient, and I feel honored for having been a part of it.

The memories of the teamwork in the department of surgery I had the honor of directing for sometimes, with all the moments of intensive tensions in the work, the spirited composure following the turmoil, and the happy rewards of success our teams shared at Saadi and Nemazee Hospitals are unforgettable, as are the unrestricted benevolent service provided to us by the hospital personnel.

The spirit in the immense friendly regards of people in that university town, and the patients' families to us as a whole, compensated particularly my feeling of owing service to those noble characters, but raised the question in my mind of what origin such sincere altruism may have other than the human affective bond that is beyond physical realms. This question haunted my thinking sphere and elaborated what is partially reflected in the present book.

The memories of my colleagues and friends that I have lost, and most of all, that of the late Dr. A. O. Lotfy, are vividly alive, as is the same feeling for so many others whose contribution to my happy memories are truly ineffaceable.

I cannot end my expressions of gratitude to those who contributed fundamentally to the fine academic atmosphere of Pahlavi University, without recalling the pioneering tasks of the late Dr. Zabih Ghorban, the first rector, who founded the tradition of modern teaching in English in the Persian community, a tradition of seeing an additional horizon of light beyond the crepuscule. The continuing progress under the subsequent royal chancellorship appointees—particularly the last chancellor, Dr. Farhang Mehr—with new dimensions in economical stabilization and spiritual renovation, was decisively helpful. The change provided more propitious grounds for broader basic scientific inquiries, of which I benefited in extending my scopes beyond practical medicine, particularly into the philosophy of life that I felt shimmering over that ancient mystic land of Hāfez, and now I express my gratitude to those who made this possible.

May the spirit of fellowship and mutual understanding, particularly for the pursuit of knowledge, be always guiding the human mind.

A. K. Amir-Jahed, M.D.

CONTENTS

Advice to Readers

Readers should be aware that neologisms used in this monogram may find matching analogues in the profuse literature of the Internet, but they reflect only the meaning explained in this book and none else.

PROLOGUE

In this book, you will face new ideas, new concepts, new words, new conclusions, some strictly scientific and some frankly philosophical and also metaphysical. I aim to show you the salient significance of the message, the essence regarding dynamism in physical and in biological realities, ultimately giving you an idea of the holistic life, physically and metaphysically. I will use pertinent scientific data scrutinized and generally supported by today's scientific and philosophical opinions. I add my own interpretation to formulate my views on creation, life, mind, consciousness, human expressivity, and metaphysics. Considering physics, material biology, immaterial biology, the nonmaterial abstract states of mind, I present all as reflecting the holistic indivisible creation.

The main point I shall make is that both physics and biology are governed by the same dynamism in the form of an identical innate principle of logic in both, explainable at microphysical subatomic quantum level with matter-energy interplay, with the message that both bear an immortality principle. The biological basic protein in genetic material, showing autonomy, is regarded as replicating a principle of immortality, same as manifested by energy in the form of photonic matter and wave. The principle seems substantially identical in both physics and biology. Detectable at subatomic level in quantum microphysics, it shows itself in organic matter-energy behavior through DNA to brain-mind complex. It is named bioforce—a functional power engendering immortality. This first postulate implicates the second and the third—the unity and the wholeness of one creation system in physics and metaphysics—but in two philosophical modalities of subatomic time indefinite and molecular to larger scale time definite manifestations.

A natural imposition then comes into consideration, which is the reality of spacetime in biology, which allows us to understand and explain all subsequent phenomena that are the results of evolutionary change with time. Psychologically oriented discussions are entertained, and the concept of biocracy is born out as the ground for the essential biological autonomy with the principle of survival, growth, and reproduction following the autonomous deep self here named *autos*. This primordial power invested in biology with the bioforce, impersonated as an autocrat, acting autonomously through tropism and dynamism initially, is the totipotent eternal power that is transferred through the genes and is the future programmer, so to speak, for psychological selfs in the developed mind to face the world. You will realize that the innateness reality of the spacetime in the biopsychon unitary life element, the biological autocrat, implicates an innate sense of time, a giant dubbed *time titan*, in the live time bearer long before the conscious realization of time delineates, fragments, and measures this giant wholeness progressing to infinity.

The interesting subject of consciousness, you will find, will be understood to be an all-in-one life phenomenon to be reduced to *nothing but consciousness*. However, this state of plain permanent consciousness transcends through biognosis with the evolution of life in steps and by stages. It goes through stages of biological gnosis including paleognosis, bioawareness, bioconsciousness, and psychognosis. This leaves our customarily called consciousness to be served by idiognosis, conferring recognition by idiomatic elements named words. The basic conceptualization, however, has been formed millennia earlier in psychognosis by semions and moveons as structural models for words appearing later.

The psychoperceptive-psychoreactive operations prior to language ability, existing and providing models as psychognostic models of holistic concepts, served survival and early social life in hominids. Thus, the mental models for expressivity have been formed earlier in our ancestors, a knowledge that will answer the puzzled query of language. With this evolutionary reality, further evidence is found for the evolving biognosis to ultimately serve the expressivity trait in biology. The stages of expressivity demonstrate a vivid evidence of unlimited expansion of the mind power in conceptualization. The free choice of networking for concepts of visual-visional types with all possible combinations of the matter (figures) and energy (time changes) in dream scenes clearly mimics the quantum characteristics and its inherent uncertainty principle that will be seen discussing dreams. Examples being experienced in dream consciousness, and defined in wakeful life, form the respectively undirected time indefinite and directed time definite conscious processes, all contributing to keep the autobiographical memory constancy through a logical psychodynamic equation.

You will share with me the passage of consciousness from psychognosis of the state of speechless mind to idiognosis with the speaking mind, on the very basis of the expressivity trait accomplishing this task. You will note a two-frame model of psychognostic thinking forming a pattern of life commandment of logic and happiness and pleasure-pain principle. On that pattern, the expressivity trait, using the innate spacetime principle, applies the mental impetus called Navand (N), with mental energy (E), and time (T), making NET trilogy of the expressivity principle essential in language formation. The realization of NET forms a solid basis to explain future language production. You will realize that the psychognostic language foundation facilitates deciphering the mechanism of speech production as explained by the Navand theory of languages.

Finally, the metaphorically used word *symbiogenesis*—celebrating an idealized ontogenesis for the birth of the twin physics and metaphysics that starts the first chapter with the first section subtitle *symbiosis in destiny*—will be seen clearly in the phylogenetic veracity of the biological life. The celebration for the twin birth, just eons late, is yet legitimate to honor logic and happiness, binding the siblings, body and mind together. The body-mind complex, with the full idiognosis ultimately enriching the abstraction with double identity of psychognostic iconic and worded concepts, realizes the metaphoric symbiogenesis and symbiosis in destiny. Thus, the biology of spacetime, you will see, will transcend from material to immaterial to the total Gnostic wholeness of energy and time minus matter.

A final horizon of clairvoyance leads to *truth-above-all* radiance from firmaments of truth examined in relation with material biology of survival and reproduction and the immaterial biology of the mind and abstraction. The puzzle of our origin from the indivisible creation of physics, material biology, and immaterial biology may ultimately reveal less undecipherable, holding physics and metaphysics inseparable, as evidence of the indivisible creation that can be explained through the formulated unified theory of creation. You may find the last chapter as a happy ending of a movie that should not all be told at the beginning. A sage said millennia ago that "the word *happy* is more meaningful in anticipation than in remembrance"—the first is to a beginning, the second from a forgone one, and this is the life.

CHAPTER ONE

BIOFORCE IN SYMBIOGENESIS

SYMBIOSIS IN DESTINY

This metaphoric title "Symbiosis in Destiny," which sounds taboo in scientific writing, and the more generic term *symbiogenesis*, are reflecting personal preferences of my views on the biological production and changes that have ended in the making of mankind with a physical and a mental life to be lived in one human life in one space and at one time. When I consider the interplay of the evolutionary changes between life and the living world, regardless of how objective I try to be, I feel an inner impulse to use the word *destiny* rather than any other qualifier. The reasons may be my hidden inclinations, philosophical sense, poetic nature, or my subconscious feeling of some scientific truth in the unscientific taboos like destiny. I hope I can overcome this personal weakness ultimately after I have told you my tale of intricate subjects I will have to deal with in this book.

Let us say that knowledge in evolutional biology has gone far beyond solving some life puzzles but yet leaves the perplexity surrounding the core questions: what is life, what is its origin, what is beyond its apparent end? To these questions, philosophers somewhat more subjectively and biologists more objectively, and both ardently, search to find answers. The first question can be defined and in some way answered in biological terms and reasoning, but yet cannot be entirely cleared fully in either biological or even philosophical or metaphysical words. The second one, the origin of life as we see it on earth continues to remain uncertain except as related to solar sources of matter and energy. For instance, how the first amino acids were formed,[1] though experimentally duplicated,[2] is still debatable at the atomic level, and as to why they eventually evolved as they did, and as are the reasons for replication, or hyperthermophilic life at 242 degrees Fahrenheit on the sea bottom,[3] or for autotrophic chemosynthetic bacteria making their own food.[4] Also, the possibility of making genetic bacteria,[5] and genetic programming with genetic instructions, autonomous action and expected results, show the puzzle of irreversible progress of life's specifications observed but not fully understood or explained. The problem in essence poses a series of repeatable whys, inherent to each step of the evolution back to its very origin. This problem can find stepwise answers in a retrograde examination, but the process stops near origin where biology and molecular science become moot. For example, when we reach back to replicating molecules of nucleic acids, the *why* that comes to our minds for the precise orderly processes that take place can hardly find a satisfactory answer, though probably hidden at the subatomic level, but even there and even if known, a *why* imposes itself. When we examine the set of linked nucleic acid molecules in genes and realize that the set is now irreversibly established and must again replicate in orderly processes of mitosis or meiosis in identical haves, we are harder pressed as to find a clear answer to that fundamental obligation which seems to separate itself from mere sequence of mechanisms. We see the complexities following simplicities in the stages of evolution which offer more ground for life and evolution to continue, but also leave more questions to be answered.

The last question of what could be beyond life, which is evidently an abstract question only in our abstract mind, will be left for metaphysical deliberation that are ultimately unavoidable and will occupy the very end of my biological discussions treating the abstract mind power.

Irreversibility in evolution, no doubt, indicates evidence for a transition in time, which can make ground for the process of evolution to answer the natural selection in new ways: adaptation to please selection, but for what aim if not for providing a balance between the individual and the environment, to secure a better survival for the individual in the inexorable continuum of time. And all this depends on *life as energy*, naturally terminable, against a time naturally infinite in which life is being presented theoretically in an infinite number of living successors as quanta.

Irreversibility is thus an indication of an obligatory somatic directional determination in line for a function. Prior to this directional determination which grants specification, a change in direction for a more significant adaptation trait in the face of evolution is still possible, but after any such change, the progress goes on to perfect the specification along determined functional lines and commensurate with foundations established. Thus, irreversibility in the stages of evolution with the developed traits marks decisive changes only compatible with further outgrowth leading to species specification in the newly determined functional line in the same species for the life to perpetuate more securely. For instance, the evolution of life in arthropods of many million years before vertebrates, though allowed fine specifications to arthropod living, did not allow a drastic change from that type of well-balanced living to a different one, say reptile, by changing physical and functional features of arthropods to new forms and dimensions creating examples of science fiction monsters. The contrast of evolutionary time for changes to occur in this genus arthropod compared to amphibian, for instance, with more elaborate system of sensing and reacting and a primitive brain, is evidence for this evolutionary principle. But other new genera of nonarthropod organisms developed in much shorter times and with much greater bodies including vertebrates. This observation suggests that the evolutionary rule of new species creation is not a simple one-theme story, and the genetic changes leading to greater soma concomitantly with greater muscles, greater mobility, functionality, nervous system, and other associated capabilities are reflecting steps not well-known for explaining everything but evident enough to show the commensurate parallel growth of soma and sensing, reaching the stage of soma and psyche at the far end of the scale. Thus a fundamental puzzle remains, which is the action of an autonomous factor, agent, or determinable force that keeps a selected direction as facilitated by the natural selection.

It may sound far-fetched to try to establish a proportionality between somatic and nervous system dimensions and functionalities to show that a symbiosis in fact can exist between the two at all level of life including the rudimentary stage of species functionality. But such possible proportionality, if shown, may reveal directionality acting on natural selection. Nevertheless, keeping with this esoteric consideration that suggests such relationship may exist, a very interesting comparison has been made between extremes of large and small bodies of biological creatures and the primary particulate masses in physics that could be cited as representing unitary elemental examples of their respective worlds in showing an analogy between physics and biology. The comparison is made between an elephant and the smallest species of ants with a difference of about 6000 kilograms to 0.01 milligram showing more than eleven orders of magnitude, roughly the same span between the top quark and the neutrino.[6] So, it seems that regardless of the size of soma, there is an innate factor in the live organisms cofunctioning in the directional selection to improvement, and in all, dynamism of an innate force is the sine qua non of their selective drive.

The other interesting remark to be made is about the fact that the natural selection can cause changes that pass from one step to another, differentiating the species in the way of better survival and reproduction (SR) without possibility to regress, and in all steps an approximate proportionality is kept between the somatic size and the nervous system elaboration along the course of differentiation. This crude proportionality in size and structure is far less significant than in functional capacity of both systems matched. The basic genetic code must therefore also secure this mainly functional parallelism that is not a strict blue print example and is subject to environmental conditioning as well.[7] Nevertheless, the symbiotic evidence between the physical and neural functions is clearly reflected in the commensurate functionalities of the two systems to culminate in soma and psyche as distinctly evident in primates and in humans.

So in essence, the philosophy of life may be reduced to invoke two interdependent states needing one another to represent life together until eventually one fails and both cease functioning: The material particle and its force, demonstrated in clearest way in physics by the positron and the electron, the matter, and wave of propagation, the energy of the photon, or the soma and the sense in the living element. A symbiosis of two separate constituents to functionally present a unity: the life.

BIOQUANTUM SYMBIOSIS

If we stop here and assemble our ideas about the physical world of which we are a part, reflecting on its constituents, processes, and rules, we find the evidence of energy and matter as a sine qua non in all items of creation. This coexistence of energy and matter with its universal constancy seems to evidence the primordial dualism in purest presentation detectable by human means. Then, when we further scrutinize the molecular to atomic level physics, we find again this constancy of simultaneous presentation in quantum physics, where the energy units in particles in motion as waves, with time allocated frequencies according to the motional speed as nonparticulate component, again show the principle of synchronic mass and energy presentation.[8] This too is a dualism, but here there seems to be utterly a live connection between the two components which shows a clear indication of all-or-none as one cannot exist without the other. The relation of the two components is such that each is a sine qua non for the other to be, invoking reciprocal causality in their mutual being in an orderly repeatable way that is reminiscent of life or is life itself in nonbiological form.

Here again we find the principle of symbiosis but specifically demonstrating reciprocity and being conditioned by it. The interesting point that reveals itself however is the fact that in quantum physics energy is confined in partitions, reminiscent of material character which can be called particulate, but its motional presentation is nonparticulate and does not invoke material character but a field of energy. The particulate component, however, is time bound, depending on the nonparticulate one, the wave, with frequencies of presentation that imply time. According to the classical string theory of quantum physics, the smallest particle represents a linear dimensional string of 10 to the negative power of 34 meter.[8] In fact this one infinitesimally small linearity particle with its pristine frequency of presentation can be regarded as the simplest representation of a nearly massless energy and time. Therefore, the principle of the subatomic quantum physics, as described, tacitly includes the extreme near-zero mass, undefined time, and multiple infinitesimal dimensions still being debated.

In other words, if our initial crude consideration of matter and energy of the creation is now revised, we can say that creation has included the dimension of time as an intrinsic component in quantum physics. If we allow our inquisitive mind to go further, we may see the symbiosis of matter and energy in physics ultimately transposed to that of energy and time in biology, a bionic symbiosis regardless of the material carrier.

With this understanding, biology must be really regarded scientifically as an autonomous chronogenic physical event, and we may conclude that living in reality is a natural symbiotic event of matter-energy confined in spacetime and having its own inherent autonomy just like the elemental matter-energy components in quantum physics. Then energy in life's material, in both corporeal presentation (soma), and noncorporeal form (psyche) in the phenotype should be regarded as a quantogenetic obligatory symbiosis, an axiomatic principle that only changes its manifestations according to the spacetime structural features of each living element.

If this analogical consideration is sound, to which in principle appears to be no apparent objection, we must accept that a basic biological unit analogous to the paired elemental quantum unit should also exist with that pristine form of the symbiosis in it. This assumption, a priori, suggests that a simplest form of body and mind should exist in an extreme biological reduction, possibly just at the molecular or submolecular level. This condition of size reduction to molecular or atomic level in biology to match the equivalent level of particle size in quantum physics restricts biological choices to DNA, or to amino acids, or even to smaller dimensional integers.[9] However, accepting the above analogical principle, we may choose the better-known such unit to be the gene with the familiar genetic biological manifestations, avoiding further search for practicality and ease but not rejecting the possibility of mechanistic analogy reaching even the subatomic dynamic levels.

In this context, the symbiosis seems to be inherent in all live matters between the material and the energy parts. In biology, what makes this symbiosis to present itself differently in the living organic forms is the evolution which imposes the differentiation of the carrier forms of the basic unit. The differentiation by evolution per se shows clear evidence denoting and proving the truth of the symbiotic pattern in the progressive complexities of the evolving life by the inherent energy in the live material making the evolutionary adaptation. Evolution only shows the steps, and adjustments in the form of adaptations by traits reflect the fine-regulated relation between the material and the energy parts for keeping the balance and allowing the life to continue according to new conditions. The time bound component of the quantogenetic origin in the living spacetime of every organism can be assumed to share the energetic responsiveness with the imposed effects from the environmental spacetimes. This intrinsic innate time element, without which no energy, no quantum, and no quantal symbiosis exists, must be considered a pillar in the theory of what can be called *quantogenetic symbiosis* or *bioquantum life*, being initiated here and exposed in further discussions to follow. All that can be briefed at this stage is that the crux of the matter lies in the binding relation between the two components of the matter and energy in quantum physics. The unknown force making the binding which I will call ***bioforce***, may be regarded the ultimate culprit, and I will expand on this and its inherent relation to life and to creation of mind in evolutionary dialogue in the next chapter and further on. This force can be called variably according to any meaning attributable to it, but as it is essentially biological, it could be simply called by the prefix ***bio***, or the name bioforce.

In the following section, I will discuss the role of phylogeny as related to the foundation of environmental conditions affecting neurobiology of the reproduction aiming the individual phenotype's versus the species' survival in connection with the symbiotic body-mind question.

PHYLOGENY FOR LIVING TO LOVE

The question of life, its bioforce and its chain perpetuation leading to the elaborate form of physical and mental symbiosis in the human species, symbiosis of body and mind, which needs more clarification, can be now examined from a different angle. We realize that the most significant and decisive evolutionary trait in the human species has occurred with the central nervous system development giving the brain, the consciousness, and the language. From the standpoint of evolutional laws of irreversible species specifications, the *Homo sapiens* has acquired these traits as a whole and unique element allowing the *Homo sapiens* to flourish, possibly even at the expense of a threat to other species of the genus *Homo* that have perished in the course of evolution like *Homo Neanderthalensis*. However, in the line of natural selections in the humans, these traits in *Homo sapiens* are synonymous with progress and prosperity both for the individual and for the society and satisfy natural selection rules. As this natural selection has produced the social living in many species and in humans, the proper place of these traits of bigger brains, better awareness of self and environment, and greater mental power of communication and problem solving in humans—evidenced by mind, consciousness, and language—to be properly studied must be considered in the social setting.

Extending our discussion on social living and its conditioning effect on the single animal member of the society, it is appropriate to remark that the fundamental principle of phylogeny is the same as the primordial bioevolutionary principle for continuation of life resumed in survival and reproduction (SR), which can be regarded as an axiom, a must-to-be-secured rule, in single and in group living alike. In any event, phylogeny must be secured through genetic transfer by the phenotypes of the species, whether by individual free reproduction or conditioned by society rules like in some insect species. This primordial rule of phylogeny then appears to be twofold and in two respects in actual application; individual versus societal in one respect, and survival versus reproductive in another. The question is then how much of the respective functions, survival and reproductive, can be independent from the society and be uniquely individual, and how much conditioned and ruled by the society, and on what basis?

To give an example, the individual free choice of reproduction in human society is in contrast to conditioned reproduction in honey bees where the task is restricted only to queens. Considering this, a more fundamental question comes up, which is why such chasm should exist in the most basic rule of biology (survival and reproduction) to present such a wide difference in some animal societies by granting versus denying reproductive capability to the phenotype.

Reflecting on these questions leads us to ponder on the principle of SR, and we may doubt if these two adaptive capabilities, survival and reproduction trait, should be considered always inseparable, as the evolutionary evidence points to the contrary in some examples like in bees. In addition, accepting separation imposes possible priority values as the two functions are related. When we consider the separability and the possibility of related priority that this separation of survival from reproduction would imply, we come to evaluate and accept these two specific essentials of life as two independent evolutionary traits but able to acquire interdependency under specific conditions. As usefulness must exist with definite value to life for a primordial functional capability to be accepted as a trait,[10] we can unquestionably see this in both survival and reproduction, but survival is useful primarily for the individual phenotype and secondarily for the species whereas reproduction is essentially useful to the species. As the species must survive, the reproductive trait must prevail to assure this task in situations of inadequacy in the phenotypes' overall functionality by assigning the task to specialized phenotypes like bee queens. This priority of the reproduction over the survival for the species when the individual phenotypes are somatically limited both in dimension and functionality for assuming responsibility and securing execution of both functions perfectly, the evolutionary course forces the adaptive changes into two separate lines of survival and reproduction. But changes to the opposite with full interdependency for both traits are evident when the conditions can be secured in phenotypes of greater somatic dimension and more efficient functional capability like in the human species. The main reason evidently is the more perfect central nervous system in humans, and its role in promoting reproduction along with other sublime mental activities.

In fact, survival seems to match value with reproduction in evolution for the more complex phenotypes able to adequately secure reproduction. But, as it is known, the shortest life spans in unicellular organisms and in parthenogenetic ones, as examples, are commensurate with their simplistic reproduction types and necessitate casting sexual assignment in monopolizing reproduction to the one most perfect phenotype of the species for that purpose. This forms solid evolutionary evidence that adaptation has been in the direction of promoting reproduction as the target to be reached and survival only as a contingent condition. So in reality, we can realize that we live to make love not for the sake of making love as our individual goal but indeed for securing our conscious autonomy to perpetuate to reach our ultimate material biological perfection and trespass beyond the material boundary.

This somewhat esoteric, witty conclusion will find some transcendental significance in the biodynamic life commandment that I will outline in chapter 6, which is based on the essential supreme logic of survival as one sine qua non in the face of happiness in life as another. The consequential ultimate species' perfection with logical abstraction and Gnostic power reinforces the interdependence between pleasure and logic. These are two inseparable fundamental conditional factors for life: happiness with our pleasure seeking autonomy emanating from the material biology and logic elaborated from the abstractive power that can be called the immaterial biology with autonomy, consciousness, and unlimited Gnostic power, appear as a biological immortality principle akin to that of the photonic quasi-immortality in physics. The physical immortality analog is, so to speak, time inclusive, and our autonomy with the abstractive Gnostic power is periodically time exclusive needing perpetual renovation, but being an immortality principle in the same meaning as in its physical analogue.

But now, further reflecting on the phenotype's somatic size and function, comparing the somatic and neural developmental characteristic examples of two socially living species like insects and humans, we can see that the task of organization in the insect society has been assigned to what can be called a collective social nervous system because the individual one is too rudimentary to carry the duty effectively alone. In contrast, obviously the structural complexity and the functional efficacy of the human brain can provide adequate control for all functions including the agreeable one of sex. The transfer of the reproductive function from its affectively meaningless status in insect examples aimed for species survival only, to that of the highly affective meaningful one specifically exteriorized by the nervous system, can be plausibly interpreted as transfer from a collective nervous system to an individual one with the flourishing affective meaning associated with it.

This observation and interpretation explains why all species with well-developed nervous system are sexually independent individually, whereas those with rudimentary nervous system (or equivalent) may be, or are, dependent to the society casting assignment for reproductive activity.

Is the great affective value of sexual reproduction pleasure a conditional factor to build a trait for increasing this same value, which ultimately helps saving the species? Can we say that creation of mind was conditioned by this trait acting in a way as a positive autofeedback increasing the substrate and the function of the pleasure sensing system? These questions may find an answer by assuming that the pleasure sensing by the nervous system may act in a one-way positive feedback as long as the balance reached with metabolic satisfaction of feeding and energy gain is secured by the same pleasure seeking, a state of affairs that can be indeed intermittently realized. We can also simplify and say that the mind of pleasure seeking essence as a trait in evolution has contributed to support the reproduction assuredly and has developed miraculously, precisely because of this positive auto feedback mechanism based on the primordial trait of pleasure seeking. This subject will find more explanation in chapters 5 and 6. For the moment, we can summarize our deductions in terms of laws of phylogeny for reproduction in the animal kingdom in the following three statements:

1. Phylogeny Reproductive Rule I. The natural trait for reproduction is the primordial biological trait that is included by the bioforce in the biological material, akin to the reciprocal model of perpetuation of the two components of the bioforce, matter and energy.
2. Phylogeny Reproductive Rule II. The trait for reproduction can be evolutionarily separated from the trait for phenotype survival and

given due priority to secure species survival by being limited to one specifically fit phenotype for the reproductive task to secure priority for the bioforce perpetuation.

3. Phylogeny Reproductive Rule III. The trait for reproduction and the trait for survival are inseparable in all phenotypes in species with functional nervous systems, based on phylogeny reproductive rule I that secures priority for reproduction for which the functional nervous system is a major adaptive support, enhancing bioforce perpetuation.

According to the phylogeny rule III, the mind should have a very special role to play in reproduction with love, which seems to be highly decisive in the human species. In this consideration comparing animal species, one may accept the tendency to compare neural and somatic functional capacities for finding a correlate for clearer explanation of what mind can do in reproduction.

The shortest way to find a reasonable make-sense explanation is to consider a clear example by starting with the Eukaryote cell, the prototype unicellular organism that has a unity of soma and no detectable substrate for nervous system but evidence of adequate regulatory organization for cellular functions including cell division: the very initial form of autonomy. The hypothetical proportionality for nervous system function to somatic function in assuring reproduction in this animal would be the unity or near 1 for the coefficient N/S representing neurosomatic functionality. We can see that N/S can be accepted as a rule for explaining the adequacy of the neural functionality commensurate with somatic functional development. Checking the animal kingdom, we can perhaps think that a progressively increasing value for the N/S coefficient through the evolutionary steps leading from unicellular to multicellular life and further onward would be a rule. However, biologically there is no reason for such presumed rule to be found until the evolutionary stage of the triune brain is approached in reptiles. Thereafter the N/S ratio seems manifestly exceeding 1 till the evidence becomes frankly noticeable in primates with social intelligence and overtly evident in humans with language and abstractive expressions. In reality, biology is served by adaptive traits in assuring useful functions commensurate to needs and not otherwise by facilitating the control mechanisms to be based on feedbacks aimed to establish balance. These feedbacks all use biologically normal, negatively directed reciprocity to misbalancing positive stimuli so that the response to stimulus triggers negative charge against the stimulus to reestablish the balance. The effect of the mind in enhancing the reproductive trait is by increasing the stimulus under normal biohormonal changes specific to reproductive arousal in animals. In the human mind, this effect, originally bound to reproduction, can surpass the other biological factors and become purely pleasure seeking exerting power on expressivity in language and abstraction. An autonomously operated positive reciprocity feedback in biology does not exist except in somatic pathological conditions, drug addiction in psychopathology, and in the chain of abstractive thinking. In those situations, the threshold for the responsiveness to stimulus can be regarded to be artificially lowered while that for the balancing response being unduly elevated. Or the feedback uses incorrect chain of intercalated steps with abnormal overall positive reciprocity result that is ending ultimately as a one way positive feedback.

Mind-related facts on the pleasure-pain principle in the human society, clearly exemplified in connection with reproduction, are of major value when symbiotic reciprocal effect of soma and psyche are considered. In the brainy human society, each phenotype is a complex structure of soma and a miracle treasure of psyche. A host of radically dynamic factors can come into play with effects conditioning physiologically the survival as well as the reproduction of the single phenotype, and thereby the species, or sociologically speaking, the individual and the society. If we take the best survival chance as the sine qua non for securing the continuation of life beyond the phenotype, we can consider it as an idealized constant inferred from phylogeny reproductive rules I and III that we can call constant P for phylogeny. We can admit that P should include both individual and social biological chances for reproduction alone if it were considered in animals without psyche and both individual and social conditioning effects of the psyche on reproduction if it is considered in humans. We could then write,

$$P = p \text{ individual} + p \text{ social} = pi + ps$$

We can imagine that P has been originally equal to pi as in the Eukaryote cell, and through eons of evolution leading to society living, ps has been added by new adaptive traits and has reached the present complexity that is operating in the mind and particularly in psyche. Although evolutional adaptation to social living forms the new ps, and appears to add to P value in principle, in fact it may devalue pi as in phylogeny rule II. We can say that phylogeny rule II may be also presented as ps>pi. On the other hand, with brain and mind that practically grant a spiritual symbiont personality to the phenotype, we can conclude that with phylogeny rule III we can also have ps<pi.

The idea of the phylogenetic constant *P*, and its individual and social constituents combined with phylogeny rules, facilitates the understanding of evolutionary variations observed in the animal scale and the recognition of natural adaptive traits and coadaptive added characteristics. If the three notions, phylogeny rules, neural to somatic functionality, and phylogenic individual and social constants are used in combination, interesting deductions on the subject of symbiotic activity of basic physiological forces of biological traits, and influences of emotional and spiritual tendencies (particularly on the pleasure-pain axis) can be gained.

All considerations in this section, so far meant to expose *phylogeny for living to love,* in fact point to all evolutionary attempts to add to the secure perpetuation of the bioforce through the material biology, generation after generations. A natural question of why this progressive irreversible course should indeed be real and to what end it would ultimately lead us is a legitimate question. The answer is not clear but the outcome seems to serve the future nonmaterial biology, ensuring mind and consciousness.

METAMORPHOSIS IN SYMBIOSIS

Phylogeny seems to represent a clear early effect of self-sustaining bioforce through subsequent evolutionary development with mental life and pleasure-pain principle that we will see in details in the chapters to come. In the course of evolution, the supremacy of ps over pi, initially evident in the primitive animal societies like in insects, tend to decrease as the central nervous system appears, and its organization increases in species through the evolutionary course. Organized specialization in reproduction like in the case of bee queens will be lost in new species with nervous system function. In this situation the survival of genotype is directly dependent on the individual phenotype, which has both tasks of survival and reproduction, following rule III of phylogeny.

A new question now comes up. Suppose that a sporadic tendency develops in a living society of organisms, which still allows the bearer to keep society membership but promotes values independent from the primordial quest to survive and reproduce, like in maladaptive hedonistic propensities for substance abuse in humans; to what extent, if at all, could such tendency develop into a character and evolve as a trait and eventually overshadow the fundamental biological instincts? The answer is theoretically never, as trait should only appear if they have adaptive useful values for the species. However, in the human society with the mind, consciousness, language, art, and science, and the extremely well-developed communication, learning tools, living facilities, and all the complex interrelations contingent with this state of affairs, extreme vagaries are possible to occur. In this society, individual members make and share a living over and above the basic biological one, their mental life, which, if naturally not able to force evolutional traits to form, can still act conaturally and cause decisive mass impressions.

The limits between such tendencies and propensities and an established trait (basically genetic) may become looser in one phenotype life span with whatever effect it can have. Raised pi value can be formed over ps in maladaptive mental and behavioral reactions not counting eventual mutations with possibility of similar manifestations. Genetic changes, notwithstanding the mental life per se, form an independent field for all sorts of psychological intents and events, sometimes not in line with any rule of phylogeny. Indeed, in humans, ps may be more vulnerable than in all other socially living species. Such mishaps can also develop collectively and ends in examples of massacres, the most inhuman types, and paradoxically, being even religious in origin.

It is always true that the fundamental biological drive is basic and prevailing in human society, yet we can see that in this same society the mental life can tend to control or suppress some biological characteristics of the human being. This new life, granted by the evolution of the brain, is a real source of wonder. It is both dependent and promoting, or occasionally demoting the physical life. The mental life needs its substrate, the brain, to keep alive and contribute to living through the individual's physiology by facilitating physiological functions, both as guide and regulator neurologically and psychologically, and as provider and balancer of needed mental feedbacks. In this society, the individual cannot be a cast following ritual traditions but has liberty of action, which he can control to reasonable extent. The controlling is the function of optimum balance between the exigencies of biological physical versus mental life with consciousness of norms being paramount in this controlling mechanism. Thus, maladaptive disorders in the mental sphere can disturb the balance and reverse the ps>pi values to pi>ps in which conscious power could fail to function properly with ultimate danger to both individual and society.

As the complexity of biological evolution has increased adaptive abilities significantly, so have the mechanisms of those adaptations. Biological basic traits and their adaptive presentations to the outside world to forms, colors, functions, and values provide capabilities radically evolved and their regulating mechanisms adjusted accordingly through the evolution. The regulating power house, the brain, provides a structured priority logic center, which, together with pleasure-pain principle, can establish the foundation of the most fundamental rule in life that I have called the biodynamic life commandment to which we will come in the chapter 6, a commandment that is the foundation, making the most fundamental line of conduct for the mind with consciousness, providing the final determinant mechanism in it.

In brief, the mind seems to include various interconnected functional organizers acting in a modular fashion in establishing communications and easing tasks to be performed uniformly and harmoniously. There appears to be a compendium of codes concerning the biological priority values of both physical and mental processes. Construction and assessment of the codes seem to be the task of preconscious states, and the choice application of codes seems to be the duty of consciousness. Priority of values and decisions for choices appear to represent basically the ethos and the pathos of self and ego and the judgmental authority and power privilege for acting characterize features of self(s). The interplay of these three functional capacities, self-consciousness-ego, interrelated and conditionally reciprocated, guides the human action with the mind holding the connections in line with the factorial effect of the bioforce impersonated in *Autos* to be described in detail in chapter 3. All these make the total conscious symbiont consisting of undirected time indefinite iconic consciousness (dream visions) and directed time definite iconic-logogenic consciousness (awake states), with the impersonated archaic organizer, *autos* to be described.

This assemblage, a fortiori Cartesian theater view,[11] is a good way to represent the facts for the readers to visualize and to grasp the connective relays to the basic idea to be explained. Here is the mental life that prevails and not the biological life. All these limitless possibilities need explanation, and in this sectional brief review of the changing faces of the nonmaterial biological manifestations (the metaphoric symbiont), the explanation seems to become evident linking the vast order of potential variations seen to the fundamental bioforce's coding in genes.

TIMELESSNESS IN TIME

Going back to our working hypothesis, in my personal view, and in connection with symbiotic principle, the most important change imposed, or I rather should say granted by mental life, is the independence from time in free imagination. Whereas all biological life and processes are essentially time bound, mental life, in comparison, is in a way time independent (time undefined in free imagination) and may act as time controller (time defined in directed thinking). Mental activities practically seem to be time free or timeless with realms of imagination extending from remote fathoms of past to unknown frontiers of future in a blink. This statement, as conventional as it appears, fits well our human timescale served by memory. Viewed philosophically, mental life seems to be a production device to allow expansion and continuation of life in the mind beside the physical life, or even an elaborated independent form of life as a carrier of the inner time component: a symbiosis with the physical oneself representing the continuum of time. This production of life has also created a paradox which is time consciousness granting a relative permanency of self with that inner sense of timelessness. Coexistence of biological and psychological livings or symbiosis of one living creature with itself in extended mental life thus realizes the time outside and inside its own being, so to speak, with some illusory inner timelessness. Metaphorically, it is as if human eyes could look into the mirror of eternity seeing the revelation of infinite time and identifying its own being as part of it. This metaphorically imagined expression can reflect that the speed inherent in neurodynamic mechanisms may serve biology to allow detecting, seeing, interpreting, and experiencing virtually timelessly. Though this imaginative explanation is prejudiced, it can stand as abstraction metaphysically extending beyond the relativity of the physical world and independent of the speed of light, making the speed of thinking its best analog.

This analogy is meaningful and invokes the spacetimes in physics to be also true and applicable in biology. The fact is that according to the old Galilean relativity, the zero point for detecting and estimating any speed of a moving object by an observer is the constancy state in the speed of the observer's space frame and therefore the observer's biological time constancy. So the biological time, essentially nychthemeral or circadian, basically represents a constant time effect, which can equilibrate with any constant speed frame of the environment moving the observer in equilibrium with it but within the speed range compatible with biological life. Einstein's special relativity based on the constancy of the speed of light as the ultimate speed would allow in theory that we as living organisms, if sharing that spacetime, would not be feeling the speed as it is constant, and our biological life could also continue unchanged, meaning that our biological processes would still follow the pace of our inner spacetime constancy. Thus, the notion of biological spacetime for the corporeally delimited life elements must be accepted, permitting the corporeal unit having its own inner energy and speed fitting its internal dynamic processes regardless of the constant speed its phenotype is subjected to. The biological neurodynamic constancy time reaching conscious self-awareness is a prominent such example.[12] Einstein's general relativity related to gravity waves[13] and spacetimes is also superposable to biological somatic spacetimes. The analogy incites macro and microcosms, a subject that will be more clearly understandable in my further discussions. For the moment, we can say that the time constancy in the mind can theoretically represent a portion of a continuum of time in biology, a state of inner eternity with an inner sense of timelessness that is described in chapter 4, dubbed *time titan.*

Getting back to biology and using a more materialistic language, one could consider the purely biological drives as essentially somesthetic and the mental ideations as primarily psychesthetic to make a distinction still in line with scientific reductionist thinking and with dualistic inclination for the sake of discussion, to remain within orthodoxy of usual norms. However, one cannot deny that the complexity of these functional factors could possibly combine interactions with resulting variable effects. As this would have to be considered and defined, a host of new events can be theoretically suspected to occur, and all will have to be explored by finding one or multiple forces to explain the causalities of events. To predict the degree of combination of component factors and to apply adequate names to the events and forces behind them would be impossible without a full evaluation of the elemental biological and psychological factors that would be beyond our scope. We can bypass such a gigantic task assuming that basically physical and psychological factors are interrelated biological effects and both must evidence rather one form of causal energy than two or more. Then whatever the nature of this predicted theoretical unique causal force would be, I assume, it cannot be categorically different from the bioforce initiating life. The nascent notion of mental force, its solid recognition, and its documented role in evolutionary neurobiology[14] also indicate that we are facing but only one category of force.

A NEW FORCE OR AN OLD ONE

The adjective *new* that should be justifiably applied to the interpretation of the envisaged bioforce in reality describes a force as old as the creation itself. In the field of mental variations and deviations from the norm visualized earlier, the most decisive factors for changes to occur is provided by consciousness and language, the greatest means for evaluation of the mental force and for the identification of extensions and limitations of mind. Examples of different mindedness can be enumerated in the normal as well as in the abnormal mentation in one and the same biological phenotype.

In this line of thinking, a whole system of argumentation has developed in the scientific interpretation of materialistic, epiphenomenological, and plasticity mindedness in the last several decades, which has drastically changed some fundamental biological beliefs on mind and continues to do so. The point made clear is the fact that mind as a nonphysical, nonmaterial actor does possess the capability of changing one's own biology including the biology of the physical brain. In philosophical parlance, the biologically created element as part of the creation, can act on the whole creation by controlling and changing the creator's act, or the brain and its output being regarded as the source, and its manifested energy as its innate force, could present the same scenario. The string quantum principle, as briefly stated earlier, could serve as analogy to mean that the mind to brain relationship actually presenting the nonmaterial and the material biology follows the same fundamental pattern of forces used by the creation in physics and metaphysics. Thus, what has been separated in Descartes' judgment could be unified in today's thinking. Mind and brain, energy and matter, philosophy and science for a paradigm could now be more firmly united, and expressions such as *mental force*, *spiritual force*, *volitional force* are used more openly by investigators.[14] Even the old warning *attention before intention and action* is finding measurable support in scientific investigations showing time differences of delays between unconscious and actual conscious actions by the causal force.[14] As added evidence for this surfacing unknown force that I have labeled bioforce giving it an extended scope, we can consider the arguments in pros and cons for interpretation of the neuronal adequacy time (NAT)[15,16]for consciousness of our acts. The experiences, as will be shown more clearly in chapter 8, show time requirement for the consciousness of our acts that seemingly follow an unconscious initiation. NAT is the time necessary to elicit a conscious sensation by repeated continuous near threshold direct cortical stimulations until a conscious impression is sensed and reported by the subject. It can be considered to act in two ways: accumulating effects of the unit of force in a nonexpanding substrate (fixed neuronal network) using only time extension or both accumulation of time extension and substrate expansion (spread neuronal network). In both situations, the time extension is the main factor that conditions the quantitation of force by stimuli to reach the level sufficient to cause consciousness. This would imply the presumed unconscious implicit memory, perhaps in the form of unit per unit of stimulus mental force, to be recalled tacitly and implicitly before reaching explicit memory and consciousness.

Indeed realities of mental characteristics attached to attention and intention, and the NAT, pose questions to ponder upon and lay the foundation for whatever relationship these mental processes may have in expressive ideations and the conscious formulation of language. I will deal more with these subjects later, but what the symbiosis of psychological and somatic biological life means and implicates that originated before consciousness was completed with conscious coding of knowledgeable items reaching enormous extension with language. It is indeed with the language, the most perfect means for expression, that the time factor from attention to intention to execution is demonstrating the effect of the bioforce.

In considering the evolutionary scale before development of elaborated mental functions, all that strikes the observer is the biological rule dictating survival and reproduction, SR, as axioms, and the organisms' drives that follow these rules scrupulously and unfailingly as if they were unquestionable directives. The course of events indicates birth, growth, reproduction, aging, and death, repeated ad infinitum by an infinite number of descendant living organisms in an infinite time and overtly in an astonishing irreversibility. Biologically, we are witnessing the mechanisms of forced transmission of units of life through units of living elements to an unending, hypothetical, unknown destination. In pure biological terms, phenotypes are only there to carry the genotypes to that unknown destination. Philosophically, we begin to sense the nature of the *unit of life* as part of the whole unknown and the common task of these units of living elements as allowing a forced generalization by carrying the inner unknown, the bioforce by the biological vehicles.

If a computer program is set to answer the obvious question of what starts this course of events and keeps it going in perpetuation indefinitely, the only answer it can give, based on information fed into it, will be some presumed causality between the two suspected unknown hypothetical ends of the process, initiation and termination. The unknown cause initiating the course of events, as reflected by its action, will be naturally a force, one or more of the four basically known microcosmic physical forces of electromagnetic, gravitational, weak and strong nuclear force types with their variable possible manifestations as a presumable cause for any observable happening. Or as alternate possibility, we can also assume a new type of force to be added to the known arsenal to represent the proposed bioforce. As any energy can be related to matter in one way or another, the presently known material constituents of the macrocosmic physical world with their inherent energy forms should not be excluded from possibilities contributing to biological matter-energy complexes. In fact, the mass-energy components of the universe mainly come presently in four broad types: the mysterious dark energy estimated as 74% of the total universe's constituents[17] that causes expansion of the universe acting as antigravity, invisible dark matter estimated at 22% and known for its gravitational effect, and the rest being visible ordinary matter and neutrinos.[6] So including all known and unknown hypothetical forces in consideration, each can be playing a decisive role singly or in combination to represent the bioforce presumed to be causal in the biological life. To sum up, the force of big bang, the force behind physical phenomena, and the force in living organisms, all can be in essence of one category with their possible biological manifestation being in one specific form algorithmically, including potential changes with the course of time.

Although the fundamental bioforce engendering life cannot be defined in any precise way, biological mechanisms of energy usage at cellular level have been defined with interesting significant clarity suggesting ideas. For example, the nanometer scale (ten to the power of minus nine meter) molecular protein machines in the cell, including various specific enzymes, interact to secure energy usage for metabolism, reproduction, and response to environment. The adenosine triphosphate (ATP) synthase as an example, a protein molecule of about ten nanometer that is regarded as the currency of cell economy, can act as rotary motor with some turning when adding phosphate groups to molecules of adenosine diphosphate (ADP) in the presence of a proton gradient and in reversing direction when it consumes phosphate groups hydrolyzing ATP back to ADP, converting chemical to mechanical energy.[18] The nonmechanical cellular energy usage also seems to follow traditional physical energy conversion mechanisms. The link between the unknown fundamental energy, the bioforce, which must ultimately trigger the orderly sequencing and the actual conversion of cellular energy for usage sustaining and continuing life, must therefore be searched in DNA in terms of a fundamental code allowing those cellular processes routinely. Or, believing new ideas for possibly simpler molecular components than DNA starting the whole thing, the two main lines of beginning for the life to be with either replicator first theory or metabolism first theory should ultimately lead to the same result.[9] The single most fundamental theme of that basic code, according to biological evidences, is unlimited expansion with time requiring growth and reproduction with all contingency events in the biological carrier forms. So in essence, the biology can be understood, by and large, as a life replica of the cosmos: A cosmobiology model containing the material biology of measurable matter-energy constituents and the nonmaterial biology of energy-time nature of the abstract mind domain.

Perhaps all known cosmological physical forces enumerated earlier are contributing to this specific bioforce as a unified factor to generate and conduct biological processes at the molecular level according to the fundamental coding replicating the cosmological creation-expansion-condensation principle.

It appears more convincing now that whatever that force would be, its analogy in binding the matter-energy changing forms in string quantum physics, named bioforce, reveals being attractively compelling. The nature of the unknown theoretical ending of this force in physical terms implying theoretical global minimization of spacetime with maximization of gravitational energy (black hole) would be another big bang and a new phase with a new vehicle for the bioforce in the biological microcosm. The patterns of processes in biology furthermore, with the overall continuity they show, are also reminiscent of physical conservation and transformation-renovation of the energy, a well-known principle in physics that could be now translated in conservation of life in the biological continuum, or one may philosophically say conservation of the phantom of time, the memory, in biology considering the principles of physics applied to biology.

THE ETERNAL TRAVELER

This attempted explanation seems philosophically acceptable on the basis of analogies described. It takes the causality to be the only logical interpretation in the course of action of the perpetuating unknown, the pristine force, the prototype energy making the physical and the biological worlds' specific spacetimes. This imaginary untiring traveler, so to speak, could either go at the speed of light in the macrocosm embodying virtually infinite energy theoretically but no mass, invoking analogy with light, or slower with mass but still with the innate pristine energy keeping a momentum with its mass analogous to the material biological organism. This eternal horseman never reaches any destination, but theoretically changes horses, and trots far distances, running unfailingly to infinity. In fact, no destination can be foreseen with any tangible explanation except another big bang and infinite series of them in the macrocosm, to believe the classical string quantum theory or newer M-theory, and to admit biological deaths and rebirths to be renovations in the microcosmic biological world.

Although the questions of a start and of an end that can be posed, and indeed have been repeatedly posed, both find no satisfying answers; these two questions, the beginning and the end, have been the worries of theology and the past time of philosophy and science, and I suspect they could remain as such for long time, whether reflected in mind or in a computer analysis. However, a light of hope entertained and discussed in the last chapter of the book, binds ultimately the physics and the metaphysics together. Advances in modern physics with string quantum principles not only provide good explanations for phenomena in the physical world, bust also hint ideas in biology, on the life and its beginning and ending, based on the primordial force that I have called bioforce. The quintessence particularity of this unknown force seems to be the mutual effects to one another of the nonparticulate energy wave and the nearly no mass string shown as *ls* in the classical string theory,[8] moving at the speed of light. This mutual effect appears to be such as to allow a wide comprehensible spectrum of quantum frequencies matching energy values between two extreme physical states, each representing the start and the end for one another like the beginning and the ending of a wave curve. The nonbiological and biological phenomena seem to be confined between these extremes that form a phase of theoretically any possible size. If we accept a modified big bang proposed by the string quantum theory, with the expanding universe to a limit and then retraction and regression back and another big bang, we can imagine the same general principle to be applicable to physical and biological life: a wave of perpetuation in series, not so much recognizable in physical forms for the human mind basically carrying the abstract imprint of time, than for the physical biological entities showing observable changes of living cycles. The other way of finding an explanation for this ***bio*** element would be to see the subatomic spinning interrelation of the positron and electron of photon denoting perpetual immateriality waving of energy between material elements as an immortality principle and considering the bioforce with its inherent biological autonomy as an analog to it.

Long will have to go our horseman through the time that he will create with his being and will evidence it in his consciousness which will create new perspective of future arenas for him to trespass in perpetuation. We will gain more concrete ideas on this line of thinking through chapters to come.

REFERENCES FOR CHAPTER ONE

1. Smith, J. M., and E. Szathmary. 2001. *The Major Transitions in Evolution.* Oxford, New York: Oxford University Press Inc. Pp 28-32.
2. Miller, S. L. 1957. A Production of Amino Acids Under Possible Primitive Earth Conditions. *Science* 117:528-529.
3. Longstaff, A. 2005. Quest for a Living Universe. *Astronomy* (April): 28-34.
4. Alcamo, E., I, and K. Schweitzer. 2001. Cliffs Quick Review Biology. *Hungry Minds.* New York, New York. 108-109.
5. McEvoy, J. P., and O. Zarate. 2001. *Quantum Theory.* USA: Totem Books.
6. Kane, G. 2005. The Mysteries of Mass. *Scientific American* 293: 41-48.
7. Marcus, G. 2004. The Birth of the Mind: How a Tiny Number of Genes Creates the Complexities of Human Mind. *Basic Books,* Perseus Books Group.
8. Veneziano, G. 2004. The Myth of the Beginning of Time. *Scientific American* 290(5): 54-65.
9. Shapiro, R. 2007. A Simpler Origin for Life. *Scientific American* 296(6): 46-53.
10. Flanagan, O. 2000. *Dreaming Souls.* New York, New York: Oxford University Press. 115-116.
11. Ross, D., A. Brook, and D. Thompson. 2000. *Dennett's Philosophy: A Comprehensive Assessment.* Cambridge, Massachusetts: MIT Press.
12. Libet, B. 1982. Brain Stimulation in the Study of Neuronal Functions for Conscious Sensory Experiences. *Human Neurobiology* 1:235-42.
13. Mallarino, A. 2005. Why Do Physicists Think Gravity Travels at the Speed of light? *Astronomy* (April): 62-63.
14. Schwartz, J. M., and S. Begley. 2003. *The Mind and the Brain: Neuroplasticity and the Power of Mental Force.* New York, New York: Regan Books. Harper Collins Publishers Inc.
15. Libet, B. 2004. *Mind Time: The Temporal Factor in Consciousness.* Cambridge, Massachusetts. London, England: Harvard University Press.

16. Blackmore, S. 2004. *Consciousness: An Introduction*, 58-60. Oxford University Press. Oxford, England. New York, NY.35(5):28-33
17. Kruesi, L. 2007. Cosmology: Five Things You Need to Know. *Astronomy* 5(5):28-33.
18. Phillips, R., and S. R. Quake. 2006. The Biological Frontier of Physics. *Physics Today* (May): 38-43.

CHAPTER TWO

THE NUCLEUS OF LIFE: A PHILOSOPHY OF BIOCRACY

NEOLOGISMS DEFINED

My intent to scrutinize the biological force in describing the theory behind it could naturally start with the force of mind noted in the last chapter, a force more accessible in studying the mind, but the definition of mind is to be first settled. Indeed, the human mind, the symbiotic superpower created by, and living with its biological substrate carrier, I firmly believe is too elaborate and too complex to be defined accurately or described comprehensibly at the present stage of human knowledge. Examples abound mainly in biological, psychological, and philosophical literature of attempts to describe what mind may be. The complexity of modularities and interconnections of anatomical structures of the brain contributing to the modularly manifested processes, and the resultant mind, its experiences and subjectivities, are simply beyond the frontiers of our present knowledge for a comprehensive clear cut conclusion. The state of ephemeral, indescribable, and often changeable feelings sensed inside oneself, also called *qualia*, which is inseparable from self(s) and the scope of consciousness including emotions, memories, noetic facts, the whole mind for short, still remains the primordial unknown to science. I am convinced that the best I can accomplish in this book, when it comes to defining mind, senses, and feelings, is to offer a summary of my own understanding of what I believe the mind could be without being able to convincingly provide testable evidence for views and theories I expose except for those with already established scientific foundations.

The explanation of matters of mind and mechanisms involved pose difficulties and biases when discussions touch upon the realm of philosophy facing pure science. The rooted understanding of the old Cartesian dichotomy of the mind-body link with its subsequent modifications that are still the subject of arguments and the more recent renovating ideas in evolutionary psychology that have emerged indeed make a ground of diversity for the study of mind that imposes caution for avoiding misinterpretation. Therefore, I will approach the subject of life, its causal force and its neurodynamic manifestations in relation to mind, using a different starting point with a terminology that I hope offers better clarity when it comes to deal with life and biological processes regarding the mind-body problem.

As the beliefs of pure scientific and philosophical deductions often seem to part astray one from the other, even if sometimes both evidence solid logical foundations, I believe a dialect to ease the dialogue between these two realms of knowledge will improve better understanding them in one another's terms. Scientific findings establishing norms, of course, cannot be changed to agree with pure philosophical beliefs, but such beliefs can be expressed to include a nucleus of philosophy in a protoplasm not too incompatible with a symbiotic life of science and philosophy or body and mind. Such ease in interpretation, even if dubious at first glance, can be taken as arbitration and can remain open to ultimate acceptance or rejection but is worth trying.

With this in mind and with the notion of bioforce already outlined, I would like to introduce three related neologisms. First is the word *biocracy* by which I simply mean the autonomous order in life processes to keep replicating endlessly. This is originated by the bioforce and conducted by it. It uses all the mechanisms in living organisms to assure uninterrupted perpetuation of genotype indefinitely through reproduction and with unfailing assurance and constancy which reflects a true autonomy with autocracy. Two other neologisms are biotropism and biodynamism. By biotropism I mean the primordial innate biological drive or directive that leads to energy gaining and detracts from energy losing in accordance with the physical principle of conservation of energy. By biodynamism I simply mean dynamic biological processes in the service of biotropism as governed by the innate autocracy in biocracy. I give the biotropism the fullest meaning of a self-oriented reactive response intended to the phenotype with no allobiotropism or heterobiotropism but only autobiotropism, in short, biotropism. Thus biotropism can be the only observable immediate effect of biocracy, and biodynamism reflecting clearly the autocracy may seem reminiscent of *selfish gene* of Dawkins reflecting its ultimate play in perpetuation.

A brief clarification regarding genetics is in order to properly define the effect of biocracy on genetic power. This effect goes beyond the simple blueprint. Biocracy gives both the outline and the outcome potential of gene action with time. In other words, the life history of the phenotype under biocracy, through biotropism and biodynamism, reflects both the initial firm ground and the later constructed edifice evidencing distinctly the effect of time and the action of environment in an autocratic way. In this belief, I fully subscribe to the autonomous agent theory of Marcus which keeps the gene responsible for both the initial categorization and the later developmental permissive changeability constructing the actual phenotype. Thus autonomy with autocracy for self-supporting and expanding seems the sine qua non condition for life to exist.

Other neologisms that will follow are autos, bioawareness, biochronogenesis, bioconsciousness, biognosis, bion, biopsychon, bioquantum, neognosis, paleognosis, and psychognosis in alphabetical order that will be described in sections to follow as they will relate to the bioforce and its biological actions concerning mind.

BIOCRACY AND MIND

Realizing that biocracy with its tacit autocracy appears to imply a deterministic meaning, I must say that no deterministic philosophy can be included in it anymore that is included in life itself. All evolutionary laws applicable to life and evolution of species are also applicable to biocracy. However, it is easier perhaps to sense such determinism more with biocracy and less with life, for life is purely a name, but biocracy, also a name, is *living* with defined autonomy orders especially with a linguistic nuance. Life is a name, but biocracy is a kind of *nomina actionis* in linguistic parlance. It is a one way directed cause-effect action by the bioforce which in fact is the strongest determinism evidenced in the product of life, a product that is modifiable by evolution. The reason I insist to use biocracy will become clearer when I reach the subject of autos, self, consciousness, and ego and will find more justification and explanation with the concepts I will develop. For the moment, let us continue with the introduced neologisms.

Today's dominant philosophical thinking departs from Cartesian dualism to find, hopefully and rightfully, a unification of the body-mind properties in establishing the foundation of mental individualism based on mind and consciousness. The mental experience regarded as manifestation of biophysical factors that are neurodynamic manifestations in the brain seems a metamorphosis of the dualistic thinking. This has appeared to be yet more plausible than the reductionist materialistic considerations implying that the mind is simply a direct consequent product of the fundamental atomic and subatomic processes in the brain. This fact is true but not so simply, directly, and so reductionistically. The modified dualistic thinking opens an avenue to seek the rational reciprocal that may exist between the body and mind, ultimately a better way of unification, a form of inclusive rather than exclusive reduction in brain functionalities.

The inclusive reductionist idea would try to find and explain the processes producing and the mechanisms maintaining the mental experience, which is regarded as completely independent from physical effects in pure dualism, and as simple direct result of them in exclusive reductionism. The biocratic philosophy, I believe, can come as a messenger in this avenue considering the mind an elemental item of the evolution just as the body is one such item and evoking a crude simplistic analogy by taking the brain as the living matter and the mind as its energy. Thinking along this line, one can argue that biodynamism certainly served biotropism in the evolutional progress in accordance with the increasing complexities of the neurobiological development allowing psychological states, all initiated by the inherent bioforce and directed by biocracy. The last and the eternal question is then: What is the power, the impetus, the decision force, or the will, and how does it form in the self to direct actions? We will come across this question again as we advance through the coming chapters throughout this book. But the question's basic fact, the time-related conditioning of biocratic autonomous steps before realization of consciousness and *the will* needs an initial attention. Let us consider this aspect of the question both from a braod angle looking at preconscious capabilities and in a more precise focusing consideration in an entertaining thought experiment.

A THOUGHT EXPERIMENT

A philosophical Persian aphorism by the Persian poet and philosopher Mowlavi (also known as Rumi) is the adage *"the unbegotten cannot beget but itself"*, meaning in extension that *"zero can only give zero, or timelessness cannot be but nothingness."* However, if we hypothesize about the initiation of life, we can take the point of initiation as a beginning of everything from nothing to mean stepping with life into time dimension that we can presume started with life or to have always existed as an immaterial nothingness now being materialized by life as a living energy coming in the pre-existing nothingness. In either instance, that start could be a *finite point* with some physical dimensions or only an infinitesimally small linear dimension as in the classical string theory, and time will still be one of the dimensions so that initiation of life steps out of theoretical timelessness into time. If we follow this living time vector as a quantum of energy in its path that could have a hypothetical end, that end would again likely be nothingness with timelessness, so that we would be at the starting point. We could then imagine a circular path rather than any other for the life as time element, a continued time segment, between two timelessness limits of initiation and termination that should be theoretically one or a hypothetical circular path of time between hypothetical starting and ending timelessnesses that would resume to be one if we ignore the possibility of multiple big bangs.[2]

Now if we try to define life's possibilities of perpetuation (and forgetting for the moment the known course of biological reproduction and decaying), we can imagine the possible ways that the finite point on the circular path, the living element with finite energy and life span (a segment of finite time itself), may take to perpetuate, based on its autocracy.

Obviously, to assure continuation and permanency, any replenishing for energy loss must occur before the life span of the living element is over. The only possible way for the living element to do the job would be to renew its full energy loss in a time-related mechanism before reaching its very end. This could be imagined to be theoretically possible using one principle in two modalities. The principle would have to be an energizing mechanism in the *finite point* on the string theory model between positron and electron of the photon incriminated as bioforce or in the bioforce living biological model to replenish energy loss continually occurring in time in equal amounts. This would imply a balancing feedback between the two subatomic elements, or in the biological model, counteracting the imposing decaying effect of time.

The modalities would be to either use a long timing to allow the largest possible feedback or a short timing with the smallest adjustments repeatedly, taking shorter time, to include two extreme hypothetical modality examples. In the first modality, if the life vector could, by chance, reach its starting point again, it would have used the longest time and could eventually have renewed its full capacity of energy by the theoretical autofeedback and could follow the path on the circle repeating the maneuver indefinitely, continually and permanently, as long as the elusive eternal time could last. However, as this living unit is confined within the general spacetime constancy where it is, for doing such maneuver without a factor of chance, it would either have to reverse the time to go back to reach its own starting, or would have to run forward, faster than light and jump over the hypothetical timeless gap in the circular path of time (a big bang?) in order to reach the beginning to renew total energy and continue. Both of these imagined happenings seem to be physical and biological impossibilities as is the hypothetical chance factor.

Or in the second modality, the unitary entity models could use the same principle, not in totality but repeatedly in part, to replace any energy loss with an equivalent gain in the shortest possible time and continue indefinitely. This pattern could well explain the physical model. In this second modality, both impossibilities of having to go back in time, or to exceed the speed of light to bypass the hypothetical timelessness are excluded as the shortest time required for the feedback could be only the reciprocal of the speed of light for one wavelength which is infinitesimally small and adjustable according to variable photonic wavelengths. The particulate element as matter (positron and electron of photon) on the one hand, and the speed of light as energy appear to be the only obvious factors to actuate the feedback, but the action of gravity and antigravity possibly appearing in this situation of infinitesimally small mass-energy change also affecting the feedback cannot be ignored.

The *finite point*, if now we remember that it could be the biological model that must decay, seems to use the second modality in its phenotype form in its living time and the first modality with its decay leaping time in its genotype renovation. As we remember, its dimensions include time, and that is the only part of it that can theoretically remain incorporated in the circle of time to continue with it in the product of reproduction. If however the unit is regarded to be *ls* of the string theory, being quasi nonmass matter (10 to minus power thirty-four of a meter),[2] it can be seen to persist by instant perpetuations indefinitely, to be eventually renewed between two big bangs or multiple ones simulating the leaping biological renovations.

Considering the photonic positron and electron, we would have to also incriminate the inherent spinning of these two elements as playing a role in the forward wavy propagation of light. This hypothetical effect could imply that neutrinos, as quasi no-mass energy quanta with spinning particularity and possibility of progressing with the speed of light,[3] would have been perhaps precursor models for the finalized setup of light with electron and positron. This thought, pure imagination as it appears to be, suggests analogical value and provides deeper significance for the metaphysical standing of the bioforce in creation. A further hypothesis may thus be envisaged by considering this possibility as the nucleus for making the bioforce by what can be called biotrons.

We can clearly see the analogy between the physical and biological paths, both representing the continuum of time comparatively that evidences an identical principle though in nonidentical ways due to different mass-energy exchange compatibilities in physics versus biology. In simple words, if biological life was matterless and massless, it could go on forever with no decay. Now, is the mind with the energetic essence of emotions and abstractive artifacts making up the abstractive world constituting a pure energy life is not that nonmaterial biological life? And is the mind's force not the bioforce, the same as the biological force, made up of the hypothetical biotrons in some theoretical ways? These deliberations will be re-examined in the metaphysical discussions viewed from different angles in the later chapters.

This thought experiment providing a hypothetical circular frame for the infinite time could theoretically apply to any mass in motion having a directive inner energy. The directive energy may be nonautocratic like in all physical bodies in motion with the speed of their space frames, or autocratic like in living elements capable to change self speed within compatible limits to their environmental spacetime. The crucial principle permitting that element to be permanent in time would be the element's theoretical capacity of autofeedbacking to keep its energy constant and undecaying with time. Such theoretical capacity should be inversely related to mass, and the principle appears to be applicable to all imaginable bodies, physical or biological, from one big bang to another and so forth—if and only if—the autofeedback could be real. The mass implication naturally limits the theoretical autofeedback to the smallest near massless physical possibility which is hypothesized in the quantum string theory. As the force of gravity can increase directly proportional to mass and decrease inversely proportional to distance between two masses of matter, it should be possibly the smallest with the quantum string, presumably allowing a state of oscillating equilibrium between the lowest gravity force and the highest speed, readjusting one another combined in an equilibrated balance, or there could be a microcosmic antigravity force as observed in the macrocosmic world. In physics, imaginable autofeedbacks between infinitesimally small moving particles are seen in atoms securing the state of equilibrium of positively and negatively charged subatomic elements, in electromagnetic fields, and in spacetime related gravity fields based on Einstein's general relativity principles. The autofeedback, presumably involved between the particulate components of the photon of light, the positron and the electron, conditioning light waves and securing light's constant speed, is a state of affairs composed of a principle of perpetual phase changing securing a principle of constancy in variable presentations whose directionality only can be influenced by the internal vectorial result of the spining and external factors (gravity effect as is well known). The bioforce principle in biology suggests an outstanding analogy with light in physics with the phenotype-genotype phase changing, autonomy constancy principle, and variable phenotype presentations in the material biology denoting directionality change in evolution of species. It has been shown in fact that electrons and ions indeed act like waves in displacement and like particles on detecting plates.[4] Thus the noted analogy in the material biology and the physical world can be regarded as a bona fide fact.

A theoretical explanation with an interesting final conclusion could be that the big bang energy of explosion, in its maximum, reached to allow ultimately a level of unitary stable form, so to speak, of energy and matter in the form of light (biotrons of neutrino origins?), with the minimum of mass and the maximum of speed, the two of them representing the two components of the string theory, transposable to photon and its energy. This state of affair occurred some 380,000 years after the big bang once an initial state of equilibrium was established after the chaotic expansion initiating the inflation, equivalent to 10 to the power of minus 34 seconds immediately after experimentally simulated big bang.[5] The initial foci of light then appeared simultaneously in different places accordingly, and the presentation assuring the formidable binding of the two elements, positron and electron in the photon making the particle, secured the speed of the particle's wavy displacement in time showing the energy. Such conclusion is presently deductible from recent experiment reconstituting the initiation of the big bang in relativistic heavy ion collider that shows the early microseconds of the big bang advent with subatomic elements interreactions.[6]

In fact, simple reasoning on the interaction of the positron and the electron components of the photon being of equal mass with opposite positive and negative charges posits that spinning of these subatomic elements giving momentum, electromagnetic field, and gravity field, could form possible centrifugal vectorial effect in a wave form for the light, also possibly aided by additional antigravity effect. Theoretically, the evading speed of the energy (photonic light), and the attracting effect of the gravity of the mass (photonic mass), one force at its assumed maximum and the other at its minimum respectively, still act on one another. In this interacting effect, three variable presentations seem legitimate possibilities for the ensuing beam of the light. The first appears to be the wave presentation, the second the straight trajectory forward, and the third the speed. Explanation to account for the wave in the most simplistic way may include effects of the spinning positron and electron and gravity effect of each combined, with forward propulsion forming spiraling depicted as wave. In simplistic explanation, the straight trajectory of the light beam may be counted for by the final evading energy overcoming any deviation more than could be expected with the permitted natural innate gravitational spiraling effect of spinning. The speed of light, however, not only overcoming the inherent gravitational effect but also remaining at its highest constant value seems to reflect the initial equilibrated constant antigravity effect that should increase eventually affecting its physical constancy.

In this theoretical viewing, the photonic mass (spinning positron-electron ensemble) assumes a straight trajectory with respect to the expanding motion of its evading energy but with allocation for the spiraling wavy appearance of the photonic quanta observed. As an interesting remark, the most surprising similarity of the most fundamental bipartite spiral presentation of the quantum particles in motion is the spiral bipartite structure of the DNA in biology that has stimulated using DNA structure nonbiologically, as in nanotechnology for information storage with the benefit of space saving to molecular sizes.[7]

The thought experiment as described, may not be of any radical revelation but does allow us to be clear on three points; similarity of the physical force and the biological force needing to be explored more, time dimension being included in both physics and biology implying spacetimes to be true in both domains, and close analogy and homology between physical and biological presentations, effects, and courses of events. In biology, this similarity evidences the fact that the modality of transfer of the time element in life from one carrier to another, and in the substance of genes as unitary life elements, is the only available mechanism for perpetuation to a hypothetical eternity. It also reveals the truth of the very character of the biological life in one phenotype phase which cannot expand the inherent time dimension except in mind processes. In this regard, it is interesting that biological material dimensions of body and brain show some parallel with mindedness and keep the neurosomatic (N/S) functionality ratio greater or equal to one to reach ultimately the real N>S with language, abstraction power, and conscious depth in metaphysical concepts in humans.

Philosophically, other suggestive ideas come out of this thought experiment. Two such ideas should be treated here, one biologically and the other metaphysically significant. If we recall the *finite point* on the time circle in our thought experiment as a living element with its theoretical dimensions including time, we may foresee and predict a direct relation between a longer life span and a greater mindedness, and such is indeed evident when comparing the brain sizes and the longevity in various animal scales.[8,9] Philosophically said, the longer the lifespan and the further away from starting timelessness on the hypothetical circle of time, the longer the living unit contains time, the constructive immaterial dimension of life reflected in mind. The force could be a theoretical energy without mass in the vehicle of ions on biological scale, or an as yet to be defined biological form of the bioforce in organic substrate DNA only evident in its coded form. The amazingly mimicking wavy spiral of DNA reminiscent of matter-energy wave relation appears simplistically analogical. It is of interest to note that test-tube DNA models used to make computing similar to biological cellular regulation, even detecting some cancer, is an actuality[10] that opens unlimited horizons of biological application and, in essence, supports the reality of the bioforce. Biologically, the limit (not so limited) is the elaborate mind of a creature, us or an unknown to come.

The suggestive metaphysical significance of this thought experiment is in the human mind's incessantly realistic searching for the idealistic reason of biological creation forming the immaterial Gnostic world. This search, related to the inherent logic of causality in the human mind, is dependent on the expressivity trait of the mind's reasoning and expression as we will see further in future discussions. It is genetically formed and constitutes an autonomous element of the human mind conditioned by the level of mind's logic searching for truth. In brief, it is likely hard wired in human psychology for the truth, a truth that is the ultimate goal in the human mind's logic to be reached but obviously for a purpose, which is a sublime secure hope of eternity. The search is incited by the wonder for the unknown as stimulus to find the unknown target as the solution justifying the psychological mind's need for security, peace, happiness, and salvation. The logic behind this search for truth is the primordial principle that contributes to the life commandment that will be discussed in detail in chapter 6. Is the human mind's searching **To Reach** the **Ultimate True Hope**—***TRUTH***—the path defining the human destiny in the corporeal material and spiritual nonmaterial life? Is this ***TRUTH*** a Gnostic reality that can be ultimately substantiated? Is it a reality supported by the anthropic principle as advocated by Gardner regarding the cosmos' fundamentally life friendliness naturally harboring physical and biological worlds?[11] Or rather it is the reality of the original **Biothymic Creation** (bios = life, thymos = spirit) being reflected in the material soma and the immaterial psyche that I will expand in more detailed discussion with the subject of Gnosis in the next chapter. These inquiries focusing on the same problem will be reexamined and discussed as I go on through the future chapters.

Let us now undertake some reflections on the innate time dimension included in biocracy, in the biological vehicle of the bioforce.

BIOCHRONOGENESIS

The puzzle of time and its beginning in biology can be considered perhaps more easily approachable physiologically than either physically or philosophically; as philosophically, its justification seems inseparable from the creation and remains insolvable with it by the ever reciprocal redundancy of the theoretical priority question of creation versus time, and in physics, hypothetical time zeroing in the black holes and running outside them cannot be unified in one inclusive theory.[12] Time can be considered therefore as both existent and nonexistent, its substance being shown by energy spent, a substrate, a body, a vehicle, a motion, elusive and changing with the substrate's changing form or location. There should be biological status change or motion to indicate time, but a very special mechanism to detect time, and a very subtle one to record it and recall it. The task is clear when accomplished in consciousness automatically and effortlessly, but how did it initiate and develop before consciousness, if it did? This question invaded the human mind probably from before antiquity to the dawn of mythology and, with the invasion gaining more land, some specific conjectured doctrine of Zurvanism has been most intriguingly associated with the myth of time giving it the ever existing self-actuated mystic aura of limitlessness.[13]

With the advent of spacetime notion, time in physics became better defined, and now it can be considered more factually in biology. Finding out how its effects could have been sensed, interpreted, and defined in the biological life before consciousness, if at all they were so sensed and defined, is a challenging task. If this problem is solved, the theory of life based on the bioforce can be also regarded as tested and proved, realizing that indeed time as one dimension in the spacetime notion, together with the other two, matter and energy, is an inherent component in the biological spacetime as well. Thus, any corporeally delimited biological spacetime, a cell or a complex creature, would possess all three dimensions, mass, energy, time, from bioforce origin in a general consideration, and only specifically from that origin, because of that force being the unique one as a model of unlimited permanency in physics and biology.

How can time become concretized in biology? In what form did it start? And what changes did it take, if any, to be now revealing the real time in our minds? These questions can be examined philosophically and scientifically. They are not answerable clearly and make essential subjects of debates that I will try to avoid for the sake of limiting my discussion to understandable biological facts only. The crux of the puzzle seems to lie in the functional meaning of the biological time span in the biological living and sensing animal considered as a biological unit.

The biological time span in the living unit includes two functionally different value times—one that is used in metabolically functional steps in all various stages and processes, used ultimately to finalize energy gain and loss to close the feedback loops and establish the needed biological equilibrium of a balanced energy state, and another one that should represent the equilibrium value time of the balanced state itself. We can therefore see that time in the biological creatures is not uniform and does not have always one effect and one impression to the sensing system except when in fixed equilibrium, both with the internal energy balance showing constancy in the biological spacetime, and with the environment spacetime. This biological, very early and very basically function-oriented time, as opposed to the equilibrium-fixed time, presents a difference to the sensing system. The first one, the function time, plausibly represents the inner inherent type of the infinitesimally small special spacetime at molecular level of cellular processes of the biological unit, and the second one, the rest time, represents that of the general spacetime harboring the very biological living unit in which equilibrium constancy matches that same pace and continues with it as long as the unit lives and keeps the equilibrium. The balance between the two systems, outer spacetime and inner spacetime, is the status quo that could theoretically be forever if the living unit could be just the string noted in the thought experiment. So in theory, biologically sensed time is initiated as biochronogenesis by differentiating the spacetime of the major unit, the phenotype, from its minor molecular spacetime, inherently different from the physical spacetime harboring the living unit and in equilibrium with it. Thus, we can think in terms of the phenotype spacetime in the same constant speed frame of the environmental spacetime, and phenotype's internal microspacetimes for the molecular, biochemical, metabolic processes with different inner speed frames. This possibly acceptable differentiation can be reasoned to become eventually detectable to the sensing system of the living organism.

The recognition of this time difference in the biological unit, based on functionally related time versus equilibrium-related time, which itself must be regarded as a biological landmark initiating time and memory, in reality introduces the time definite scale in the time indefinite subatomic operations. This allows a fundamental extendable principle governing living organisms, the material biology, and specifically concerning ergogenetic processes in the animal life. As the difference can be *sensed* only in terms of related sequences of functions in ergogenesis, its absence can be noted at equilibrium with the status quo situation with plenitude, the happy time for the brainy *Homo sapiens*!

We can thus accept that early time dimension in the biologically living unit is recorded in the implicit memory as sequences of internal events. However, this principle of identification by sequences is in reality episodic, and if operating at the very early evolutionary times (which seems reasonably probable), we can presume that it could have initiated the implicit memory in the form of internal episodic modality. Or we can talk of internally related episodic memory functioning only as implicit memory in the beginning and externally related episodic memory with also semantic function memory at the later stages of evolved life with mind and consciousness.[14]

An effect expected from this biologically inherent time detection is that the realization of zero point, the time constancy of the uniform pace by the biological living unit, makes it also possible for the unit to recognize time sequences in the environment. Thus, a system of comparing and gauging becomes operable in life for measuring time in sequences of functions before consciousness. It is interesting to remember that the inner function time recorded by the implicit memory is not consciously realized as well as the rest time of equilibrium, which is also a sense of permanency without being consciously evident to the biological unit. In contrast, the external environmental changes also appearing in sequences can be sensed implicitly and can eventually become explicit and conscious when other evolutionary conditions are also propitious. In other words, the zero point of metabolic equilibrium in ergogenesis is practically the calibrating standard base in biochronogenesis for the real time changes of events in the environment. Thus, if we accept any value for the balanced equilibrium time, it can only be a sense of infinite constancy or permanency by the sensing system of the living creature. This sense of permanency will be further elaborated when we talk of the time titan in the next chapters.

BIOGNOSIS

Biocracy with biotropism and biodynamism, acting to prolong life obviously in an autocratic way, and for the benefit of the integral unique living entity also raises the questions of *how* and *why* of the mechanisms involved in organizing the biological processes especially if we consider the simplest unicellular biological life form. Regardless of these questions being answerable or not at the level of bioforce with inherent hypothetical autofeedback considered applicable in biology, and the reciprocally enhancing model we hypothesized, additional difficulties can be expected at the level of biophysical and biochemical reactions. This dilemma in reality reflects the *hard problem* of sensing the sense at its inception in the complex biological unit on its way to become a mindful living organism toward acquiring consciousness and will. These difficulties do not negate the analogy we observe between the accomplishment of the sensing-regulating system of the simplest unicellular living organism like an amoeba, for example, and the most elaborate soma and mind of a human being in terms of the essential universal task of SR of all living creatures including these two extremes. Therefore, this fundamental biological sensing-regulating act with its essence of analogical functional significance between the simplest and the most complex sensing-regulating systems suggest a biologicl recognition and regulation power that I have named biognosis.

In the primitive biocratic life, all we can believe to be present and to activate and direct biotropism and biodynamism for adjusting molecular reactions in the metabolic chains and to keep the balanced status is presumed to be the bioforce appropriated and integrated in the individual unit as an organizer acting to register the protocol on its own model, but with the limited biological ground, and commensurate with the organism's biological possibilities. The organizer apparently disposes of only two agents, the biotropist and the biodynamist ones. In spite of this simplistic system, observation leads to detect some endowed directional capability in securing survival and facilitating reproduction. In objective interpretation, the system has a program and a programmer, similar to a mindless computer-originated intelligence without an autonomous will, or a mindful-sensing organizing setup with it. This state of affairs invokes an early simple form of gnosis, which, strictly functionally, is doing the same as the mindful human instincts do unconsciously. This power of changeability of the sensing system from the simplest to the most complete form in the fundamentally two-way relay of detection and reflection is in fact a trait, the origin of expressivity inherent in the living organism, engendered by the bioforce. If we do consider the unicellular life as the biological unit that evolves and forms the multicellular life, we should similarly consider this state of affairs as forming the unit for biological Gnostic function, which can develop and ultimately expand into the state of consciousness as reflecting expressivity to environment along the evolutional time. The chain of events not being tangibly testable in morphologically evolutive animal examples is yet reasonably detectable and plausibly logical to incite thinking of a paleognosis to go on to a neognosis in the service of relaying the phenotype spacetime to the environmental spacetime. We can therefore confidently theorize that evolution of mind started with what can be called paleognosis and reached what can be labeled neognosis in the more complex state of Gnostic conscious realization, with biognosis being invariably the unfailing ground to support evolving differentiations.

Explanation of the regulatory organization of functions in biology by invoking behaviorism allowed easy steps in discussing and comparing differences in observed animal functionalities. But pondering on an unconscious animal tending to a task or on a conscious will in attention-intention processes finally behaving in a particular acting way with a purpose are reasons that form the nucleus of the *why* and the *how* questions about consciousness. The uniqueness of the purpose directed to invariably serve the autocrat in biocracy is intented only and irreversibly to benefit the biological unit. If anything, it is evidencing a selected unchangeable direction, indicating a Gnostic basis for that selection. I will use this term *biognosis* for the initiating cellular (single or multiple) organization states in biocratic life that will undergo the exacting adaptations conditioned by evolution for that purpose but a purpose that keeps moving forward on its own by its inherent bioforce and in an autocratic way. Discussions and expansions of the subject will follow in sections and chapters to come, but looking now into some examples of progressively more complex biological organizations will allow us to grasp the idea of biognosis and will pave the way for realizing stage differences justifying further expansions. So in short, in the philosophy of biocracy, biognosis should be regarded as the initiating Gnostic ground assuring autocracy.

Some functional evidences are needed to be shown briefly by living organisms witnessing the philosophy of biocracy in its purest autocratic manifestations, and I will start from the simplest going to the most complex examples. For the simplest example, we can take the prokaryote or the eukaryote genetic protein or the well-organized deoxyribonucleic acid, abbreviated as DNA, which demonstrates a clear biocracy showing both biotropism and biodynamism in one unique act of replication at time of bacterial or cell division.

Philosophically viewed, the primum motto of life in the gene is to keep alive by continuing the essence of its existence in time. At this stage of living, biotropism and biodynamism that are the sine qua non of autocracy regulate mitotic division showing the sole observable biological manifestation: the cellular division. Is there any mind behind that action? Neurobiology cannot be affirmative in any way, but philosophy is more permissive. For scientists affirmation needs scientific proof, but for philosophers, the effect implies a cause even if undetectable or only analogically permiscible as an evidence—as the aim justifies the means, and justification per se implies choice, and choice implies reason and relativity, binding the end to an initiation, the causal force. We sense regulation in invariable biological stages in constantly repeated processes, a regulation that justifies recognizing an inception of basic analogy to mind. Respecting this very fundamental fact of biological regularity in purposeful function, the word *biognosis* seems the right choice to explain it.

Next to DNA molecule and genes are three well-known organisms that show a gradual degree of physical complexity, demonstrating a proportionally more elaborate nervous system and functional specialization. The first one is *Entamoeba histolytica*, which is a unicellular well-known parasite for *Homo sapiens*. Shortly called *Amoeba*, this one cell organism has an elaborate system of biochemical factory yet some membrane fragility and shortcomings in its borderline capabilities to cope with environmental changes. This single cell structure reacts swiftly to external stimuli of mechanical, thermal, electrical, or chemical nature distinguishing between life sustaining or enhancing (nutritional) and consuming or endangering (lethal) stimuli and reacts to the stimuli being attracted or repulsed. The animal's biocracy is beyond doubt, and its biotropism is evident as is its biodynamism. Now again the same question may be posed, but a clearer answer can be given this time based on the known conditions of the experiment. Here too, the principle of biocracy is governing the animal's action through its biotropism and biodynamism to safeguard the animal's life. In this example, the animal's motions, biologically permitive somatic reaction, can be interpreted to be reactive responses that constitute a short adaptation attempt to gain energy or to prevent losing it any more than necessary. The action is directed and not random, being always a positive biotropism to gain energy and always a negative one to lose it. The interpretation by both neurobiologists and philosophers cannot neglect to include the aiming of the action at least in terms of evolutionary conditioning to secure survival.

The second example, more advanced in complexity is that of Caenorhabditis elegans, which is a roundworm, a multicellular organism, that at maturity should have about nine hundred cells including about some three hundred nerve cells located in an invariably constant distribution in all members of the species, making about eight thousand neuronal connections.[15] The complexity of multicellular organization appears to have endowed this more elaborate animal with a more specific functionality with responsiveness to external stimuli, which shows some gradation of response commensurate with the intensity of the stimulus indicating a more appropriate adaptational mechanism. Here, biocracy exerts the same principle of self-sustaining and is not using only a directed but also a graded biodynamism.

Again the consideration of a more elaborate reflected and measured response, invoking similarity with a correspondingly better organized neuromuscular reaction, cannot be overlooked. Is this a form of primitive brain or mind function? Are we justified to speak of bioawareness or bioconsciousness? Faced with this stepwise progressive complexity, one has a compelling tendency to use the analogy with the brain and mind suggested by the questions. Indeed, in terms of proportionality, aiming, and causality, there seems to be enough evidence to justify the analogy.

The third example is that of Aplysia californica, a sea snail with twenty thousand neurons.[15] This animal shows something more in its biotropism, which indicates frankly observable short and long term adaptations. The animal reacts to repeated external stimulations by decreasing its withdrawal responses. In terms of biocracy and particularly autocracy, the animal controls its biodynamism by controlling its biotropism. The reason behind this control is in a finer distinction served with the more elaborate nervous system separating a gradually but surely adaptable environmental change from a sudden and apparently nonadaptable one that would dictate complete withdrawal. How would this function be possible if no measuring mechanism that supports comparison, and thereby some sort of memory assuring sequencing, were not available to the animal. Here the principle of biocracy firmly controls biotropism, which shows itself vividly with the essential element of early implicit memory based on more complex neuronal organization. This now allows the living organism to judge the environment and only use the biodynamic energy needed for coping with the environmental change and keeping the equilibrium for sustaining life. This is simple in appearance but has complex neuronal circuitry and function behind it. It reminds us an organized brain with a reflective mind. Can this be regarded as bioawareness or bioconsciousness or by any other name for gauged sensing and graded responding system functions? We can judge how to find answers to these questions if we compare these examples to some more complex ones in animals with more organized nervous systems and brains.

An excellent description of animal lives considering how animals direct their daily living, showing undeniable degrees of reasoning and acting can be found in the work of Griffin.[16] Examples from that work are clear evidences that, were it not for the dominant behaviorism in the mid-twentieth century, definite ideas on animal minding could have been crystallized much earlier. These examples can be categorized in terms of elemental psychological characteristics evidenced by the actions observed. In this categorization, the constancy observed is that of the unfailing biological force behind every action for safe SR in an autocratic way. However, in all these examples, we can see nuances that hint to elaborate directives and handlings that demonstrate memory, feeling, gauging, planning, and perfecting adequate change with time if needed. In other words, indications of choosing and intentional acting are quite evident, and frankly, some awareness and perceptual consciousness can be clearly observed. I rather briefly describe instances of intentionality and planning in line with survival and leave aside examples of biological impulses with reproduction.

A very interesting example of communicative intention and advice is that shown by scout honey bees perfecting special dancing to show the direction and location at distances they have checked for new sites for honey combs to be built. This has been studied and found to demonstrate an accurate means for informing the community of bees to find the spotted location. Another communication example showing specificity is that of the queen bees showing at least two different types of distinct sounds for two different purposes.

In terms of calling for alarming other animals of predators being noted, birds appear to produce special sounds with specific meaning. However, in the communication to convey the meaning, the three different alarm cries of vervet monkeys, each distinct from the other, for leopard, martial eagle, and python, compose an act of clear advising intention. The interpretation of each alarm call and subsequent specific escaping of vervet monkeys to the highest branches of trees for leopard, lowest crevasses in the earth for martial eagle, and standing up and looking for python, leave no doubt about the specification of meaning for each call. Communicating the meaning and the urgency by these calls is vital and indicates a semantic capacity and distinction by the brain. This level of perceptive, judgmental, and executive acting certainly has a mind behind it, and it is unjust to deny the role of an efficient perceptive function with the emotive subjective reaction manifested by their minds.

In terms of group actions in a planed cooperation to reach a collective goal, there are different examples of group hunting of predators whose tactic in conducting the hunt leaves no doubt about reflected intentional aiming, organization, and acting. A very clear example is the group hunting of lions in which the land conditions and visibility judged in a projected way by the predators estimating as well the visibility available to the prey and possible path for it to escape are clearly observable. In this organization, the predators divide the assignment of tasks among themselves to the attacking group and reserved expectantly prepared group members according to their fitness or hierarchies—the runners chasing the prey, and the observers watchfully awaiting in the appropriate positions to catch the escaping prey. This maneuvering indicates clearly a high-level of intentionality and preplanning that demands mental work in a well-functional mind.

Many more examples could be cited, but in all, the problem is not lack of evidence for mind but the overlooked abundance of it, and the beauty of mind capabilities commensurate with the living necessities and phenotypic means, no matter what somatic size is the animal, from the very primitive to the very elaborate of the nervous system functions and for the constant aim of surviving and reproducing. There is a signature of biocracy on the screenplay of evolution in this motion picture. The force is the bioforce inherent in biocracy, the supreme order is autocracy for SR, and the degree of organization is dependent on a form of mind. This mind is unquestionably present in all living creatures but commensurate with soma differentiation the mind's functional scope varies; it can have from the simplest functionality to the most developed one as in the *Homo sapiens*. The intent is to last in time in both quantum physics'scale of matter-energy reciprocal feedbacking survival or in phenotype-genotype's continual evolving survival in biology. *Philosophically, it seems as if creation has condensed genesis in both cosmogenesis and biogenesis, through a uniform principle, and in phasic renovations forever.*

Now if we try to accord structural definitions to the nervous system functional control in the examples we have reviewed, we can surely admit the biognosis to be present in all, covering from what we can call paleognosis in the first example to neognosis in the last and can plausibly talk of bioawareness and bioconsciousness in the intermediary examples. However, it will not be possible to precisely separate these two words to be applied to precisely distinct neuroanatomic and neurophysiologic structures and functions in all intermediary examples that are too numerous and too varied. Nevertheless, a crude separation line can be considered based on five levels of functional reactivity:

1. Unicellular-pluricellular reactivity and automaticity: paleognosis
2. Sublimbic reflexive-perceptive functions: bioawareness
3. Limbic-sublimbic perceptive-reactive functions: bioconsciousness
4. Limbic-supralimbic perceptive-intentive functions: bioconsciousness to psychognosis
5. Supralimbic-Neocortical associative functions, attention-intention conscious acts of psychognosis-idiognosis

The first is simple cellular reactivity denoting the most archaic origin of reactivity, here called paleognosis. The second is the fundamental nervous system reflex reactions with neuromuscular coordination and aimed locomotion to represent bioawareness. The third is the same basic somatic structural type with perceptive-reactive functionality and early implicit memory with early limbic and supralimbic functions to include bioconsciousness. The fourth is the complex nervous system structure with perceptive-intentive controlled actions with developed implicit memory and clearly distinct limbic and some supralimbic functions to still match bioconsciousness but also what I have called psychognosis that will come up for further discussion in chapters 5 and 6. The fifth level represents neocortical logic and abstractive functions that include all noetic capacities and uses full literary communication and semantic memory. This level includes psychognosis and what is named idiognosis in this book that will be detailed in chapter 5 and thereafter as needed.

These artificially staging through the continuity of increasing complexity of neuroanatomy and neurodynamism, we should not forget, serve the relay between the spacetime of the bio-organic unit and that of its environment. The accomplished task in final analysis at the ultimate stage is serving the expressivity trait, allowing the discharge of emotionally originated energy serving the old fundament of survival and reproduction. Up to this stage the expressivity uses implicit, episodic, and limited explicit memories essentially in concrete presentation of body language and action. The ultimate stage of transfer from this concrete material form to abstract will be the fifth and final stage of the prelingual and lingual (idiognosis) minds whose manifestations will be covered as needed in chapters 6 to 11.

BIOPSYCHONS

In the philosophy of biocracy with its hall mark of autocracy, biognosis as seen in the examples we reviewed, represents the unitary basic mind status that runs throughout the autonomy engendered by the bioforce from the very initial somatic unit of the living material for ensuring SR. This same sensing-acting singularity is then transferred to all subsequent replications. Contrary to this unique functional character and significance included in all creatures with biognosis, somatic differentiation introduces new neurodynamic specifications to the living soma with biognosis, thus, all living elements forming extraordinarily varied species enjoying some form of soma with biognosis can be called by a generic term reflecting biognosis. The term "biopsychon" facilitates speaking and discussing these living creatures as often used in this book and also serves as a reminder of the observable autocratic self securing and replicating principles inherent in life. To be able to converse on postulates of neurodynamic features of the five levels of mindedness discussed earlier, I will use the word *bion* for the functionally bound unit of biology invariably enjoying biognosis and the word *psychon* for the added unit of mind. Thus every gene can be potentially a biopsychon as can all other creatures regardless of somatic size and shape. These life forms, the biopsychons of single or multiple cells, have the autonomy based on self-conservation and replication, modeled by the bioforce in quantum biology, guided by the inherent biognosis, using from biotropism and biodynamism to sophisticated mental undertaking, art, literature, music, mathematics, etc., still on the same biocratic principle with autocracy (reshaped in human society) and on the same ground of biognosis with much refinement.

Reconsidering the fundamental significance of biognosis as a sine qua non for SR and continued sensory-neural progress, the hypothetical ratio of N/S that we have seen in the last chapter now can be confirmed to keep remaining above 1, and the changing capacity of the cerebral functions in any and all mind expansions in biopsychons of all types can never normally fall below that level. This fact further demonstrates the elusive character of the affective sensation basically related to survival and reproduction that is autonomously regulated by biognosis. This brings up again the question of the state of mind called *qualia*, which is essentially the sensing of usually some pleasant and occasionally unpleasant reminiscence of a passed relived moment, automatically produced and not recalled by will. This much of such an intense, often happy feeling or occasionally nostalgically tinted one, that is also of implicit memory origin as is biognosis, is yet not capable to alter in any way the basic biognostic regulation. Biognosis in the human mind is not conscious except when manifested in the emotional accompaniments with affective sides of the human biopsychon living. *Qualia* is consciously felt but cannot be categorized in a measurable way; it is a gut feeling as nicknamed by Gazzaniga[17], a feeling in a short ephemeral instance and qualified as psychognostic in my explanations to come.

In today's estimates of materialism and functionalism, neurobiological findings of the brain have come to establish clearer relationship between the structural parts and their corresponding functions so that analogy of function generally reflects similarity of anatomy and structures and vice versa. In this regard, one can assume that the observed results in biopsychon examples enumerated in animal acting discussed earlier present a significant analogy to both brain and mind and suggest a primitive mind to be the regulator of these observed responses. The biognostic state as the foundation for that mind, and bioawareness, bioconsciousness, psychognostic, and idiognostic functions as added specifications to it will be discussed in future chapters.

Biosensing and bioregulating capability with varying degrees of these added specifications could be firmly proved to change from unconscious to conscious. Indeed, as we can believe the work of Libet that clearly shows the initiation of an attention-intention mental act to be always unconscious,[18] we will realize that before our conscious mind becomes aware, what has been considered our unconscious mind in fact encompasses paleognosis, biognosis, bioawareness, bioconsciousness, and psychognosis. That mind has been taking care of our animal life all along and again takes care of our decision making presently as it always did before we realized it with our will. I will discuss this interesting apparently enigmatic autonomy of our sensing-acting system more fully in chapter 8.

THE FORCE OF MIND

One can argue that if a force of mind is shown to be real, as demonstrated by Schwartz and Begley by the will in reconditioning the mind's conduct and correcting obsessive-compulsive disorders (OCD),[4] no reason would exist to deny that a similar force did cause each of the events in observations on biopsychons listed earlier. This fact leads us to think more fundamentally to search for the organic original inclusion of this force showing the self-sustaining function in the organic biological material evolving to form the mind's will. This force's effect is in reality a precise replication of the self-renovating function of the theoretical positive feedback principle in the bioforce constituents (particle, wave, expansivity, and their interrelations) at the subatomic level. How this function was established in the organic material to engender and conduct metabolism, nutritional, and reproductional conservation all condensed in the genetic material is the puzzle of biophysics, biochemistry, and biology to remain a mystery for quite a long time. But some facts are available for examination to shed light on the main issues.

To begin, we can remind ourselves of the subjects of our conscious will, our unquestionable autonomous SR principle, our adamant self-sustaining biotropism and dynamism of biocracy down to focus on the nucleus of these uniformly unfailing biological principles, to the genetic material, and even further to the subgenetic level, to bioforce. This unknown, unexplained drive as a real force behind the production of organic molecules to reach autonomy, for example, as the causal drive of DNA replication is bound to organic material substrate of the same atomical element forming the physical world but functioning in structural self-sustaining autonomies as live matter. These amino acid materials have been formed experimentally similar to the forming blocks of life, in appropriate admixtures of elemental materials, electron current and adequate temperature called primitive soup.[19] There, the only forces provided were thermal and electromagnetic of classical physics and seemingly the inapparent physical forces of gravity and antigravity in any given spacetime in equilibrium, possibly also at microcosmic quantum scale.[20]

The problem can be approached considering two possibilities, each including alternative choices. First, the possibility that the hypothetical self-sustaining force as exemplified in the photon is indeed the bioforce so far considered the origin of the biological material with its self-sustaining autonomy. The alternatives in this possibility are that this force can either use any of the physic's elementary matter in the Mendeleev's periodic table of elements to form organic material with self-replicative autonomy or only those selected elements compatible for the purpose. Second, the possibility that the bioforce in organic material is different from that in the light but similar in function by securing self-sustaining autonomy and forming biological organic material with elemental atoms in the same way as with the first possibility. Although both possibilities may cast some hypothetical plausibility at the first glance, the first possibility has the simple basic advantage that it excludes two major elaborate structural setups (new force and atomic compatibility) for one identical self-sustaining functional effect realized by just one such setup of force, the bioforce, in macro and microcosm. This basic principle of logic seems indeed to work best in creation in which the function governs morphology, notably in biological adaptation with evolution. For the bioforce, it is also the functional expansivity characteristic of it that seems to be the actual modality evidenced in material and nonmaterial biology.

Further points in favor of the first possibility consideration and its second alternative applicability can be evidenced reviewing the story of the origin of creation of the bioforce in more detail. The laboratory recreation of the big bang by mini bangs, as is said for the replicas of big bangs from Relativistic Heavy Ion Collider (RHIC) observations using heavy atoms like gold,[7] has provided clear measurable scenario timing the big bang events in infinitesimally minimized material scale. It is shown clearly that from zero second to one microsecond the explosion temperature cools from initially ten to power thirty-two degrees Celsius, an inimaginably infernal heat, to six trillion degrees Celsius, and the liberated quarks and gluons of stupendous haphazard jumping speed start slowing to be bound into protons and neutrons ten microseconds after explosion. About one hundred seconds following the explosion, elemental nucleosynthesis takes place like that of helium atom from hydrogen but yet at the extremely hot temperature of one billion degrees Celsius. In contrast to this short time, it takes 380,000 years for stable conditions to be established for the first neutral atoms and the stable positron-electron to be realized in the photon.[6]

From then on, the photonic radiation is practically able to survive in space indefinitely, radiating permanently if not intercepted by obstacles. This permanent stability and ultimate speed evidencing the firm interrelation of the particulate and wave forms of the light constitutes a specific functionality that is unique in engendering botanical life. Furthermore, the origin of life on earth is not proved to be other than from the sunlight and the well-known photosynthesis theory.

Regarding the second alternative of specifically compatible elemental atoms making the organic biological material, the theory seems plausible but needs confirmation on the basis that a selected number of elements are used, that heavy atoms and radioactively decaying ones are excluded, and that most elements used are of rather smaller atomic number more closely following the hydrogen atom in the periodic table of elements. Nevertheless, adaptive choices are not restricted in the face of environmental conditions. For example, in the initial earth's conditions, astronomically high mantle temperature produced the propitious ground for the change in sulfuric atom, which eventually gave the hyperthermophilic life with a nonoxygen dependency,[21] and subsequently, the same occurred with carbon atom in oxygen containing atmosphere and oxygen dependent life.[22] These actual realities suggest appropriate carbon atom compatibility being pivotal in the biochemistry of the presenting material biology. In this connection, it is of interest to know that the carbon monofoam with ferromagnetic properties can be formed under the effect of laser light.[6]

In spite of all debatable points considered, once we are faced with the autonomy of the formed organic material, we have to explain that autonomous will that is the essence of life. The crux of the matter therefore seems to lie in the force that must be both legislative and executive, so to speak, organizing and executing in line with self-sustaining, but with limitation to use only materials able to carry this dual function making the biological existence a possibility. In other words, we are faced with a compatibility problem choosing the right material and are forced to consider a cause-effect relationship with a cause for both dynamism and direction and a matter securing this causality and the possibility of autonomy and perpetuation to coexist. The energy, being potentially invariable and permanent, but not the matter that is changeable and decaying, necessarily axes the biological autonomy on replication. We can remark the same principle of positive autofeedback effect for the energy to be permanent but in interrupted quanta (phenotypes) as the matter cannot be permanent. We are led to believe that the pattern of the self-energizing feedback of the force is repetitively operational in its products through metabolism, thus creating the elementary unit of the self-perpetuating biological material and its autonomy. This thinking binds the bioforce to biocracy extending biological somatic and mental life in parallel, but the progressive course is necessarily phasic in the material biology.

It replicates in essence the quantum physics where the energy-mass interaction is also basically phasic but in significantly smaller scale. Dissimilarity to be noted is in the immortality principle in quantum-regulated light that is inseparable from its particulate carrier but is separable from the biological material carrier in order to secure perpetuation through indefinite number of phenotypes. Thus, the material biology, can be said, is facilitating phasic renovation for what can be called the immaterial biology as a permanent principle of immortal autonomy through reduction in the transferable code by meiosis. Accordingly, the autonomy in biocracy may be incriminated to be intrinsic in the bioforce and to follow antigravity, with the one-way directional conduct that allows material changeability in expansion to the limit of necessary decay, but does not affect the immaterial autonomy with the bioforce, which can continue similar to its physical counterpart photonic unlimited continuation in phasic gametogenesis.

Another facet of the problem of energy in biology including the force of mind is the apparent applicability of the principle of conservation of energy by biotropism and biodynamism that is more openly observable. The effects of these mechanisms lead to usage of energy and its conservation in accord with optimum aptitude for living and reproduction securing the perpetuation of living in time. It is also remarkable that conservation of energy in biology, although applicable in principle, is not an exact copy of what occurs in physics. Although remarkably homologous to it, it has two clearly different modalities worth to be mentioned. One is that bioforce stops acting in biology in the phenotype long before atomic decay occurs. The other is that gametogenesis in biology appears to mimic decaying in the reducing process of phenotypic presentation to the genotypic one and more grossly or obviously so in the example of actual death of some casts of arthropod males at gametogenesis with copulation. Thus, decaying in physics to allow transformation is always total and complete at the molecular level, but in biology, it is both at the molecular level with organic decay after life and at the living unitary biodynamic level in life. These facts indicate that transformation in biology is normally only a pseudodecaying form by gametogenesis in living time, perpetuating the unitary biodynamic living element in uninterrupted continuum of phasic ontogenetic-phylogenetic life.

The missing link between the force and its directive in biocracy, the autocratic will, appears to be simply the DNA code order formulating biotropism and biodynamism and is the point where determinism can be either accepted or rejected. It will be inappropriate to predict any definite value for either of the considerations before the exact nature of the bioforce is known. However, both can have intrinsic values that may not appear extrinsically equal from an observing eye with biased scientific or philosophical viewing. This is not to say that we are back at the classic dualistic point for many scientific findings have contributed to bring us to the present primordial crux. I will take the position in saying that if the nature of a force includes an intrinsic effect to feedback itself adequately, keeping constantly in equilibrium at the same value in a changing physical environment, only that force alone can be permanent. This principle of self-sustaining in a one and unique energy type by the bioforce can be only suspected, believing the photonic constitution and function. However, the principle is also active in psychological autofeedback effects and by autosuggestion that demonstrates the power of the force of mind.[4] On that basis, the permanency of the hypothetical bioforce in biology, the bioquantum type of energy, can find explanation through the changing physical environment without needing any other explanation for its manifestations in the different forms it may take in its carriers. On the other hand, if the force lacks such intrinsic effect, it would certainly need an acting regulator to keep its value constantly unchanged in variable physical environments to confer permanency to it, and this consideration adds another problem and demands another solution, the very basic theme of all religions.

Looking at the bioforce through an explanation of autonomic equilibration of its value for permanency or hypothesizing the need for a regulator to direct its equilibration and secure permanency is to solve the problem with an intrinsic power, an included contingent key solution, or change it to one with an extrinsic schematic solution, doubling the puzzle. The only logically sound solution is surely the first choice. All quantum data indicate that possibly we can dispense of searching for an independent regulator as the regulator is already in the energy itself initiated at the big bang event and continued in the modalities of energy interplays, namely, between the electron and the positron in the photon conditioning wavelength versus frequency in bonding the energy partitioning. This takes us to the alternative of a beyond-big-bang possibility to be also considered,[23, 24] and not without the very probable possibility of the prebiotic organic molecules recently discovered in the astronomical observations backed with molecular spectral studies.[25] But to keep things simple, this subject will be further discussed when we reach the last chapter.

REFERENCES FOR CHAPTER TWO

1. Marcus GF. 2004. *The Birth of the Mind.* Basic Books. Perseus Books Company. New York, NY.
2. Veneziano G. 2004. The Myth of the Beginning of Time. *Scientific American.* 290(5):54-65.
3. Motz L. 1987. Jefferson Hane Weaver (Ed). *The World of Physics. A Small Library of the Literature of Physics from Antiquity to the Present.* Pp 17-27. Simon and Schuster, New York.

4. Schwartz JM, Bagley S. 2003. *The Mind and the Brain*. Regan Books. Harper Collins Publishers Inc. New York, NY.
5. Frank A. 2005. Seeing the Dawn of Time. *Astronomy*. August. Pp 34-39.
6. Riordan M, Zajc WA. 2006. The First Few Microseconds. *Scientific American*. 294(5):34-41.
7. Shapiro E, Benenson Y. 2006. Bringing DNA Computers to life. *Scientific American*. 294(5): 45-51.
8. Seeman N. 2004. Nanotechnology and the Double Helix. *Scientific American*. 290(6):64-75.
9. Bonner JT. 2002. *The Ideas of Biology*. 109-141. Dover Publications, inc. Minneola, NY.
10. Cairns-Smith AG. *Evolving the Mind. On the Nature of Matter and the Origin of Consciousness*. Pp127. Cambridge University Press.
11. Phipps C. 2006. A New Dawn for Cosmology. What Is Enlightenment. *Enlightenment*. 33:42-48.
12. Hawking S. 1998. *A Brief History of Time*. 10th Anniversary Edition. Bantam Books. New York, NY.
13. Oshidari J. 1992. *An Encyclopedia of Zoroastrianism*. Nashr-e Markaz, Tehran, Iran. (text in Persian).
14. Baddeley A,Conway M, Aggleton J. (Eds). 2002. *Epizodic Memory: New Directions in Research*. The Royal Society. Oxford Universeity Press.
15. Zeman A. 2002. *Consciousness: A User's Guide*. Pp 47, 252-255, Yale University Press Publications. New Haven and London.
16. Griffin DR. 2001. *Animal Minds. Beyond Cognition to Consciousness*. The University of Chicago Press. Chicago and London.
17. Gazzaniga MS. 1995. Gut Thinking. *Natural History*. Vol 2:68-71.
18. Libet B. 2004. *Mind Time. The Temporal Factor in Consciousness*. Harvard University Press.
19. Smith JM, Szathmary E. 2001. *The Major Transitions in Evolution*. 28-32. Oxford University Press Inc. Oxford. NY.
20. Bousso R, Polchinski J. 2004. The String Theory Landscape. *Scientific American*. 291(3):79-87.
21. Longstaff A. 2005. Quest for a Living Universe. *Astronomy*. April. Pp28-34.
22. Kasting JF. 2004. When Methane Made Climate. *Scientific American*. 291(1):78-85.
23. Shapiro R. 2007. A Simpler Origin for Life. *Scientific American*. 296(6): 46-53.
24. Choi C Q. 2007. New Beginnings. *Scientific American*.297(4): 26-29.
25. Dorminey B. 2008. Did Molecules from Space Seed Life in the Cosmos? *Astronomy*. 36(4) : 50-55.

CHAPTER THREE

PILLARS OF THE MIRACLE

SITUS SOLIVAGUS

Situs solivagus is a Latin expression meaning "posed solo" or "solitary being" and is a good symbolizing epithet for the hypothetical early biopsychon. The pristine biopsychon, if anything, is above all a separated living creature from all else living and nonliving beings. The mere fact of separation, putting a limit between it and its surroundings, poses the necessity of managing its small spacetime frame wherein it must exist. Therefore, immediately two factors come face to face into actual function in this unit of life; its potentially free independent energy usage faced with its potentially limited dependent energy supplies realized between its *self* and its *environment*. Its biodynamism must match its biotropism, keeping with its biocratic autocracy for the balance. The principle of SR then is ipso facto in action and keeps the organism along the extent of time that biological conditional allowances permit. At this stage of solo being, the bion that can be simply called autos to allude to its solo autocratic nature and somewhat in connection with the meaning used by Esther Harding[1] is governed by the innate autocracy, which keeps the independent *situs solivagus* with autonomous biologically regulated mechanisms, equilibrating energy gain and expenditure. The gain must provide for both survival and provision for a new spacetime frame beyond the one that is already reaching its biological dimensional limits, necessitating a replication that would allow that change. The principle of renovation repeats itself in physics and biology—from matter to energy and energy to matter, and from genotype to genotype, and phenotype biopsychon to biopsychon—and thus keeps going indefinitely. Is there any need for consciousness? The answer is certainly no. But is there any ground potential for this extravagance to be realized? The answer is definitely yes, and the miracle must be expected.

The miracle to occur and that indeed occurred was the appearance of a no contact communicating system, a prewireless relay, between the biopsychon and its environment beyond the limited contacting system, the surface boundary of the organism. In fact, the primitive contact sense organs from the ectodermal (skin derivatives) and endodermal membranes (gustative/digestive derivatives) can only communicate immediate contact changes from the environment, and the necessity for receivers informing the animal of more distant changes is paramount and pressing. Evolution worked inexorably, forming and perfecting olfaction, audition, and vision (principally from the ectodermal lining) and ultimately granting the miracle of wireless communication: mind, consciousness, and speech.

In the production of mind and consciousness, basic logic dictates that the building block should be the sensing capacity of the biopsychon on which further structuring to increase its functional power shall be constructed. The axis for conduction of processes adding the structural completions must necessarily be oriented to support the biopsychon, the autos in its biocracy according to the pinnacled autocracy and with its biotropic and biodynamic functions. This axis should guide as a directive to answer the exigency of evolutional rules, otherwise the species shall fail in evolution and ultimately extinguish. The lesson from evolution, teaching our thinking empirically to see directionality in the evolutionary course, demonstrates secured progressive improvement in the expressivity of biopsychons from sensing to conscious dealing. The sensing capacity as the fundamental factor in that expressivity is its observable evidence. All other biological functions for keeping metabolic balance to sustain life show also refinements through evolution, but refinement in expressivity, evidenced by progress from material to immaterial, psychoproductive sensing, percepting, imagining is paramount in supporting directionality logic of evolution. In other words, a particularly natural selection seems to favor the sensory over the somatic refinement suggesting special directionality toward metaphysical abstraction. This component of evolutionary improvement seems to be additional to the natural selection and seems to be innate in biology in the immateriality essence of the bioforce being maximally energy and minimally matter. This understanding also tends to suggest the basic Lamarckian postulate along Darwinian rules.

On a purely theoretical basis, a hypothetical self-sustaining force vector essentially of nonmaterial type, as the bioforce along its path in time in our thought experiment, if not subjected to decaying with energy loss, may acquire hypothetical increment the faster it can theoretically go, adding the incremental effect of immateriality in relation to time (relative energy increase) with its theoretically expanding spacetime. Such possibility, though hypothetical, can be theorized that would imply an innate potential for energy increase by favorable enhancing effects of the environment spacetime conditioning (antigravity effect). This hypothetical possibility based on mere time progress, helping specialization with more propitious functionality and less energy expenditure, if actually realized, will add a second positive feedback to the one already innately present in the biology's bioforce; the easier the way to go, the faster the course, the greater distance gone, and the greater the incremental gain to be turned into actual result in terms of gaining time and adding to the sensing power. Thus, the potential ground for the increment of the hypothetical force vector implying sensing seems to be real, and an automatically increasing sensing power with time seems an inherent possibility in biology, other conditions permitting. Thus, the ground for increasing neurofunctional capacity to lead from material to nonmaterial biology, evidencing the creation of memory seems naturally included in the bioforce.

At the early stage of evolution, we can only visualize and accept a focus of neuronal centralization in the biopsychon that regulates autocratic living and mechanisms for SR. Examples of earlier rudimentary nervous system like that of *Caenorhabditis elegans* described in the past chapter are of this category. We cannot yet talk of any nervous system beyond a primitive centralized neuronal organization for coordinating the autocratic processes at this stage. The essential work of this centralized neuronal group is serving the autocrat biopsychon and is reinforcing itself along a constantly progressive direction securing the potential for possible second positive feedbacks. I will restrict the word *autos* to this self-oriented agent, being formed as an autocrat, in whatever proto-mind can be considered at this level of presumed archaic sensing-acting that is biognosis based on its very root of paleognostic development.

In this pristine elementary biopsychon, the fact that is definite is the animal as a single unit separate from all other beings surrounding it, in a stage of real uniqueness, a justified physical and biological self. This self, or interchangeably now called autos, is an autocrat as defined by its biocracy with its inherent living mechanisms. It has no elaborate conscious mind, but its autocracy serves its purposes as a live organic being that recuperates its damages and regains its energy losses and replicates by renewing its unique spacetime frame. All this is the attribute of biocracy in this autocratic self that in reality has been part of the bioforce from the start. The pattern of events indicates a basic model on which more elaborate forms can be built through the same principles. The analogy of governing principles between this simple primitive self, Id, or autos to that of today's variably defined self(s) and between this clear-cut single autocratic behavior and today's modularly complex mind acting is astonishingly similar if analyzed without getting lost in contingent details.

Creation of self in its psychological definitions does not pose an undecipherable problem in appearance until the self is realized in the biopsychon as self along with autos. With autocratic biotropism and biodynamism, the self of the autocrat is truly identical to its autos and is the same in its presumed structure and function and is traditionally unconscious; but in a fully conscious creature, the self is not identical to autos, the very autos forming the most primordial inner being. We can say that the self in the evolved conscious biopsychon can be regarded generally as similar to autos in the capacity of representing autos, but certainly not as identical to it. Consciousness has complicated the problem in functional value changes of autos and self so that what can be regarded as self in the conscious biopsychon has functional similarities with autos but also differences as opposed to autos. Interpretations of self in the conscious context also has introduced variable opinions to even deny such entity as a unique self and to propose multitudes of self(s).[2-5]

Indeed, the naturally appearing course of phylogeny of *situs solivagus*, evolving from the primitive status of the simple unicellular bion with one unique autocratic self, the autos, to that complex one of man, is the most natural continuation and ending that could be expected in the evolved and complicated nervous system. It is then in the sphere of consciousness that the puzzle of self comes out in philosophical, psychological, and neuroscientific thinking. What was the foundation of the autocratic autos, now transposed through autobiographical memory to the thinking biopsychon, has created the belief of an undisputable unity on the one hand (the same archaic autos) and multiple possible functional values of the same unity representing self(s) on the other. The interconnection through the autobiographical memory has preserved the unity and autonomy sensed in oneself, the autos, but also in associated self(s). This pattern of processes following an assumed evolutionarily related phylogenetic course that I have alluded to earlier seems naturally acceptable. The genesis of self-reflecting cognitive realization of the being of biopsychon by its own recognizing neuroperception seems to be the initial step in the stepwise evolution of mind through neurobiology to ultimately end in the complexity of consciousness.

It is fundamental, however, to consider and discuss Gnostic powers of the mind before consciousness, if such power, which seems incontestable, can be called Gnostic. Because of lack of tangible evidences reducing all arguments to hypothetical levels, this consideration and any ensuing reflection should be regarded possibilities only. But I should remark that I continue to use the word *conscious* in its usually understood meaning until in chapter 5 and thereafter when given further explanations will allow me using the more appropriate words specifying different levels in consciousness states. The word *gnosis*, unlike its general meaning for the mental capacity of cognition, which I have considered in this writing, is often used in a metaphysical meaning assuming full knowledge of one's own inner life and separate outer life. Gnosis of self alone, or ipsism, simply cannot stand to be possible. Logically, Gnostic vision must have views on one's own physical and mental being, the inside world simultaneously with views on the outside world, otherwise the theoretical unification of the two, which is a logical impossibility, denies gnosis altogether although such unification may be psychologically allowed in identifying projections. In descriptive terms, the gnosis cannot be only autognosis or heterognosis, but must be both and must include abstract materials that should consider abstract inner component for autognosis in addition to the concrete one; thus, holognosis may be considered as an all inclusive defining word. Holognosis is used here to provide a generic name for all Gnostic categories of concepts, concrete physical and abstract metaphysical, which can include the two essential types, autognosis for self and heterognosis for other than self (concrete and abstract) and possibly for all other legitimate additions fitting in that group.

AUTO-HETERO-GNOSIS

The reality of a unique self for us thinking creatures—seeing every one of our actions and reactions as reflection of our personal acting supposed to be based on our will, not solicited or provoked, and consciously, and for no one else but us—is natural and not questioned. Our autobiographical memory provides us with the belief that our unique self at each moment and extended moments in continuity is the ever present actor that is no one else but us. But when this memory is disrupted, for short or long time, the self in us is affected momentarily or for good. Examples of fugues and multiple personalities reflect such disruption to variable degrees. The very continuity that saves the uninterrupted autobiographical memory in us gives us the feeling of selfhood. This state of sensing can be regarded as the basic ground for *qualia* and is possibly the most archaic *qualia* in the experience of the more elaborate neuronal complexity in the sensing biopsychon, a sense of oneself at each instant in time without any other definition. There is no consciousness as such in this sensing biopsychon, but there is the foundation of some stereotactic sensing self, some prototype Gnostic state that could be in the category of bioawareness or bioconsciousness described earlier. The closest idea that I may interpret to probably invoke this description is that of a protoself,[6] but I would insist on uninterrupted time element, its continuity, adding to the basic sense of recapitulation of self to end in making the self a sine qua non in the more elaborate biopsychon.

In my view, the purely biological pristine self, or in reality autos, reveals the very essence of the bioforce acting in biology with its autonomous self-sustaining autofeedback for securing permanency. This pristine self is modeled on the primordial pattern of the autonomously self-sustained matter and energy. In the primitive biopsychon, the nucleus of biocracy, the autos is endowed with this power with its inherent time dimension. Its manifestation is summarized in the repeatable invariable chain of autocratic biotropism-biodynamism leading to the continuation of life. The principle (in the bioforce) and its mechanisms form a reciprocal pattern in biology between two factors (sensing and acting/psyche and soma) necessitating and adjusting one another for a purpose. This is analogous to the two components in the photon that conjoin to express the light in the cosmos and the analogical abstraction in the mind as no-mass energy, one in the material world and the other in the nonmaterial biology. Studies of plant and animal biology provides multiple such examples with equilibrating feedbacks demonstrating both reciprocity and purpose in maintaining the balance and allowing the progress to continue with time, carrying the energy effect as life. The element of time is the essential ground that must be kept uninterrupted, and its continuity that is secured by the repetition of the reciprocal processes, as in all various biological feedbacks, is basis for autobiography generating the self. This constant identical time effect in the primitive biopsychon is its own *implicit memory self*, which equals autos in autognosis and needs no further assertion except by natural contrast to heterognosis. If any other recognition is needed for its inner self, it cannot take place without recognition of the sequenced time itself. It is the identical repetition of the inner biological time sequencing that episodically occurs in biological balances that can provide and reinforce a self-recognition in the unitary domain of autos at some stage of evolution with more elaborate nervous system functions.

The philosophy behind the biological meaning of memory as a whole is, in reality, the most crucial in terms of its value in abstraction and metaphysical realms of the nonmaterial biological entity that is the mind. If in fact we hypothesize that a memory registration can represent a unit of neurodynamic energy spent for time registering an event for a recall process, we are saying that the memory is in reality coursing through time to the extent of making a permanently disponible imprint that could be reduplicated. This hypothetical unit of memory in the simplest form that is the saving of the imprint in the hypothetical unit of time will express the primordial meaning that stands for the forming blocks of abstractive knowledge.

Thus, realization of the very initial gnosis prior to consciousness can be seen and hypothetically explained to depend on the nucleus of memory with the self-enhancing effects of two factors elaborating the memory—one being the automatic stereotyped repetition of life-sustaining processes in replicating sameness in maintaining life and the other being the uninterrupted continuous time allowing this invariable sequencing to be kept continually connected and to be facilitated for automatic recall in the developing central nervous system theoretically ad infinitum. In this pattern of invariable re-experiencing processes, the first factor (the experience) proceeds in its automatic consecutive steps, permitted by the second factor (the sensor), until its unconfirmed nature to its sensing organizer eventually becomes confirmed giving the organizer, the autos, self-recognition. This realization occurs by repeated stereotyped metabolic processes (first factor effects in time) mainly in similar consecutive sequences in the continuous time ground (second factor effect using n number of memory units) forming the nucleus of the episodic memory. Thus, the automatic unconfirmed autos, we have to accept, shall become a confirmed realized autonomous power, reconfirmed to itself simultaneously but separate from all nonautos recognized entities.

This empirically reconstructed state of the initiation of nascent gnosis can be regarded inseparable from the initial memory formation and can be formulated to ultimately show the relationship of the two factors in a schematic medium formulated as:

1. GNOSIS = SENSING THROUGH TIME

This simple formulated form of gnosis must be considered in the broadest meaning of cognition including the unconscious to conscious cognitive states. It can be regarded as the ground for all theoretical cognitions to be, regardless of any carrier in terms of matter harboring such cognition.

With a definite link between instantaneous reflex sensation-reaction processes establishing immediate balance at the initiating step and perception-reaction mechanisms with delayed neurodynamic response and balance, we can think that the link should be the initiating working memory, the very first extended time effect for the realized sensing to provide cognition over time. Such a timed initial gnosis can be regarded as the result of the fundamental unit of neural energy being used to provide a fundamental cognitive value, which should be interpreted as an automatic reflexive response or as a perceptive-elaborated neurodynamic conscious act. Either one of these responses necessitates energy as working memory, an invariable quantum of neural energy that would be autonomously used in repeated neurodynamic functions that may vary from unconscious in the reflexive-reactive to conscious in the attention-intention type on the complex end of the scale.

In this formula, gnosis is in its general sense of auto and hetero capacity, and the sensing factor can represent autos or self(s) versus nonautos and nonselfs or both at the same time. The time factor can be regarded as the component of the spacetime for the living environment of the biopsychon, and that of the biopsychon itself, so that it can be regarded an integral part of the organism realized in the formed memory. This time factor is immaterial per se, and certainly unreal, if not inherent in matter and energy extension.[7] As an established fact, this physical interpretation for the time as contingency of matter and energy forces a sharp contrast to its mental interpretation that forms an essential conditioning factor of it, which imposes itself to be considered in discussing Gnosticism. Indeed in philosophical Gnosticism, no matter and no matter-related energy are in fact noticeable, but logically assumed, in contrast to time that is openly figuring as a major conditional factor.

Gnostic sense of self, as a unique self or autos, is therefore biologically established and is based fundamentally on the animal's metabolic needs regulated and covered through one constant metabolic process differentiating function time from rest-time that repeats itself, and this mere repetition in time per se is basic priming that reinforces any rudimentary autobiographical memory. Such a state of neural functionality with even the most primitive autobiography implies some awareness of environmental matters not being self. Again, a primitive archaic sensing, a paleognostic state, or the fundamental biognosis with differentiation into bioawareness or bioconsciousness, may be accepted to lend credibility to the presumed recognition of self and nonself in animal species. This stage of self-recognition against nonself is the initial step in the realization of an inner autobiography, biologically defining the biopsychon's separate entity from its environment. This archaic, paleognostic sensing, is basic in the animal's autos, but only making nonself entities known to it in a functional way without abstraction. It is the biopsychon's unaware autos or unconscious functional self that becomes the autocratic inner agent universally safeguarding SR. The change from this functional self to a mental self with abstraction occurs with additional neocortical associative capacity of the brain forming the conscious self(s) but on the account of the unconsciously functioning autos.

It would be futile to try to establish any timing in the evolution of the nervous system to indicate the beginning of sensing of the selfhood, the initiation of the animal's mental self in becoming ultimately conscious in distinction from autocratic unconscious master that is autos. Examples described in the previous chapter have given us an idea about differences in animal actions showing what can be accepted as behavior and what can be suspected as more than just behavior, and finally, what can be regarded as more complex and probably or certainly representing elaborated mind, consciousness, and therefore self or self(s). To cite an example, recognizing their own bodies in the mirror by apes is interpreted as distinguishing oneself.[8,9] This mirror image recognition is documented only in some chimpanzees and tamarins.[10]

Notwithstanding this mirror image interpretation, it goes without saying that in the wildlife, sensing selfhood is invariably enforced facing nonself in the environment. Environmental images and senses as exogenous stimuli are all perceived as nonself, outside of the reach of implicit memory. However, the unconscious autos, the central archaic figure of the animal's autobiographical functional slate ultimately recognizes self from nonself on implicit iconic memory that must transit to some type of explicit memory by added sensory identification (auditory or olfactive stimuli) in addition to vision. In fact, all that is needed for biological life to continue under environmental hazards or benefits is recognition of nonself in environmental conditions to face and appropriately deal with. This much information will be made available to the animal with some paleognostic capacity or with modalities of the fundamental biognosis, bioawareness, or bioconsciousness. At this stage, if holognosis is present, the living entity only needs heterognosis for gaining information on nonself contrasting with self and only in the concrete physical world. In fact, the biological Gnostic power remains limited to the physical concrete for millions of years. If we recognize a Gnostic mind in animals, it should be the mind that can be holognostic in terms of both concrete auto and heterognosis as in humans. The metaphysical holognosis must await literary consciousness and extendable abstract conceptualization with the unimaginable eons of time behind it.

This subject will come up further, however, in order to emphasize the basic difference of Gnostic function of reflex or reaction type, regarding auto and heterognosis in relation with autos, self, and degree of discerning, three extreme hypothetical situations may be presumed for the physical concrete Gnostic capacity in nonliterary and one for the metaphysical abstract Gnostic extent in literary minds as shown enumerated below:

1. Bion = autos with implicit autognosis and heterognosis of reflex concrete type
2. Biopsychon = autos, self, implicit and early explicit autognosis and heterognosis of reactive concrete type
3. Literary biopsychon = autos, self(s), explicit semantic memory concepts of created heterognostic type in addition to 1 and 2 above.

These schematically summarized categories of gnosis are obviously crude examples shown to expose possible skeletal structures for types of Gnostic models that may be used for further reflection and debating. We can be confident that states of variably refined gnosis of preconscious perception of concrete type existed in mammals and hominids and early metaphysically oriented foundations developed in psychognostic speechless mind as will be described in chapter 7 and eight. But, certainly, metaphysical complex Gnostic thinking only flourished after language developed and recognition of self became solidly established psychologically, acting as an agent of autos in perceptive-directive functions.

In the conscious mind, the self has to assure conscious recognition of both inner and outer worlds and therefore must recognize the autos if only realizing its unexplained inner suggestive presence as the unique separate being from all other nonautos, nonself, external existences. This understanding leads us to regard the self primarily as a mediator between autos and everything else rather than just a mere solo representative of the inner hypothetical self-autos against all noninner self-autos. Even such hypothetical self-autos entity, if it can be imagined to form initially in the process of recognition of autos, it cannot be regarded as an invariable permanent decision maker. With this understanding, self becomes immediately an agent for assimilating and appropriating images or effects of the outside world for the autos to review and evaluate, accept or reject as assimilable or rejectable. Rejection is not kept long in the autobiographical memory, but assimilation and appropriation operate in it recalling the self to autos relationship. Speaking of self to mean autos in conscious biopsychon leads to confusion as each one (self or autos) is a totally different functional entity. The autos is the essential primordial element, not eliminated by the advent of consciousness, not subordinated by self, and is the ultimate power in autocracy that can be regarded as a pristine natural impetus concerned with SR.

The *situs solivagus* is pivotal in determining energy exchanges to benefit only and singularly the biopsychon, to force dichotomy of the gnosis, and to expand the expressivity trait at the later stages of the neurodynamic development as will be exposed in chapters to come, extending lingual expressivity in relation with the centrally positioned autos shown by the egocentrist conscious self. Admitting the existence of self as a primordial pillar in psychogenesis alongside autos thus becomes inevitable, but only as the mediator and not as the ruler that is autos. Thus, recognition of self as such, with its roles played in the more elaborate mental functions, and its evidence as a purely abstract psychological definition, reflect its practical value in discussing the more complex autos and its functions that present an unfathomable psychological reality, difficult to precisely define or study with unfailing adequacy.

HOLOGNOSIS AND METAPHYSICS

In approaching the subject of Gnostic sensing, I necessarily made allusion to a legitimately imposing idea of total gnosis of which autognosis and heterognosis granted our solidly established understanding of selfhood facing nonselfhood. Having served the purpose, however, definition of holognosis regarded as ultimate all inclusive gnosis including potential extent of Gnostic abilities remains somewhat unsettled, and the crucial question of what in reality may be called holognosis, or defined as a state of total Gnostic recognition to be, must find a clearer answer. Evidently, description cannot neglect the physical and the metaphysical worlds as millennia of philosophy and science have proved that these two preoccupations of the human mind are realities of logic, have solid reasons both separately and more so together as they have not been shown in science to be mutually exclusive, and finally, as their eternal unique question in the mind, *what is mind's relation to metaphysics?* must have an answer. Difficulty to answer this question is mainly in the fact that no tangible testing method common to both physics and metaphysics can be found and examined.

Obviously, neuroanatomy and neurophysiology deal with the testable brain structures and functions but have not detected any materially testable element except memory and intelligence somewhat arbitrarily representing mindfulness; and to summarize all states of mind, any of the affective, abstractive, imaginary states, or the ephemeral *qualia* as an example has no measurable quantitative representative in terms of known units. Though these states are qualitatively demonstrable to some extent by functional MRI (magnetic resonance imaging), positron emission tomography (PET), and single photon computerized tomography (SPECT) of the brain, they have no grade precision in terms of quantities to be tested. In other words, the meaning of *quality* and *qualitative* seems to imply time indefiniteness as an innate peculiarity of the matter that cannot be quantitated. This fact further indicates that in contrast to the nonquantum physical world where time is a definite and a defining contingency of energy manifestation, in quantum physics and in immaterial biology of metaphysics, time is a primordium either as indefiniteness or as wholeness respectively.

The crux of the puzzle can be eliminated if we only consider the energy alone behind all these immaterial states (zero mass) regardless of their material (mass) carrier (zero or near zero in bioforce), not as quantitative measurable energy, but as infinite energy undefinable and immeasurable except as related to the inherently indivisible carrier that is the time, making with it one inseparable reality as "energy-time". In fact, in the previous section, I regarded gnosis as sensing through the time. Scrutinizing this crude formula would reveal the meaning that the neurodynamic energy spent for gnosis by the brain substrate can keep going with the bioforce applied through the time at one given time or at repeated times in continuation theoretically ad infinitum. Consequently, if we keep with the totalitarian concept of all creation (past, present, and future), the theoretical global Gnosis of metaphysical character, can be regarded as infinite energy through infinite time. Once we grab this concept, we can see that to find the answer to holognosis, to the global Gnostic realization by the human mind, the big bang theory will allow us to assume applicability of the concept to the hypothetical spacetime of the whole creation, the spacetime that would include infinite energy, infinite matter, and infinite time. This idea can then gain ground in a simpler presentation if we consider the metaphysical world as a type of a spacetime in which there is zero matter but infinite energy and infinite time. That theoretical metaphysics idealized on the immateriality can be formulated in:

2. GNOSIS = (ENERGY) (TIME)

The formula can be regarded to sructure a ground for all units of cognitions-to-be, regardless of matter harboring such cognitions, or theoretically without material carrier. This formula, theoretical as it appears, does include the veracity of both factors of energy and time needed for the Gnostic use of memory and can be applied to a hypothetical holistic Gnostic state, which englobes abstractive realization of both physics and metaphysics. This inclusion, irrespective of forms taken by its included presentations, must be regarded as showing infinitesimal unities of subatomic energies that have reached equilibrium of constancy in physical forms without independent autonomy and in biological forms with such autonomy. This philosophical idea summarizes the creation in a unified holognostic principle independent from mass but also applicable to it in any form. It can be regarded as a purely insubstantial mind, a mind powered with expressivity creating both the physical and the metaphysical worlds hypothetically from the subatomic scale to no scale limit. This can be linguistically condensed in God, but a God of infinitized energy and no matter: a concept of infinitesimalism as in subatomic quantum theory making the bioforce the only representative prototype expression of such God concept. This philosophy asserts the metaphysical reality of infinite truth, which applies to all forms of beings, in the microcosmos at quantum level of nearly nonmaterial energy or macrocosmos at actual matter-energy bodies. As it is in fact representing GNOSIS irrespective of matter, it can be taken as an orthodox metaphysics of infinity essence if our philosophy leaves in oblivion the artifact of human made mathematical infinities and admits only the constant natural integrity of the creation. The fact is precisely clear in the truth of infinitesimals of calculations revealed in microcosmic physics, supporting the uncertainty principle of Heisenberg that allows unlimited possibilities that are ultimately revealed only in integral physical and biological created forms. The crux of the matter is in reality passing from the subatomic undefined time to the larger-scale-defined time understanding physics and material biology versus remaining in undefined and infinite time in metaphysics. Thus the theoretical unit of gnosis in any mind can be regarded as representing the autonomous bioforce forming abstractive nonmaterial concepts in what can be called immaterial biology.

In this reflective presentation, we moved from materialistic physical world of reasons to immaterially known facts in reasoned construction, representing the deductive reasoning from a premise that accepted immaterialism of the Gnosis recognizing a metaphysical immaterial substrate. This hypothetical immaterial Gnosis can be theorized representing an original hypothetical creating model, as a Gnostic state of mind at that infinitesimal subatomic level, an immaterial pure energy making a universal mind of unknown origin but with known facts comprehensible to explain the overall pattern of creation. This hypothesis assembles physics, biology, and metaphysics as natural constituents of a holistic indivisible creation that can be regarded as ***holognosis*** based on the innate Gnostic power in the bioforce autonomy ultimately reaching consciousness through biology. The old word *thymos* of Greek origin attached to meanings of courage, spirit, and soul can be used to represent the biological pristine autonomy responsible for reaching conscious states as biothymic, meaning a state of directional energy representing energy-time principle transcending from material to metaphysics in immaterial biology. In contrast, the physical world can be regarded as representing the energy-matter principle including material biology as biophysics in essence. Thus the hypothesis considers a universe that is autonomously and intrinsically organized from the very inception to the very end (if there is any), in terms of phasic alternation, existing between the states of being and nonbeing by interchanging mechanisms by and large but following an immortality autonomy principle in two manifestations of biophysics and biothymic that constitutes the physical and the metaphysical worlds respectively.

If we consider this hypothesis assuming the theoretical pure energy initiating the immaterialistic Gnosis presented in the formula 2, we can argue that the idea has realistic probability as it applies to the subatomic nearly nonmass energies and the quantum physics'uncertainty principle. Thus the biophysic and biothymic realities appear primarily applicable to biological creation for both material biology with scale differences of soma and psyche structures and immaterial biology forming abstractive Gnostic concepts of metaphysical nature.

FROM ONE AUTOS TO MANY SELF(S)

Gnosis involves self, and reviewing the subject of self through the ages, even as briefly as I will outline here, provides interesting information that helps understand this pillar of consciousness more broadly. The oldest documentary evidence on the notion of self comes from the *Gathas* of Zarathushtra in the second millennium before Christ.[10,11] It reflects on the inherent selfhood and its associated free will of man in choosing good or evil in a religious context. The self in that context is the human decision maker regarded as an independent power, able to ascend or descend in morality, facing the unity of God, Ahura Mazda, the Ultimate Goodness. Here, self is potentially changeable, indicating a hidden plurality, and is ego oriented, witnessing complexity in that archaic psychological sphere. From that earliest monotheistic thinking presuming one God, one soul, and one self, which really reflects autos, but with tendencies to vacillate between good and evil, the notion of soul and its components including self dominates the antiquity, influenced both ecclesiastically by a religious proto-ego at that time, and philosophically by further Plato and Aristotle's later definition of soul. The notion of soul then becomes the playground of subsequent philosophers but mainly Descartes with the self as the unique being and the responsible thinker. In Descartes' expression, self is I—*je pense donc je suis* (I think, therefore I am)—and its sameness with autos in Descartes' mind is crystal clear as we can realize Descartes' intention in retrospect today.

A significant contrasting view on the soul and on the uniqueness of self appears with Hume in the eighteenth century, who denies any single self.[12] James in the nineteenth century questions both notions of soul and self seriously to deny any real value for either one as a defined and recognizable solo entity but accepts self in its multiple forms.[13] However, subsequent philosophers and psychologists entertain the same subject, talking more in terms of mind with self, ego, and consciousness. Jung was especially influential in promulgating self and particularly self-ego relation as a firm unity decisive in regulating human psyche.[14-18]

Tracing the events to find out the reason for this pluralistic thinking on the subject of self, we find that the simple natural way of assuming our individuality in our inner self, our autos, is one way of reflecting on the problem and is indeed the simplest way. Another way which imposes itself comes into consideration when we try to attach material values of some kind to our mental experiences at a time when we feel we are thus affected or inclined. Trying to concretize the immaterial sensation in ourselves like feeling happy or sad, content or discontent, in pain or pleasure, and so forth or especially sensing that particular indescribable qualia in ourselves, we detect variations moment to moment. In these states, we may question whether we are really the same and whether we are right to assume sameness at all time. This is the reason we may see the self in us inconstant and somewhat an unreal uniqueness, and this is the cause for eventually rejecting the self consciously but preserving tacitly and subconsciously our unique inner feeler of our different states of minds that is us, our autos. In this second way of thinking, if self is not totally denied, as it is in Buddhist philosophy and by Hume,[15-19] it is changed from its singularity, which in reality reflects the ownership, the state of autos and not the property, to plurality that defines properties, self(s), to a single owner that is us, autos, with or without any solid unique self. The common sense in this context, however, rejects ***being oneself (equivalence with autos) and not being oneself (equivalence with selfs)***. Then what? Interestingly, this puzzling thought, if pondered upon inquisitively, shows a truth which is found in the perplexing conclusion in Schrödinger's cat example that, as is well known, can be alive or dead at the same time by virtue of the uncertainty principle of the quantum physics. So, the puzzle in appearance, though perplexing as it is, is assuring the reality of the uncertainty principle in biology in the mind's elaborating processes that in a way personify the quantum field and reaffirm the bioforce behind the will.

The natural course of beliefs based on the archaic religiously defined self from the proto-ego of Zarathushtrian monotheism evolved in philosophical thoughts, and together with the idea introduced by James for a definite notion of self but in almost as many forms as we can have mental experiences, laid the foundation for Parfit to group these categories in two classes as ego theories and bundle theories of self.[16] However, though these two categories are apparently quite distinct, the separation cannot be regarded rigidly established. In ego theory, self is firmly attached to the personal being of the individual but reflects a conductance of the individual's ego(s) that may condition the individual's mind in varying degrees. In my interpretation, the person may still sense multiple selves, but each one is more or less affected by the person's ego(s) and in turn, affects the autos in the course of perception and ideation. The proprietor is invariable and is the owner self, in reality the autos, based on the autobiographical memory. Being oneself is firmly established and unlikely to accept changes. In bundle theory, selves are multiple, essentially independent from ego(s) influences in the experiencing of the subject, but the proprietor and the autobiographical memory are unaffected as in ego theory, and being oneself is less firmly immutable. I do not rigidly separate the two theories as the ego definitions and impositions may vary significantly in everyone's mind with uncontrollable variations in selves. In my belief, self should be firmly distinguished from autos and should be given the value of the mental processor for providing presentation of any impression from nonautos sources, ego or else. Self is not equal to autos in the mind but is its potential conscious equivalent. It can bring many images to it to leave impressions but has no power to imprint them there permanently, or at least without hard mental conflicts for a change in pi or ps relationship discussed in the section of phylogeny in chapter 1.

Most important of all is the autobiographical memory, which establishes the selfhood. What is it like to be a bat? Reflecting Nagel's discovery question[17] is most appropriate to tackle this question of selfhood and the feeling it can give to its proprietor. Indeed it does not feel to be anything particular without having previously experienced that particular feeling as such. Being anything and sensing it repeatedly in time will provide recognition of that sensation stored in memory and when perceived again, and that is caused by the well known facilitation effect in the nervous system, which is a bona fide priming. The factors of continuity in time and sameness of the experience repeating and leaving its trace in the primitive nervous system of early biopsychons, living and reliving the autocratic life mechanisms, is the essence of autobiographical impressions. This lays the foundation of the self, but a self that functionally recognizes the independent autos and can reinforce the autos, yet remaining separate and not able to unify with it in any permanent way. Self is sensed as autos when constancy of status quo is ruling. But it serves the first notice to indicate any change to autos to reiterate the cognition in autos reacting to accept or reject the change. It absorbs the change until its correct position to be saved as property of autos or to be rejected is defined. Psychodynamic function of the self appears to be decisive and unlimited, both in the cognitive separation of autos from all nonautos elements, and in controlling attention-intention acts in modular functioning of the brain and helping to induce conscious reminiscence of various *qualia*.

Autognosis by self appears to be the first stage in the course of evolutionary progress of the nervous system to consciousness, prepared by as yet unknown neuro-psychodynamic processes. The theorized main result of self role in autognosis seems plausible and reasonably acceptable. Other effects of self in the capacity of the main agent acting as a relay between ego(s) and the archaic autos or as mediator in modular brain functions facilitating basic conscious processes and ultimately leading to complexity are potential possibilities subject to usage with possible extension. Some such modular processes will be dealt with in discussing the qualitative versus the quantitative conscious states in the last section in this chapter, but some are beyond the scope of this writing except for the involvement of self in language processing that I will have occasion to revisit in later chapters on expressivity and language origination.

If phylogenetic examination of self in the autocratic biopsychon is based essentially on reflections, hypothesizing analogies gathered from ontogenetic development and evolutionary evidences, a propitious field of study instead compensates for such abstract examination and that is infant psychology. This field of study is in reality that of extended brain development in the human from intra- to extrauterine life with its early observable manifestations. This evolutionary serendipity has facilitated gaining information of considerable importance. To be brief, I can say that pioneering works by master psychologists in the early and mid-twentieth century essentially differentiated two stages of development in continuity for the production of self in human psyche that I prefer to call stage of somesthetic and stage of psychesthetic self(s) formations.

According to this view, self is the internalized serene neurosomatic and neurovisceral unconscious sensation of the embryo in the uniform intrauterine environment. In the late intrauterine period, notwithstanding the real auditory perception that is well functional and the possible but questionable visional ones related to auditory sensation, the body surface sensation is overwhelmingly homogenous and acts as an ego internalized in self with its calming inner sense. In this period, the bodily contact and somatic sensations are judged as effective imprinting egos for the inner self. At birth, sources of harsh contacting external stimuli are interpreted in psychology norms either as alarming and unwanted senses with corresponding internalizations, meaning death, or pleasurable reinforcing ones, meaning libido. These are in reality the earliest rejectable or acceptable ego types to the shaping self. The so called oral and anal phases of development in early infancy related consecutively to breast sucking with the utmost sense of pleasure and to defecation with its mixed sensation regarded as relief versus loss are old examples reflecting active periods of the early functioning prominent self(s). Self(s) then develop, differentiate, and multiply as the brain rapidly develops, and language starts making defined conceptual self(s) realizing various sources of egos. This is very concisely the essence of the old classical views.

Psychesthetic stage develops on innate emotional ground with abstraction producing meanings for concepts. Self(s) in this stage are agents for trading between the egos and the autos, quite often spoken of as agencies of the mind and are practically taken to mean autos, which is wrong. These two stages of soma and psyche are overlapping continua that are not really separate, and the somatic predominant stage is preparatory for the psyche to be elaborated that in turn should progress theoretically to infinity if somatic substrate could also persist indefinitely.

The gap between the extreme simple bion and the extreme elaborate biopsychon is thus filled with the fully developed mind going through a continuum of progress of a substrate of material biology, soma and its function leading to nonmaterial biology the psyche, evolutionarily and developmentally. The substrate is the biological matter formed by the bioforce, and the function is the expressivity modeled by its energetic expansibility. They are inseparable and show continua of overlapping existences through evolution that may induce the erroneous impression of a gap between the observed soma and psyche at phenotype-genotype exchanges. The bridging is completed by consciousness with the intermediation of self, acting between the more stable archaic autos inside, and the less stable or continually variable egos outside, internalized by self(s) and through perpetual cultures

AUTOS - SELF - EGO

There has been a numerological tendency to consider ternary assemblies sometimes in the disciplines of knowledge, and I take one that applies to the subject of self from its inception and in its changing faces with consciousness. The very archaic self in the primitive biopsychon is the inner autocratic agent, the autos, more precisely the functional autocratic power, which is the single separated biological selfhood with no conscious self-recognition and in the total absence of any other recognition that could affect its autocratic behavior. This bona fide autos is its own self. With biognostic stages progressing and consciousness appearing in the primitive minding biopsychon who also recognizes self and nonself, mental images of the presenting entities to self with any comparative attribute inciting biotropism, will necessarily act on autos through self in terms of positive or negative feedbacks. The foreign powers of these images act as egos and autos through self can only choose to accept or to reject them, in trust or distrust, for uniting or fighting with, repeating its fundamental archaic formula of biotropism and biodynamism.

Once the stage of mythology and God fathers in the evolution of the human culture has passed, the self can eventually recognize its autos functionality as its inner source of power facing the unique most powerful entity, the God, the superior outer source *of* power surpassing all others. In this situation, we can assume that the conscious mind will find itself in the face of that supreme power blatantly helpless and therefore must unite with it. Autos with its ethos and pathos, the metamorphosed basic biotropism and biodynamism, naturally adopts the rational conduct, guided by the self in conscious realization and understanding of the benefit of this unification. The primitive autos must incorporate the supreme power in itself by identifying with it. At this time, self serves as an agent in the service of autos, which gains more reality as a proprietor, whereas self becomes more of a property but with limited proprietorship. The addition goes where it belongs, that is in the conscious property space of self(s), although ultimately appropriated by the unique proprietor that is autos. In the primitive unconscious bion, autos and self are one and the same, but in the conscious biopsychon, the identity sensed and initiated by self may exteriorize, but not necessarily reflecting the autocrat, the autos in its hierarchically dominant position. Autos is in the position of the old autocrat and the proprietor at all time. Self, working always as mediator, serves to identify nonautos and presents it to autos to be accepted or rejected. But self is not incorporated in autos so as to unify with it in inseparable union. In fact, we often notice this separation when we make a decision against all logic of our conscious self, but in line with our wish, and act in a way that in essence reflects the supremacy of our inner *autos* over our *self*.

To visualize the positions of autos, self, and ego, a figurative example would be to view the autos as an inner source of light and the ego as an outer one, facing and shining to one another, both with varying brightness, and to view the self as a transparent screen glass panel between them, containing a network of radiation-absorbing embedded lead, patterned to function selectively (modularly) to different wavelengths of light. When the brightness of the light sources shines on the leaded glass screen, a two-way passage from one to the other lights can result, which depends on the functional modularity of the screen and the brightness intensity of the lights. The variation may actually go from zero to full light transmission through the screen. Thus, metaphorically, self can represent brightness or darkness with subtle nuances outside in or inside out for a possible graded effect from flat rejection to full assimilation by autos. Appearance of self as modulator functionally means neurodynamic balancing between autos and ego until ego charges, either assimilated, rejected, or adjusted, will be absorbed in the mind equilibrium.

In terms of variable ego figures, earlier I used the exceptional figure of God as the best-known archetypal ego[18] and assumed full identification with it by autos through self, which is of impossible occurrence. Failure leaves autos without any definite self property other than established by biocracy in the autocrat, which, I believe, is the metaphysical self of God, the belief that God exists, a sense of conscious possession in the form of ego-self of half emotions, half logic. In all other more often observed situations, the identity of ego-self mental possessions also remains understandingly vague, but autos remains clearly the same and reflects its own archetype, hierarchically the oldest one: the autocrat of the bioforce. This conclusion can be verified in actual observations of the statistical norm of predominant invariance in mass psychology in religions contrasting with variance of sporadic disorder cases of unstable, shaky personalities and beliefs.

A clear example of distinct separation of autos as the master proprietor from self as the subdued property is the figure of ego-self axis of righteousness abhorring lies when, however, necessity in a self-defense induces the most righteous person to tell lies. In this situation, no matter how strongly the ego-self axis can be supposed to be structurally fixed, the supremacy of autos is unfailingly predominant and gives the ultimate verdict in favor of the falsehood. But why should this be true?

In my view, the answer is because of the archaic hierarchy ruling of SR priority implying that the ***autocratic-only truth*** must prevail. This safeguarding of the autos in face of a negative feedback in principle is basically what would happen in the unconscious biopsychon with its only mechanism of defense in its biotropism governing its biodynamism. In practice, the same model is also used in principle in the conscious biopsychon. This example sets a firm conclusion that consciousness is primarily a tool developed through evolution to basically serve the same old principle of SR in the face of evolved biopsychons interacting with one another, but it developed additional sublime functions as well. We find the phylogenetic rule of values described in chapter 1, again asserting the ultimate superiority of the innate individual pi over the social ps.

To simplify more and retouch upon the thought experiment of the previous chapter, I would entertain once more the analogy of personification of the time segment in the infinity time circle and would say that consciousness is now a mirror showing autos to itself often with beatification by various ego-self makeups and avoiding ugly changes and preventing an eclipse of the image at all times and by all means—all these are being made possible by idiognostic language coding that will be discussed later. Philosophically reflecting, the time dimension has finally accommodated the evolutionary psychogenetic material spacetime with its nonmaterial psychognostic spacetime ultimately asserting its continuity with the ultimate immaterial infinite one.

REFLECTIONS ON BASICS

Approaching the subject of self in a different line of thinking but with the same aim of understanding the *raison d'être* of self as a concept in consciousness, we can envisage the subject from two different angles—one from the somatic standing and the other from the psychological one. Some of both views have been so far briefly sketched, but fundamental aspects need additional considerations.

On the biological side, with establishing biocracy, a bordering separation between these two approaches to the self, somatic and psychologic types, appears a must for better understanding the psychological formation of the self. Presuming the basic biopsychon to be normal in every way, except for a hypothetical absence of a limiting boundary, elicits the immediate idea that such situation eliminates the integrity of phenotypic spacetime and microspacetimes and reduces the physical material realities to subatomic levels, to regression reaching hypothetical real nil state without possible cognition to be formed other than by the hypothetical metaphysical immaterial being. Such thinking could be entertained on logical basis giving an antithesis of immateriality rejecting materiality. So to come to the realm of reality, if metaphysics could be imagined to be possibly real beyond the frontiers of the mind as it is reflected by the mind, it would have to have the Gnostic sense in a substrate that must be necessarily biological, but abstract, with both automatism and direction with self-sustaining, or any theoretical nonbiological substrate with automatic self-sustaining (computer) but with an outside control. We have already seen in the past discussions that the logical argument supporting the first alternative prevails. Therefore, biological phenotypes, phenotypic spacetime, microspacetimes, and environmental spacetime must exist to grant the possibility of organized sensing through the ever-persisting regulator sustainer for the biocracy that is the autocrat animated by the bioforce. Thus, the separate identity and uniqueness of the living autocrat is the sine qua non for its forms of the basic phenotype spacetime, the microspacetimes, and the biochronogenesis based on contrasting function time and rest time, as discussed in the last chapter. In brief, the unique identity of soma leads to the unique autocratic sensor that is autos, with early implicit pseudo-episodic memory for internal affairs and bioconscious realization of the balance with the environmental spacetime facing external affairs. Thus, the primitive self forms and separates from autos possibly through bioconsciousness and psychognosis gradually to assume more precise functionality in consciousness with developing and perfecting explicit memory, language, and abstraction. The very essence of the autos will not change, but some intermediary mechanism of defense would be formed by the conscious self to absorb impacts and allow the inner autos to continue in better adjusted autocracy.

Psychologically, the effectuating agent must be primarily in the service of autos, independent from the outer forces but able to sense them, must be swift acting, changeable and not immutable, and must be effaceable or reenforceable if needed. This agent should act as an informer, alerting the inner autos of outer influences. It should act as a messenger and facilitator to ease the recognition of the outer world to the inner autos, thus making gnosis possible. It should make relays between the implicit memory of the autos and its psyche with its conscious meditation using the explicit memory. Self appears to be the only one of the three psychological entities—autos, ego, and self—that fulfils these characteristics. Psychodynamic value of self is this action of informing and alerting the inner side of the biopsychon that is facing changes on the outer side and transforming the boundary between the two spacetime entities to a checking gate, providing the means for communicating in the language of gnosis of time and space between the two sides.

We can say that self is the agent of gnosis or is gnosis for autos. Testing this hypothesis psychologically can be best performed in cases of schizophrenia when unreal causes are invoked by the patients for phenomena of real or illusory character, or unreal characters are assumed by the patients in their inner self(s). These patients' recognitions return to normal once their treatment succeeds to unify their multiple pathological unreal selves back to the closest one to their archetypal self compatible with their basic autos.

Modularity of self(s) is the psychologically posed question that may be seen clearer, keeping with principles of somatic development with related neuroperceptive conditioning. This conditioning is primarily internal, emotional or affective, generated by the various sensing of the internal needs in imbalanced energy feedbacks versus balanced situations. They seem to be all archaic and use the archaic periaqueductal gray (PAG) matter of the midbrain.[19] They form modular self-representation of visceral origin as models that, I believe, can form the basis for further modular self(s) with mental acquisitions in art and science subjects. These are generally regarded modular self(s), but let us remark that PAG is frankly part of the limbic anatomy and reflects the archaic autos in function. So these modular self(s) must be really regarded as true representatives of autos psychologically speaking rather than the prefrontal self[20-21] of reason, which is a metaphorically more autonomous intermediary between autos and egos.

I take self as the original evolutionary agent laying the founding pillar of ultimate gnosis starting in bioconsciousness and as the base ground on which to edify consciousness. I believe, the ontogenetic creation of the nervous system from the ectoderm that is the outer boundary of the animal soma to its environments, and not from the inner mesoderm or endoderm, gives a hidden message that the ultimate gnosis-to-be will naturally originate from the boundary layer, separating the animal from the outside source of its information access to provide ultimately the connection of the inner individual gnosis to the universal holognosis. The messaging by self, bringing recognizable information to autos prior to full language development and conceptualization, must have been practically the basic gnosis of acceptable versus rejectable conditions in the environment. In other words prior to language formation, conceptual needs must still have been served by gnosis through self to advise the autos for ultimate action. This idea can be extended to mean that gnosis by self in reality laid the basis for concepts to be expressed by the future language-to-be. I should not miss to indicate that the early self dynamism, even before the conscious self appears, could indeed make the building blocks for the language to start on an unconscious ground, and I will discuss this subject further with psychognostic consciousness in chapter 6 and further on in connection with language development.

CONSTANTIA AETERNA

To end this chapter with another exotic Latin expression is not solely to harmonize the end with the start with an artistic symmetry, but to also indicate an additional reason for doing so. *Constantia aeterna* means *eternal constant*, which I am using to emphasize the functional constancy of autos, regardless of apparent variations with evolution ending in modular self(s) consciousness. Autos in its journey from the unconscious biopsychon, through bioconsciousness, to the fully conscious one in the human mind has left eons behind like autoecious parasites infesting many varieties of organisms without ever-changing character. In reality, although consciousness has garnished the essentially unconscious autos in mind with various investitures, self, ego, and modular specificities in many mental functions, it has not changed its autocratic essence. All the neoformations described to date in the anatomic substrate of autos, the limbic system, and in its additional functional possibilities in the mind confusingly replacing autos by self must be precisely understood in order to realize autos constancy. Although the very basic principle of autocracy, so far explained, is sufficient to support this constancy in the primordium of autos, which is based on the unfailing rule of SR of biocracy, a tangible example is in order to demonstrate reasons for the confusion. I will describe such an example and follow it with discussions related to autos and self, mainly to try to correlate the neuropsychology of functions of autos and self with their neuroanatomy. Understanding mechanisms in descriptions to follow may not be so clear to nonprofessionals, but conclusions reached are in clear understandable terms.

Reviewing the studies of split-brain patients in whom the two brain hemispheres were surgically divided (transecting the corpus callosum for treatment of epilepsy now no longer practiced), having lost all direct communication, we can usually see that each hemisphere functions normally but independently from direct control of the other. For instance, presenting distinct different pictures separately to the right and left visual fields without allowing cross visibility and asking the patient to pick up a corresponding picture from a series presented so as to match the target picture shown freely to both eyes prior to testing, the right hand picks up what matches the right visual field that corresponds to the left hemispherical action, and the left hand does the opposite indicating the action of the right hemisphere as shown by Gazzaniga.[20] This duality of action taken as a contrast against the unity usually accepted for the decision making self incited curiosity, though the perceptive separate right and left hemispheral neuromotor coordination explained the difference.

Another more interesting experiment reported by MacKay in 1987 demonstrated a more sophisticated degree of independent autonomy of the two hemispheres in surgically made split-brain patients,[21] raising the question of possibly two independent minds being the decision maker rather than one. To find out if the split-brain patient is really one or two persons, the experimenter taught the patient JW to play a twenty-number guessing game guided by the experimenter's saying up or down or OK to lead the patient in reaching the right answer on picking up numbers.

Answers by the patient, if given orally, indicated the function of the left hemisphere and, if pointed by the left hand picking up numbers on cards, showed that of the right hemisphere, so that the answer could either represent the function of the left or the right hemisphere totally independently. Next, JW was instructed to use his own hemispheres to play one another. The playing was successfully carried out and indicated two independently functioning cortices that were even capable of reciprocating the play. In spite of the obvious independence of the two hemispheres and the greater complexity and autonomy observed, the experimenter's conclusion rightly was that there was still no compelling evidence for two separate persons in the patient, and in fact, the patient also felt being the same person at all times. The explanation by the experimenter was that there was only one self-supervisory system to determine the priorities in the conscious brain even with two independent hemispheres regarded as two persons acting separately. It becomes of interest to see if the same supervisory system is functional in unconscious acts or in mixed conscious and unconscious situations like in musical composing and performing.

The answer can be possibly found in reviewing more complicated examples of mixed situations, but even in this single experiment, findings indicating separated cortical halves do not imply a full neuropsychological separation of the total substrate of autos but its added refined conscious functions. As a matter of fact, the anatomical hindbrain (archencephalon) and midbrain (mesencephalon) in split-brain patients are untouched and fully functional and are still connected to the forebrain (telencephalon), which is also undisturbed in each hemisphere, but the hemispheres are not connected directly. The important limbic system of the midbrain and its direct connections with either hemisphere and all the relays of sensorimotor types of the peripheral perceptions and central motor responses to each hemisphere are remaining intact. In other words, the impact of the many million years of autocratic evolution on autos, now in its best demonstrated representation in self, is still vividly detectable with the constancy of autos.

The limbic system can be said to literally represent autos with all emotional and feeling characteristics related to the archaic autocrat. Connections between the limbic system and either hemisphere must be symmetrically similar so that in the split-brain patients in whom the autos is still in firm communication with the right brain half with synthetic holistic capacity (like face recognition and musical tone synthesis) and with the left brain half with essentially analytical coding capacity (like language and scientific symbols), these functions follow neuromotor rules correctly. So it is naturally true that the unity of the primordial autocrat in the limbic system will remain intact in the split-brain patients and still perfectly capable of controlling the two hemispheres. But if each hemisphere can act independently as shown by MacKay in split-brain cases, and yet the patient feel being the same person, the unique conscious self must be somehow in both right and left brain hemispheres, or as equivalent autos-self(s) in both, or as a whole of cortical and subcortical limbic integrity. This situation could pose an important question as to exactly where could be the abode of a unique conscious self. The question itself is of course not without flaw since firstly, the self is presumably represented not only consciously but also with the influence of autos to it carrying some affective unconscious feeling, and secondly, global limbic affective function only opposes the neocortical logical reasoning function in the right prefrontal area. Thus only in terms of conscious logical self localization there seems to be some uniform agreement.

Regardless of this basic reasonable distinction, the answer may be possibly found in the nature of self-imposing arguments for its constructional multivariate structure.[22] If self is presumed to present either solely reason or affect, and not both, the definition and understanding for self would be clear. However, even such theoretical presentation of either of these two mental functions and not both by the self seems not to be precisely so clearly separate. In fact, self is sensed to represent autos and all nonautos effects of ego types starting to be more and more defined with linguistic conceptualization. All reasoning, planning, and decision making reflecting also some inapparent affects are decided by the conscious will and exerted by the conscious self. If self is regarded as such conscious exteriorization presumed to represent the sole decision maker as both judge and executor, then it is holistic in its function of representing variable affects and reasons of state of minds as contributing to a finalized conscious will. It is therefore more limbic but refined in its solo conscious finalization to reflect more reason or more feeling. In this definition, self in its purely conscious manifestation, regardless of its components and considering its synthetic essence, is plausibly the state of balanced limbic autos with the added right prefrontal lobe specification for logic and left parietal lobe specification for spatial orientation. The right prefrontal function for self, regarded as self's only final consciously formulated exteriorization, may be accepted because the fundamentally analytical function of the left hemisphere theoretically excludes synthetic elaboration of a reasoned decision but of course does not eliminate the contribution of the left-sided work, meaning that at any time both right- and left-sided contributions are involved in making a unitary representation of modular self modalities. Furthermore, these analytical left-sided functions such as language and science idioms have been developed after formation of the self and not before. Therefore, if any place could be presumed for the logical judge self in the brain functions, it should be in the right hemisphere both for reasons of priority in connection with the limbic system as a whole and concordance of functional activity compatible with the right hemisphere where synthetic holistic functions are the norm. In addition, the newly acquired scientific noetic aspects of the self that cannot be neglected impose self(s) categories or nonemotional modular selfs representing left hemisphere functions more than right. Furthermore, as self is always conscious, even in dreams with recognition of admixture of visional types, the real comprehensive self should be regarded as an integral unity, reflecting the wholeness of autos. It should be a functional unity but with several modular functions, both emotional and noetic, with cortical representations not in just one location.[21] Thus the coherent integral autonomy of what is called self should be really considered cortical reflections of the limbic autos working in harmony from different cortical sites that represent the modular mind with modular self priorities.

Further significant facts about modular self(s) concern the evolutionary timing and mechanisms of self(s)'s realizations associated with neurosensory reactive functions or neuroperceptive mental functions. For example, the unquestionably abstract mental functions of more recent times relating to logical evaluation and judgmental decision are located in the neocortical right frontal lobes.[23] However, the more archaic egocentrist positional self specification in the environmental spacetime that is based on vision orientation seems to be in the left parietal lobe connected to the occipital visional foci.

The issue is important and represents the best example to understand how consciousness can act through self and in a holistic way, an act that is in essence originated by autos. This order of the course of action from autos to self, unconscious to conscious acting will be discussed elaborately in chapter 8. However, practical understanding can be helped by an actual example of typical artistic production originating from an ego that we will see here.

Reproduction of a musical theme by a player's self, from the composer's original creator producer, an autos-self, as an example, involves characteristic transfer of psychodynamic significance of the chain of autos-self-ego-self-autos to the player and further similar transfer to autos of the listeners. The composer's artistic power in structuring the melody and embedding the emotional effects in it with the special *qualia* of the composer's internal world of experiences while creating the melody (autos-self production forming an expressed ego) is all beyond the power of the player (self-autos receiver) to mimic and the listener (a different self-autos receiver) to grasp the affect of the melody (autos-self-ego effect). So far, the passages of autos-self-ego 1 (created by the composer) to—ego2 (perceived by the player)—self-autos to—ego 3 (perceived by the listener)—self-autos, in this example, have completed the transfer of a state of mind, a *qualia* of artistic nature originated by autos and transferred through selfs that probably reproduces as close a *qualia* in the minds involved as it could be artistically reconstructed by different players quantitatively. Qualitatively, however, in some cortically damaged conscious musical artists with left hemispheric hemorrhagic stroke and aphasic problems, there is always one self, and the actual compositional percepting ability produced by that self is not different and certainly not destroyed as I have observed in such patients. Furthermore, musical reproduction by electronic devices that are supposed to produce exactly the same quality are equally well-estimated by these damaged brains as by undamaged ones. Consequently, the limbic modular musical autos-self can be regarded as a typical model of affectively based modular self with purely synthetic qualitative capacity[24] in contrast to analytical modular self(s) with scientific capability. The purely holistic qualitative effect of music as an evidence for this type of emotional modular self can be noted in every virtuoso that performs with a quasi-absence of consciousness in the reproduction of the play, which must be done by the holistic self as the actual representative of autos. The same is true in every listener that can grasp the essence of the emotional effect of the reduplicated music but not with the same *qualia* in the composer or the player.

What can be concluded on the subject of autos and self may be emphasized by briefly recounting their evolutionary course. Indeed, from starting as the autocratic unconscious self, the autos, in the primitive biopsychon, to ending as the ultimate director with an agent, the self that represents the conscious selfhood in the human mind, both autos and self, have functioned through chronologically older and newer brain structures. The archaic autos resides in the archaic brain structures where it continues to be in the structures known globally as limbic system. The conceptual mental self, by acting as the limiting separator between inner old autos and outer world and using functions of the newer brain additions, synthetic holistic reasoning and logic, must be functioning mainly in the right prefrontal cortex with secured access to the left cortical speech centers. Dynamism of self appears causally decisive not only in gnosis of space and time, but in multitude of information bits as concepts, primitive or elaborate, archetypal, ordinary, or nuanced by ego(s), or by any other external stimuli, or internal drive, and in the uncontrolled production of qualia of all kinds. Egos that are evidently variable and can include both natural figures and concepts and conatural educational or symbolic elements and metaphorical abstract types are the hardest to decipher, categorize, and attribute to modular functions of the neocortex.

REFERENCES FOR CHAPTER THREE

1. Esther Harding M. 1963. *Psychic Energy. Its Source and its Transformation.* Second Edition. Princeton University Press. Princeton New Jersey.
2. James W. 1890. *The Principles of Psychology*. London. Mcmillan.
3. Baars B J. 1997. In The Theatre of Consciousness: Global Work Space Theory, a Rigorous Scientific Theory of Consciousness. *J Consciousness Studies*. 4(4) 292-309.
4. Ramachandran VS, Blakeslee S. 1998. *Phantoms In The Brain*. London. Fourth State.
5. Strawson G. 1999. The Self and the SESMET. *J Consciousness Studies*. 6(4) 99-135.
6. Damasio A. 1999. *The Feeling of What Happens: Body, Emotion and the Making of Consciousness*. Heinemann. London.
7. Gallup Jr GG. 1977. Self Recognition in Primates: A comparative approach to the Biderectional Properties of Consciousness. *Amer Psychol* 32:329-38.
8. Suarez SD, Gallup Jr GG. 1981. Self-Recognition in Chimpanzees and Orangutans, but not Gorrillas. *J Hum Evol*.10:175-88.
9. Hauser MD, Kralik J, Butto-Mahan C. 1995. Problem Solving and Functional Design Features Experiments on Cotton-Top Tamarins, Saguinus Oedipus Oedipus. *Anim Behav*. 57:565-82.
10. Mehr F. 2000. The Zoroastrian Tradition. An Introduction to the Ancient Wisdom of Zarathushtra. 42-45. Zoroastrian Benevolent Publication.
11. Boyce M.1992. *Zoroastrianism. Its Antiquity and Constant Vigour*. Mazda Publisher. Costa Mesa, California.
12. Hume D. Referenced in Blackmore S. 2004. *Consciousness. An Introduction.* University Press. Oxford. New York.
13. James W. Referenced in Blackmore S. 2004. *Consciousness. An Introduction.* Oxford University Press. Oxford. New York.
14. Jung CG. 1990. The Undiscovered Self, with Symbols and the Interpretation of Dreams. Revised Translation by R.F.C. Hull. *Bollinger Series XX*. Princeton University Press.
15. Blackmore S. 2004. *Consciousness. An Introduction*. Oxford University Press. Oxford. New York.
16. Parfit D. 1987. Divided Minds and Nature of Persons. In: Blakemor C. Greenfield S. Eds. *Mindwaves*. Oxford, Blackwell.

17. Nagel T. 1974. What is it like to be a Bat? *Philosophical Review* 83: 435-50.
18. Edinger EF. 1992. *Ego and Archetype*. Pp 154-156. Shambhala. Boston and London.
19. Carter R. 2002. *Exploring Consciousness*. University of California Press. Berkely, Los Angeles.
20. Gazzaniga MS. 1992. *Nature's Mind*. London. Basic Books.
21. McKay D. 1987. Divided Brain-Divided Minds?, in Blakemore C. Greenfields S. Eds. *Mindwaves*. Oxford. Blackwell. 5-16.
22. Zimmer C. 2005. The Neurobiology of the Self. *Scientific American*. 293(5): 93-101.
23. Feinberg TE. 2004. Not What, But Where is Your "Self"? *Cerebrum* 6(3):49-62.
24. Peretz I, Zattore R. Eds. 2004. *The Cognitive Neuroscience of Music*. Oxford University Press.

CHAPTER FOUR

THE KINGDOM OF TITANS

THE PSYCHOZOIC HORIZONS

Neurobiological processes creating memorable registration of facts to ultimately reach and extend the consciousness in the human species could not have just been formed and remained a simple memory ground with fixed concrete concepts and without further change. The inherent plasticity of the sensing-acting process naturally expanded its horizons into imaginations. In fact, memory and autobiographical reminiscences could only expand in time with the implicit register acquiring greater extent and stability registering sequences of the time toward explicit realization. This fact of seeing both the changes and their carrier times as well and registering the enforced actuality of events made the realization of conscious sensing firmly established. Enriched with observations of environmental and interbiopsychons' interactions and renewed with time and ultimately having reached perfect conscious auto-hetero-gnosis, these impressions had original roots certainly real but unknown to consciousness. In this chapter, I attempt to eventually expose these unknown origins through retrograde scrutinizing of their known results in actual consciousness.

A host of intrinsic mental concepts, based primarily on concrete observation of surrounding events and forces and the later use of language, was gradually overshadowed by newly imposing abstract concepts and ideas. In the primitive mind, mental images mainly represented human and animal figures in both forms and functions, and the greatest initiating ego concepts in the primitive consciousness no doubt elicited constant effects on the sensing-acting mechanisms to be more refined in terms of comprehension and concept reformation. Based on these refining effects, figures of concrete external power presentations as egos eventually transformed into some fantasies forming the foundation of myths. The myths of superhumans and superanimals were created in the conscious minds and defined in reproducing forms once they could be categorized with stable specific names. Mythology thus was created, and with it mental images of titanic superhuman egos that lived with us, our self(s), and our autos from prehistory, continuing in some cultures even today.

Transition steps from the unconscious biopsychon with autos and admittedly without conscious self and ego, changing to form the mind with autos, self(s), egos, though presumably not known, can be nevertheless imagined on analogical basis. We have some ground on which to build coherent ideas about the basic neurological regulatory function at the start without consciousness and its corresponding similarities at the end with it. The facts reflect one major example—that of biocracy, which is undeniably real and is shown in what the evolution has shaped consciousness with its observable course and results. The final result of edifying a skyscraper by consciousness on the ground of biognosis, itself based on the archaic paleognosis and the progressing increasing functionality through bioawareness and bioconsciousness, are in fact astonishing evidences for biocracy and its presumed roots of consciousness, roots that can be investigated and possibly identified.

The early life figure, to reiterate briefly, is an unconscious biopsychon in the customary meaning of consciousness, with biotropism and biodynamism safeguarding its autocratic living, albeit subject to environmental conditions and forces. This organism has some regulatory agency inside it that responds adequately to types and intensities of external stimuli representing environmental forces: a paleognostic capability. I have named this agency's director or regulator "autos" as you are now familiar with it, somewhat equivalent to *Id*, an unconscious inner self introduced by Freud. Other than autos that presents itself as a necessity for the biocracy to go on with self-regulation, the unconscious biopsychon shows no evidence needing any regulatory organizer, neither structurally nor functionally. Therefore self(s) and egos are nonexistent, and autos is practically the only organizer in the unconscious but bioconscious organism.

In the conscious biopsychon, however, the regulation of mental forces and tendencies and the needed balance have been achieved with the collaboration of autos, self(s), and egos that as hypothetical entities are yet essential to explain psychodynamic mental processes. If therefore we argue for these entities to be functionally necessary to explain keeping the balance in the conscious biopsychon, we can look to find eventually elementary models for them in the unconscious autobiocrat. The focus of our attention to find such hypothetical unconscious models should be the organization of biognosis and its autos, which represents the central agency for regulating the organism's simple life mechanisms in the unconscious biopsychon. In fact, internal mechanisms in regulating energy supplies and usage in the unconscious biopsychon is on the same basic biological feedback effect for equilibrium as used in conscious biopsychon, including neurodynamic plays between autos, self, and ego. Therefore, protomodels for self(s) and egos can be assumed to be involved with the archaic autos.

Our efforts to possibly find a plausible explanation for supporting the hypothetical concept of autos and biognosis extended to prototype models for self(s) and egos may be based on three considerations. One is the pattern of psychological course of generally accepted interplay of autos, self(s), and egos in consciousness. This pattern includes an innate autonomy for the autos in the trilogy of autos-self-ego that implies an independence from the logical self considered in the standard self-ego explanation. As I already explained with modular self(s) elaborated from two categories of limbic and neocortical functions, the first category closer to mimic limbic emotional forces and the second mirroring more the noetic cortical concepts are more plausible to present two separate autos-made self modular types rather than self(s) of one type that cannot be equally limbic and cortical and modularly so different. The other consideration forms a twofold reason, theoretical and practical. Theoretically, the bioforce engendering life on quantum principle is so far the most plausible explanation entertained, compatible with the physical and metaphysical thinking, and further shows the evidence for the priority of limbic ground in the structuring of the future consciousness. This point will be thoroughly discussed in chapter 8. Practically, the change from subatomic uncertainty of time indefinite to conscious certainty of time definite type indicates possible graded precursors in differentiation, imposing roots in earlier times. Finally, the third consideration is the unity of the phenotype's spacetime that logically should not allow two forces of different self(s), but rather one of the autocratic autos of the limbic origin, preventing modular varieties of emotional and logical types and chaos in the unconscious biopsychon, suggesting only undifferentiated categories of prototype roots. Such roots as models for the future egos necessitating self(s) can be presumed to only relate to the biopsychon's basic metabolic balance affecting autos.

The components of the bioforce, the perpetual energy and time, are thus both affected and affect autos in its function in biotropism and biodynamism in response to changes imposed. These external effects causing biological changes and tending to imbalance the biopsychon's metabolism in biognosis, centered on autos, are akin to ego effects in consciousness impressing self and triggering the psychological interplay mechanisms to reestablish the mental equilibrium. Autos is the solo actor in its sensing and acting and naturally uses variable amounts of energy and time according to demands depicted by biognosis in the unconscious state. It is accomplishing what self is doing in consciousness.

As self acts to establish the psychic equilibrium by psychological means at different levels of conscious Gnostic states, autos does the same for somatic equilibrium by biological means in biognosis. In consciousness, egos could be variable and numerous, eventually converging to give the conventional ego models of idealized oneself, but in biognosis, egos can only be basic primordial ones affecting the functional components of the bioforce, energy and time. In this reasoning, the archaic prototype egos appear a plausible theoretical reality to be considered. These presumed ego models, imagined as such, indeed can become interestingly enlightening when their theoretical functional possibilities will also come into consideration.

At this stage of archaic functioning, based only on biotropism and biodynamism, invoking an early sensing-acting mechanism as biognosis with a central autos may appear unnecessary and unrealistic to those who prefer behaviorism instead. However, some grading in the sensing-acting domain that is poorly served with only conscious versus unconscious states but seriously needed to understand mechanisms involved can be exposed by behaviorism only confusignly. In addition, whenever a precise idea needs to become a concept for frequent use, more neologism becomes unavoidable to prevent confusion. The prototype egos that can be regarded as models at the stage of biognosis for future egos to be formed in consciousness will have to be defined according to their hypothetical ground. In this connection, the hypothetical foundation of all psychological agents like self and egos, introduced by the pioneer psychoanalysts in the twentieth century with the neologisms attached, though still remaining hypothetical, have survived because they tremendously facilitated discussing the psychological disorders. What I will introduce here is also given in the same spirit, but possibly on a more objective ground based on analogy with the established psychological norms.

THE FOUR ETERNAL TITANS

It is customary to ascribe egos to the conscious mind, and it is further normal practice to categorize the word *ego*, regardless of various meanings and interpretations attached to it, as mainly representing a wanted figure, character model, desired value, an enhancer of self-esteem, and in short, not a despised category of anything to be rejected. Although this view is correctly applicable in the psychology of the conscious mind, it is not tenable in the archaic sphere where any hypothetical sense provoking external stimulus could presumably trigger reaction by autos, both physically and neurologically, facing theoretical pseudo-egos or autogenous egos. Hypothesizing a state of paleognosis, including different functional levels of biognosis with biowareness or bioconsciousness, will necessarily allow only biological judgmental sensing to be accepted, but in potential duality of somatic and sensorial types. This would mean sensing biocratic mechanisms in terms of energy gain or loss and metabolically balanced time gain or time loss. Thus any external effects on the biopsychon, tending to cause imbalance in this dual equilibrium, must be regarded as an archaic ego type acting on autos.

Here, both good and bad effects, positive with gain and negative with loss on either energy or time, will form what can be theoretically called prototypes for egos in conscious sphere. Thus, there can be only four prototype egos or proto-egos theoretically, two for positive or negative effect on energy balance and two in similar relation with equilibrium time, the time of peaceful living in balance. However, unlike the quantum force with the equal reciprocity effects of the energy versus matter, or in reality, energy versus time in the hypothetical spacetime relations without decaying matter, this equal reciprocity is lost in the biological substrate with decaying matter, and the quantum principle in biology would be reduced to energy gain or loss, meaning longer or shorter living times. So in short, theorizing for models of proto-egos in biognosis, one could consider four prototypes—two for positive and negative effects on energy and two similar ones on the equilibrium time. However, considering all as hypothetical possibilities, one should retain the two prototypes affecting energy gain or loss as likely the primary models of future egos in consciousness but the other two, time gain or loss, as secondary prototype egos that are dependent on energy changes. In fact, the balanced biological time as discussed earlier, is strictly dependent on the energy gain or loss that affects the metabolic working time, which establishes the equilibrium-balanced time. Furthermore, the equilibrium-balanced time as a theoretical time span, could persist if the energy changes would keep the balance constant. This theoretical time span of the balanced biological time when the metabolic processes keep energy gain and loss in equilibrium is reminiscent of the time dimension in the bioforce. In this analogy, it could be shorter or longer according to negative or positive energy status and could be interpreted directly as negative or positive biologically balanced time change. Thus, comprehensively, the four categories of positive and negative effects on energy and time must be included in the hypothetical protomodel types for egos-to-be.

If we consider any directive of the bioforce as outlined in the end of chapter 2 in relation to its autofeedback balancing, we can visualize it in the form of an autos-ego agency that is concerned with maintaining the energy and time components in their downhill course of biological decaying, preventing any additional negative effects of environment on that pace. We can refer to the effect on energy as ergotropic and on time as chronotropic. The ergotropic directives should include energy changes (gaining or losing) causing theoretically chronotropic effects with lengthening or shortening of the balanced time of the status quo, which is the time in equilibrium hypothetically equivalent to the time dimension of the vector in the thought experiment of chapter 2, meaning time dimension in bioforce.

In summary, ergotropic effects controlling energy balance in activating energy-gaining mechanisms by being ergosynthetic, or losing ones and being ergolytic, can be regarded as archaic proto-models of egos inciting autos to act accordingly in the anabolic regeneration of the biopsychon. The ergosynthetic prototype ego here is as one hypothetical ego-autos that will have to do with all endogenous nutritional metabolic and exogenous contact and radiation energy gains by the biopsychon, and the ergolytic one is the one that spends energy in biological processes of the biopsychon including metabolism, locomotion, defense, and reproduction. These hypothetical proto-egos can be regarded as ego-autos. They do indeed incite the idea of a model for future egos to develop in the conscious biopsychon on the same basic principle of directives acting on autos but through self.

These assumptions do justice to the time dimension of the bioforce by recalling its biological form in the balanced biological time span. They also confer credibility to the biological ground for memory formation with biochronogenesis. Indeed time as the bearer of changes occurring in it can be considered as an abstract capacity that, if permitted to be unlimited, can include all changing events in an endless continuum. And memory in the conscious mind, in a way, theoretically expands this abstract time capacity and exposes it unlimitedly. In the unconscious biopsychon, hypothetical chronotropic effects, chronosynthetic or chronolytic, may affect the inner time dimension of the bioforce, which is the permanent limitless time for the implicit memory. In fact for the unconscious sensing-acting system, there is only one limitless time, which is the balanced biological time, the time that reflects the indefinite time dimension and is not consciously sensed. In the conscious biopsychon, however, biochronogenesis that has evolved with clearer recognition of past and present time by explicit memory makes interpretation of the measurable time a reality. The limited time thus appears to be then realized outside of the limitless inner time.

These hypothetical proto-ego figures that can be called autogenetic in biognosis in the unconscious biopsychon, as opposed to psychogenetic egos to be formed in conscious mind, are all established on the theoretical basis of enhancing gain or loss effects of the metabolic energy component on the autos. The effect would be sensed to act as stimulus on the bioforce in autos and the regulating mechanism would specifically ensure reequilibrating the energy-time component of the bioforce to keep the biological-balanced time as an inner time constant. Such time, sensed infinite, is in fact a whole unlimited and eternal figure of time bound to the implicit memory. Although the hypothetical foundation cannot be proved for either of these proto-egos, the basic theory appears facilitating the comprehension of steps and modalities of future ego formation in consciousness. Thus, theoretical proto-egos of unconscious types are fundamentally addressed to the autos and affect the observable balanced time of the biopsychon that is not spent for any activity other than consuming itself—the free time to enjoy life—representing the truest meaning of biocracy.

THE INNER TIME TITAN

To start talking about the colossal giant in us that I have called time titan, I would like to first describe a childhood experience I had in a marriage ceremony when, as a youngster, I was fascinated seeing the reflection of candles in the mirrors particularly of candles between two mirrors facing one another. The repetition of light reflections in the two facing mirrors was infinite and beautiful. It gave my novice mind an impression of immense enlightenment and serenity. I immediately thought that if I could be in the place of the candle, I would feel being infinite as my own self, in an infinitely expanded time and space, seeing interminable views of my dimensions: a feeling of encompassing the infinite past and future in the present time. In later years, in courses of physics, studying light and optics, I had occasions to ponder on this memorable phenomenon, and it attracted my imagination that if there was no light placed between the two facing mirrors, the mirrors still would theoretically reflect into one another infinitely, but showing infinite emptiness, a continued uninterrupted empty time frames only. I started to feel the notion of eternal time in a physical being, in any space occupying matter, grasping the analogy of an endless time continuum that becomes sensed when an event takes place in it and separates a frame of it, making a spacetime like the virtual endless spacetimes between facing mirrors.

Now, coming back to the subject of the time component of the bioforce, I find the same old analogy and believe that time can be regarded as a continuum of infinitely present spacetimes not noticeable unless at least one of the two other components of the spacetimes, matter or motion (energy in any form) is detected, breaking the undetectable continuum. The point of breakage is the actual realization immediately extending into past and present. This situation is real in the conscious sphere with self and explicit memory, but not in the biopsychon with autos and implicit memory. For autos, time is holistic and is nothing but a continuum in which biocratic processes are automatically carried out in a chain with repetition and no interruption but phase change for transfer of genes. Indeed if any sensing of time in biognosis could be imagined, it would be a sense of sequencing in continuity, keeping the same pace without a break. Dividing and segmenting the time would have to be realized by imposed impressions of different pacing or meaning of time category from the environment, like in urgent fight-or-flight situations. That would be from a different source of time, the external spacetime changes, as against the innate sense of inherent sequencing metabolic processes. The stereotyped repetition of the uniform biotropism and biodynamism in the unconscious biopsychon in repeated chains of steps can form the nucleus of engraving as a ground for initiating implicit memory in which time is holistic and eternal that can accommodate sequencing of life processes in a continuous chain. This gigantic time dimension of all-or-none type, virtually unlimited, permanent and eternal, here called time titan, is innate in the biopsychon and realized by it when sensed as the balanced equilibrium time. This created concept of the time titan personifies the eternity for the living organism as long as the biological life span of the living body is allowed to last, akin to the light between the mirrors that keeps the infinite unending reflections of the endless spacetimes to be visible as long as it shines.

The foundation of autobiographical memory in the biopsychon is also based on the sequencing recognition of biological living steps in the endless time titan. If the subliminal stimuli of biocratic processes reach and exceed the primitive neurological perceptive threshold repeatedly, facilitation in the neurological pathway can be effected to eventually shorten the interval of perceptions in that connective chain and form a primitive recalling capacity, which will recognize sequencing as an implicit memory of proto-episodic type. It will not be difficult for this primitive memory to become as extensive and as complex as in the developed human mind, based on extensive networking, modularity, and effective transmissions of engrams to recognize external time coding with the explicit memory expansion in consciousness.

The basic evolutionary value of memory with its role throughout the mental life, taking origin from the biognostic archaic sensing with the implicit memory, justifies calling its original endless time dimension the time titan. It also ultimately grants the sense of conscious mastering of time to the biopsychon identifying itself with that titan. If considered in an unconscious or early conscious biopsychon with the hypothetical interplay of basic ego models of ergosynthetic and ergolytic character, this sense of interminable source of time would assure a guaranty of an ultimate equilibrium. An infinite source of time that can dispense of all hypothetical chronosynthetic and chronolytic time effects as independent from them; a source of infinite time inherent in the bioforce and endowed in the biopsychon seems thus to be virtually inculcated in biology. It is the permanent present time at each instant, a real invisible titan with no beginning and no end.[1] In fact, we do have this undefined inner feeling of eternal being in us as noted and alluded to also by other writers.[2] On the other hand, the fact that is not clearly realized and recognized before conscious mastering of time is the relay the consciousness allows and whereby facilitates the mental conceptualization and unlimited regression and progression in time. This achievement is the result of coding in consciousness with the language codes being the communicative prototype that is served in mental ideation using chronologic logical bonds. In this situation, the biopsychon titanic personality is reigning by using the titan's unlimited capacity in the form of memory in repeated segments of codes with languages. These are possibly the simplest codes, making unities of ideas that initiate concept formation in the conscious mind. This aspect of time dimension, evolving from the unconscious biopsychon to the conscious one, will be further and more appropriately treated with the Navand theory of languages in chapter 10 on idiognosis, discussing the basic monodimensional present time. But now, we can depart from the kingdom of the time titan, leaving it for exploring another metaphoric land to sketch a comparison between a giant autocratic master biopsychon at early conscious stage and a more-evolved one with full consciousness as relates to egos.

THE CYCLOPIAN VISION

The mythological story of Ulysses (Odysseus) and the giant Polypheme (Polyphemus) has been told by different translators at different times and in different versions. I remember a French version of it that is particularly appropriate to show an autocratic conscious titanic master.

In Homer's *The Odyssey*, cyclopses are anthropomorphic giants with a single frontal eye. Ulysses, the Homer's hero, with his companions, are held prisoners by Polypheme, a cyclops, in the giant's cave to be devoured by him sooner or later as he has done so already with some of them. They have no hope to escape as Polypheme has closed the mouth of the cave with a ponderous gigantic slab of stone. After Polypheme devours another one of them for a meal, Ulysses tries to please him and appease his wrath by offering him wine from their reserve, hiding his real intent. To Polypheme asking Ulysses's name, he calls himself *no one* and offers him more wine, which pleases Polypheme, drinking plenty and falling asleep. Ulysses and his companions find time to sharpen a big timber making a pointed heavy rod with burning point and blind the giant's only eye with it, and they all hide. To Polyphemes's cries, cyclopses come for help, asking, who did that to you?

They only hear "no one, no one!" for response. They ultimately leave closing the gate with the same colossal stone. The prisoners eventually escape by hanging themselves under the bellies of Polypheme's thick-fleeced rams they had tied together by straps of bark so that each three rams would cover one man hanging underneath. In the morning, the blinded Polypheme moves the stone, stands in the doorway, stretching his arms and checking the back of his rams on their way out, finding nothing.

The story has its own rights to be interesting, but for my purpose, I feel that Polypheme is the best metaphore of the giant biopsychon titanic form with the sense of power and eternity, a pure autocratic biopsychon who has consciousness and language but is inefficient in his world vision, a linguistically developed mind without benefit of psychogenetic egos and with only autogenetic egos in the service of autos. This metaphor represents a conscious autocrat with a unidirectional vision of the world that only shows to him what makes up his living needs and the ways for providing for them irrespective of any other consideration. Even the mastered language does not go beyond the strict concrete. The absent depth in his monocular vision matches the same deficiency in conceptualization, which is almost totally absent beyond the tangible facts. Polypheme is probably the only full metaphoric example that can be found to show an anthropomorphic humanized autocratic biopsychon with consciousness and can be used to typify reinforcement of biotropism and biodynamism. Indeed, the story invokes anomalies that could alter significantly autognosis and give rise to profound disturbances in self(s) and in autos, if they were related to a *Homo sapiens* today.

The interpretation of *no one* by Polypheme indicating the culpable, and no culpable by the other cyclopses, and its use by Ulysses in its antonymic sense in his planning scheme maximizes demonstration of steps in concept formation from the strictly concrete stage to the extended mental scope that is made possible by the language in consciousness. The total inability of correct reinforcement by cross-examination when the most important detector of the environmental change, the visual detector, is deficient is also the other obvious important reason for misconceptions. The paramount necessity of perfect wired internal and wireless external contact with the environment to provide the inside system regulating autocracy with total information must develop in parallel with evolutionary changes granting more refinement with better developed neurological system. This in fact has been observed, times and again, in neuropsychological studies.

On the other side, for Ulysses and his companions, maximized negative feedback from Polypheme, shown with the threat of death, is the justification for self and therefore for autos to destroy this superpower ego figure. For them, conscious biopsychons with depth of vision and structured conceptualization, the saving enterprise must be a mental strategy to either destroy the destroyer or escape from him for good. It means that the threat's elimination, for the autos to remain undisturbed, is instantly needed one way or another. The biopsychon's permissiveness cannot allow acceptance of and identification with the mighty dreadful figure of Polypheme (a prototype ergolytic-chronolytic ego) in any way and in any length of time beyond an immediate decision making to reject the ego figure for the biopsychon autos only survives if the threat is eliminated. This situation affects the evolutionary allowance for survival in an *all-or-none* imposition, an allowance that can otherwise usually permit adaptation causing the ultimate fate of the biopsychon to be affected by more permanent adaptable environmental forces, the psychogenetic egos, to which the unconscious and the later conscious biopsychons could adjust.

Psychological interpretation of adjustments has used explanations of interplays between egos, self(s), and autos. In my description, I am limiting autos to the fundamentally same autos of the primitive unconscious autocratic biopsychon, and treat self(s) as representing agents of autos in the stages of autognosis-heterognosis and thereafter that check information from all sources other than autos forces, all considered as egos with variable polarity, permanency, and authority. Here, I include egos, from the one to be rejected instantly on the one hand and therefore only included in the category of egos for their common superpower character (like Polypheme and his threat of death) to all those that have had shorter or longer impressions on autos in the scope of evolutionary time. In this approach, to be comprehensive, all four protomodels of egos on energy and time for the unconscious biopsychon, and all externally and internally originated concepts of the conscious biopsychon that could be regarded as egos shaping the psychological interplays between autos and self(s) impose consideration and are included.

However, egos in the orthodox psychological understanding are all formed concepts that define value models to be accepted or tolerated by autos or Id through self(s). The quintessence of standard psychological egos is that they are formed in conscious minds and can be clearly factual or potential in real time. However, I do believe that we can benefit from considering what I have called autogenetic egos characterized as premodels of psychogenetic egos in tracing the background processes to unconscious governing autocracy that is a biological reality. They could facilitate a more profound understanding as being functional without being conscious and as can be reflected in behaviorism without being evident.

PSYCHOGENETIC EGOS

Processes through the evolution to provide consciousness have been gradual, and not knowing the exact stepwise-detailed evolutionary course of the progressively developing consciousness does not exclude its grossly evident change in becoming more complex. Therefore, its origin from a simpler archaic model based on principles governing its functions appears to be a legitimate reality that I have considered in invoking the archaic ego models for development of the psychogenetic egos. Possibility of such archaic models in the autocratic unconscious biopsychon is further supported by the behavioral human newborn's unconscious acts labeled instincts.

The psychology of intrauterine behavior in late pregnancy, mainly supposed to be perceptual in terms of hearing sounds of human voices and music and rapid eye movements indicating possible dreaming notwithstanding, the evidences fully observable in the newborn are undeniable facts denoting reactions to external stimuli interpretable as egos. We can realize the contact effect of the caretaker—her body warmth, caressing acts, emotion filled assuring tone, and of the impact of the environment of warmth, lighting, and music—all being factors to form egos. These are indeed adamant directing forces of the environment for the developing mind of the new born. It is basic to compare the contrasting effect of the intrauterine state with that of the immediate extrauterine life on the newborn infant. The loneliness of the intrauterine life, darkness, and absence of the external stimuli, which provide a uniform protective, calm, and comfortable environment, is the ontogenetic model for night reclusion and sleeping into an unconscious refuge. The sudden extrauterine change into brightness and shower of external stimuli imposed with variable effects of pleasant and unpleasant nature, a model for day break and all unexpected events, is the infant's first ego experiencing. The very initial external bodily contact especially in the newborn's sucking reflex searching the breast, which represents the archaic fundamental positive biotropism and biodynamism, is now showing the observable forms of the hypothetical archaic ergotropic (ergosynthetic ego) and chronotropic (chronosynthetic ego). These reflexes, named instincts, are said by Freud and his adepts to be the early ego representation in life. Even the term *libido* is used at this early stage of neuropsychological development for the instinct of life against death. These interpretations on Freudian terms represent ego figures that I would regard as archaic autogenetic egos of ergotropic and chronotropic types.

If therefore we consider egos in the simplest descriptive terms, as the somatosensory or neuropsychological representation of external forces causing good or bad effects on us, we can see that the principle is essentially the same as manifested by the autocrat in autocracy using only positive and negative feedback mechanisms. In fact, egos in psychology of the conscious states are equivalent of directives in autocracy by autos, organizing the regulator mechanisms of biotropism and biodynamism, either directly or after changes through interaction of archaic ego figures with autos. It seems crucial to clearly outline internal regulation of the unconscious autocratic biopsychon in comparison to conscious state of minds with theorized distinct autos, self(s), and egos.

In the unconscious biopsychon, autocracy entirely functions to maximize positive feedbacks and minimize negative ones in regard to life with keeping a balanced constancy unfailingly obeying autos. By definition, in the pristine simplest unconscious autocrat, from bacteria to unicellular organisms, there is an internal regulation by biocracy with tropisms and dynamisms. In a plasmodium to a multicellular organism considered to be an unconscious autocrat, the internal regulation still shows a simple but multifaceted task necessitating harmony in the mechanisms involved. There is one autos in the autocrat, there is no self(s) and no apparent egos other than the hypothetical archaic prototype egos. All forces of the environment (egos-to-be) are adjusted by biotropism and biodynamism directly from autos.

Contrasting with this simplified plan in which autos is faced with the presumed archaic egos in material forms, we find the more complex plan of autos, self(s), and egos in the conscious biopsychon in conceptual forms. Here, the egos are defined as independent agencies facing self(s), and complete separation of egos from autos and from self(s) (the agent of autos) is clearly evident. The real separate nature and the multiple conceptual significances of egos, showing in the conscious sphere of mind, grant them independence and authority reflecting the force behind their foreign origins, a force that once assimilated enhances autos. In both situations, the egos are ultimately influencing autos, but in consciousness, the ego, though clearly elaborated by conceptualization, is yet devoid of any direct relation to autos and any similarity with self(s) and is totally foreign to both before being assimilated or rejected.

Appearance of consciousness with autognosis exposed the biopsychon's conscious mind to a great number of environmental impressions as well as inner imagined ones, mostly representing forces out of control, recognized by the biopsychon only to know how to cope with them going undamaged. This recognition made the biopsychon use conscious mental resources to support and reinforce its biological means of adaptation for survival. Facing the far superior powers by the primitive hominid was then an early preoccupation of conscious thinking. Consciousness also allowed exchanging ideas between the peers and, no doubt, was determinant in more open expressivity being aggressive and exercising force, or being submissive and obeying it. Advent of language and concept formation introduced the figures of powerful superpowers of natural environmental and human forces collectively and individually, giving rise to myths, god kings, and numinous figures, to account for all that was superhuman. Psychogenesis of egos is indeed inseparable from the myths. Legendary extent of imagination of the primitive minds, perhaps mainly influenced by the autogenetic ego models, allowed extraordinary imaginable sketching of stories of the superpowers and their superhuman lives and deeds.

The story of conscious ego formation is interestingly revealing in passing from concrete to abstract stages involving metaphysics. The period of concrete ideation of the human mind probably extended from the early hominid thinking about animals, material environmental changes and objects, to concepts of superpowers, natural phenomena, god kings and their deeds and meanings agonistic and antagonistic to human expectations—a polytheism tending gradually to favor one figure more than others. This era of Gods for all natural events and effects on the human mind culminated in Mithraism in the Aryans, before separation of the main creed in subsequent geographic divisions or shortly at its beginning, to face ultimately the monotheistic Zoroastrianism.[3]

In this period of progressively refining ideas, the crux of the impetus in the abstract concept forming mind was to gain power to exert out or to protect oneself. This was the quest for the all wanted superpower, projected by the psyche to form an independent entity to which identify and with which gain comfort. Creation of an immaterial image, unreachable, unassailable, and of unsurmountable power and value in the myth of the society's beliefs has been a necessity and has in fact taken place probably at a very archaic time. The starting was probably well before language became ubiquitous and concepts refined, but exaltation of it should have been the product of full language development and actually recounted the early history of mankind when writing appeared and allowed information to reach us. Today, we are witnessing the exhibition of a compendium of egos in the human mind, mostly undefined, and a continual metamorphosis of investigational ideas on interpreting their intrinsic effects on self(s), or more formidably on autos.

ABSOLUTE KNOWLEDGE AND POWER

Psychogenesis of egos is inseparable from mythology, for both myths and egos are products of imaginable figures—the former too unrealistic to face autos and the latter nearer to a wishful improved state of oneself and ultimately possibly compatible with autos. Any psychogenetic ego could be regarded to have the background of one or more of the autogenetic ego models.

If we study the mythology of east and west of Mesopotamia and their private sectors of Indo-Iranian, Babylonian-Egyptian, and Greco-Roman lands, we can find innumerable vivid examples of archaic character egos that fit one or more of the four autogenetic model types that I have described. However, first I take the example of the abstract concept of absolute knowledge. This choice is justified and valuable for three reasons—firstly, for the abstract character it is supposed to present, which can be knowledge in a substrate or by itself as name for an immaterial entity; secondly, for englobing the ergosynthetic model character with enhanced mental gaining; and thirdly, for implicating the chronosynthetic model for extending the mental value of time. The classical story for the quest of absolute knowledge takes place in the Garden of Eden.

God places man in the Garden of Eden, according to Genesis, telling him that he could eat fruits of every tree but not from the tree of knowledge of good and evil, for if he does so he will die. In the meantime, God creates Eve from Adam's rib, and the serpent tempts Eve to eat the forbidden fruit because not only would she not die, her eyes will be opened by it, and she would see clearly and would be like God, knowing good and evil. Adam and Eve eat the fruit, see clearly their nakedness and cover themselves with fig leaves. God discovers their disobedience and says, *Behold, the man has become like one of us, knowing good and evil.*[4] God expulses them from the Garden of Eden before they could find the tree of life, eat its fruit, and become immortal.

The story includes three important points—the initial one is the innate immortality that can be supposed for Adam and Eve in the Garden of Eden, which is later lost; the second is the tale of two trees, one for life and one for knowledge, unknown to them at the beginning, hidden from them by God; and the third is the oracle by God referring collectively to *us*.

Metaphoric significance of the story is shining with the initial unconscious biopsychon autocrat with its innate eternal time dimension and the presence of one autoregulating power of unconscious life and none other with any conflict by concepts of good and bad. The story shows the change from unconscious to conscious mind, realizing the value of time and threatening a possible change of Adam and Eve to deities if they access the tree of eternity and eat its fruits.

The myth is also significant in its representation of autos and the four subtypes of the archaic models of autogenetic egos to become later conscious. One can interpret the creation of Adam as autos to indicate the ergosynthetic ego presentation of the power God gifted to Adam, the secondary creation of Eve from Adam's rib to depict the ergolytic loss for possible libidinal gain, the eating of the forbidden fruit to indicate ergosynthetic and chronosynthetic models by adding to energy expanding the time, and finally the expulsion from Garden of Eden to show the clear loss by ergolytic and chronolytic types of the hypothetical archaic autogenetic models.

The plurality of Gods and their supposed anger and greed in expulsing Adam and Eve and keeping the tree of life as their sole divine property is finally to be emphasized indicating the conscious importance of the ego of immortality: a coveted wish of unsurpassed value even for Gods! The plurality of Gods, other than indicating the human fear of loneliness, also seems to indicate quantification in the coveted quality ego of mastering time and immortality through projection in a generalized common conscious mind. Based on these hypothetical archaic ego models, interpretation of psychogenetic egos in presuming correct direction of changes is more important than finding clear-cut similarity or analogy between the psychogenetic and the autogenetic ego types.

The quest for immortality, the envied eternity in mind, can be regarded as a gift of timelessness given to the conscious mind. We may barely notice the titanic time giant in ourselves in some earlier young times of our lives without realizing it consciously. This quest for immortality has occupied the sphere of consciousness throughout the history of mankind. Richness of mythology here too can provide interesting legends to match the ego of immortality that takes its ultimate form in consciousness and not before when the giant titanic time reigns in the unconscious autocrat. If we study the subject with objectivity, this crucial hypothetical time titan appears in consciousness as an autogenetic ego model that forms the conscious ego of *being* in the permanent present time expressed in words, which will appear with the initiation of the early Indo-European root languages in the form of the phoneme /a/ as suffix to names or pronouns, as I will describe in chapter 11. This ego, a bona fide chronosynthetic autogenetic ego type, seems to constantly show up in words to reject conscious fear of death that can be regarded as an autogenetic chronolytic ego. In fact, archeological studies excavating ancient cites of early civilized human colonies have invariably shown needed objects for daily life buried with bodies. The interpretation is that for these early humans death was not the end of life, and that life would continue beyond physical death, a tacit acceptance of the inner immortality. This acceptance is in fact clear evidence for enhancing the chronosynthetic ego and categorically rejecting the chronolytic one.

Excavations extending over lands that cover from proto-historic to prehistoric and historic times throughout the living sites of early dwellers from China to

Europe show signs of fundamental human beliefs. They extend to include many findings of Mesolithic to Neolithic times and thereafter from prehistoric to historic eras.[5] Extent of findings show the importance of the hoped immortality that has appeared in the human mind and has continued so long overwhelmingly. Scrutinizing this belief, we can find two salient features to it. The first one is the rejection of the fact that death is real, in fact rejecting the strong chronolytic ego, and the second is the conceptualization of a soul and projection of immortality onto it, a psychological defense against losing immortality, defending the strong chronosynthetic ego. This solution to secure immortality by identifying with an object that could legitimately harbor immortality then appeared clearly when polytheistic religions gave way to monotheistic ones and the concept of the soul as an inseparable part of the unique immortal divinity was shared by the mortals. Thus, religions are in reality rationalized versions of older irrational mythologies, themselves originating from autogenetic ego models, and the essential of mythological beliefs continues to live in religions under disguise.

In the older polytheistic beliefs of more mythological character, notably in Mithraism, we can find Gods of variable power and ranks, some mighty and ruling, others peaceful and tolerant, some praising righteousness, and some promoting deceit, some controlling specific natural events like light, fire, wind, rain, storm, darkness, disease, death, and still many other deities of second ranks with more mundane activities. In this chaos of so many divine characters, a model of a crowded Pantheon, a God of time is hard to be properly placed in its correct position without damaging the system and is in fact totally absent in the position we would expect it to be, i.e., in the summit and with the lasting veneration he/she should deserve. This God, named *Zurvan* [6] that appears more clearly later in the Persian mythology as a replica of the same deity in old pre-gathic myths, is called *Zravaana (or Zarvaana) Akaraana* meaning the unbounded time and is also named *Abun bun bunomand* in the Pahlavic terms meaning *the unbounded of no boundary*. He is then placed in a high rank and considered as generator of Spanta Mainyu and Angara Mainyu, the good and the bad spirits.[7] In these terms, Zravaana simulates the trees of life and knowledge of the Garden of Eden in one personification. Somewhat similar to this deity are Tiamat and Apsu of Mesopotamia and Purusha of Vedic origin. These deities are also the origin of Titans in eastern mythology.

Comparing the story in Genesis with that for Zravaana Akaraana, what is blatantly different is the unequivocal separation of the sources of knowledge and immortality in two trees in the Garden of Eden as opposed to the unique source for both knowledge and eternity in Zravaana Akaraana, the one God of eternal time and good and evil. In contrast to this difference, tacit polytheism appears to be visible in both stories.

However, Ahura Mazda who is the only one God in the established monotheistic Zoroastrianism is said to have created Zravaana, and therefore, the real importance of this God of time and his position are basically considered secondary in the Zoroastrian Persian myths. Thus, it appears that seemingly, both biocratic autocracy and Persian mythology have not recognized this God of time as autonomous. But interestingly, the situation is the same in the Greco-Roman mythology in which Cronus (Roman Saturn) as the God of time initiates ruling over the other Titans but is dethroned by his son Zeus and thrown down from Olympus and nothing else is heard of him. One cannot avoid sensing the analogy between these examples of the God figures for time when religious thinking has been more or less polytheistic than monotheistic at the origin of myths. A simplification of the ego archetypes is easily observable in these examples.

The explanation for the implicit and tacit characters of the main chronotropic ego in the polytheistic religions without complete ignorance of its mythologic figure, the deity figure for time, is understandable considering the inability of correctly positioning the God of time in the Pantheon. In monotheistic thinking, on the contrary, all deity powers are incorporated in the one and unique almighty God who is the creator of everything. God grants life, a life that conscious biopsychons are supposed to enjoy in their physical material life and in their souls when they continue living their eternal nonmaterial life after death. The chronotropic ego in its inapparent position in the conscious mind is thus reinforced as a dominant structural part of autos that now has gained soul for eternity. All additional perceptual and conceptual egos in consciousness make up the psychogenetic ego figures to fulfill the protomodel autogenetic ego functions. The chronotropic ego and its chronosynthetic and chronolytic models can be seen to have become inseparable from the ergotropic ego models as implicit parts of autos. In the unconscious form, the autogenetic egos act through a permissive gate of undefined mechanism to incite autos action directly. But the psychogenetic ego figures, showing more concrete faces than the autogenetic ones, act on autos through self. The feeling of permanency in time in the unconscious biopsychon is innate and inapparent to autos. Only consciousness detects and conceptualizes time in the form of the actual real time out of autos and archaic ego models for time. Egos for self-esteem, superiority, aggressiveness, despising fear, risking, fighting, and destroying can be explicit examples of inflated archetypal egos[8] that can be assimilated to one or more of the archaic autogenetic models. Studying them can ultimately evidence relatedness to one or to another or to several aspects combined of the four model types that I have described as autogenetic egos.

An interesting conscious presentation of the ego models witnessing the archaic safeguarding of autos with autocracy is depicted in Spanta Mainyu of the Zoroastrianism mentioned earlier. The ergosynthetic and chronosynthetic models of autogenetic egos consciously personified in Spanta Mainyu, the expanding spirit that is regarded holy, is basic to promote expansion of life for autos irrespective of possible opposite effects by Angara Mainyu, the contracting spirit, less significantly considered. Here comes an important consideration for the conscious morality as to how far this expanding trend for autos can be allowed to continue if it could damage other beings in its progressive course. This primordial conscious moral question simply does not exist for the unconscious autos bound to the principle of autocracy but is unavoidable in the conscious mind with autognosis and heterognosis creating various defensive mechanisms between egos and self(s) to establish adaptive balances for autos to survive in consciousness. The answer to this question is interestingly evident as will be seen in discussing Zoroastrian Asha concept in the next section.

EGOS IN THE QUANTUM MIND

From Freudian psychology to the present time, ego as the agency of self between the Id and the superego has acquired an established personification in the mind in our present stage of knowledge. As such, ego can act in the human mind both consciously and unconsciously. I have regarded as probable the possibility that ego in its yet unconscious form could be acting in the unconscious and early conscious biopsychons so far discussed with the prototype models, possibly as directly aiming autos, advanced by the hypothetical but yet attractive autogenetic nature. I have based this possibility on the observed behaviorism and on the functional analogy that theoretically exists between the living mechanisms of unconscious and conscious biopsychons.

Thus autos is in reality the nucleus for this variability of relations with egos and selfs in the unlimited permissiveness of the time titan. In essence, autos in its position is the interchanged ability factor in quantum physics between energy and matter, an analogy that cannot be overlooked. Indeed, the same inherent characteristics are operating both in the quantum physics and in the conscious mind. Shrödinger's probability, Heisenberg's uncertainty, de Broglie's matter-wave correlated variability, and Bell's theorem,[9,10] all pointing to a state of uncertain choice in probabilistic possibilities in quantum physics, also indicate changing potentiality of autos facing autogenetic egos in the unconscious or egos through selfs in the conscious mind. The mind seems also to function on the same evident basis making this analogy significant. In fact, it has been demonstrated that every conscious act is in reality started unconsciously, and the conscious will, noted as such, is a *wont* rejecting an alternate possibility presented to the consciousness.[11] Viewing egos with this understanding in the analogous quantum mind can show the extent of possibilities less well-appreciated otherwise.

Ego in the conscious mind undergoes changes along the learning and assimilation of new knowledge in society living. It is a mental value mostly to be assimilated, increasing the self-esteem, a gain for autos, and rarely a threat to be rejected like that described with Polypheme. Fundamentally, the mechanisms follow the four autogenetic ego directions to use ego-self axis to ultimately involve autos, but for the mental life to prosper sublimation is the rule. A sublimation of ego into a superego can occur in the concept-forming mind by identification of an existing ego to its newly elated form.

The very solid archetypal superego that has ever been formed in the human mind is that of the supreme power of an imagined almighty God representing personified eternity.[12] Mythology witnesses many such changes accorded to lower class ego figures endowing them with an ultimate sublime character, but the one concerning the all potent, all autonomous, and imperishable power of the almighty God is the most ineffaceable example in the transition from mythology to religion. Changes have gone from multiplicity to unity, increasing the potential power, and from one all-inclusive concretely originated to abstractively perceived figure.

These God's attributes, the superego's functions in evolved theological doctrines, are considered to be essentially theocentric but of two types—one that engenders creation and controls creatures and their acts and destinies, which is a fatalistic creation allowing no free will, and the second that engenders the creation but leaves the creatures free under the laws of causality, which is a deterministic creation allowing conditional free will. Philosophical interpretations, with theological orientation or without, have encumbered the subject greatly and have caused subdivisions in theophilosophical thinking. The one doctrine that has approached physical reality and kept the essence of theological inclination on a solid logical basis compatible with autos and autocracy is the original gathic belief in Asha (truth).[12]

This doctrine's theoretical interpretation concerning metaphysics will be briefed in chapter 12, but its practical applicable interpretation to life concerns our present purpose. Asha in the Zoroastrian gathic writings in practical interpretation means the path to absolute purity of the Anaghra Raochanghaam (infinite lights), greatest power, the almighty Ahura Mazda. Its practical interpretation poses grades of purity as opposed to impurity, and thus Asha establishes a basic relativity in goodness. It is of interest to note that many religious myths and rituals have followed the same principle in belief and conduct to purify the mind, and thereby the soul, to approach God and acquire eternity in identification with God. In gathic belief for achieving this, one has to choose the path of Asha by one's own free will, a free will clearly reminiscent of autos tending from uncertainty to certainty. And the individual conscious mind, with one's own conscious judgment indicating the ultimate choice that can be Asha, meaning truth, correctness and justice in life, or its opposite, Draogha and Aka, meaning lie, corruption, and injustice,[13] offers the choice between the two. This is a clear example of conscious interaction of two opposite poles of power, both in ego forms, acting on the autos-self(s) axis following a quantum model. This interpretation of Asha exposes its abstractive aspect based on abstract goodness as the dominant power, which, in my opinion, is opposing its concrete aspect that bases dominance on the energy hierarchy in the physical world.

Now in the structure of this superego, one can see that the abstract concept of Asha as an example in Zoroastrian philosophy is theoretically representing both trees of eternity and knowledge of the Garden of Eden showing unification akin to the monotheistic Zoroastrianism. Thus, the mechanism of conscious analysis and synthesis for egos and superegos forming new concepts is essentially based on cultural foundation and depends on persistence of archetypal beliefs on the one hand and spread of new knowledge on the other. In addition, the mechanism uses a stereotyped two-step process in which the first step is to assimilate ego in self, and the second step to serve accommodate the self-accepted concept of the ego with the supremacy of autos.

From the early primitive encephalic development without elaborate consciousness to the fully developed stage, egos start as primarily concrete concept types and not really imposing persistantly on autos. This is an intermediary stage between that of autos and ego models at the bottom of the scale to that of imposing egos and super egos with self(s) and the use of well-developed adaptability processes and defenses between the self-coupled egos and autos at the top. The ultimate stage is in culturally developed societies where moral egos are of imposing character and can press upon the autos heavily. These moral and spiritual egos increase the value of ps over that of pi as described in chapter 1. However, the natural dominance of pi over ps is basic and fundamental by the natural potency and supremacy of autos in all normal equilibrated mental lives as long as the balance is kept in favor of autos or just at the equilibrium level for a particular social environment of life. Unbalancing the equilibrium in favor of ps by the restricting effects of strong ego forces on self(s), educational, severe abstinence, unbearable oppression, deprivation, separation, and so forth is well known to result in anger, depression, regression, isolation, or eventually persistent complexes to neuroses.

Ego sublimation on the other hand can also result in mental value gain through mental productive work. The results have been shown through artistic, literary, and scientific achievements. Here, the ego may or may not be distinct as a model, but the mental elation serves as an impersonal stimulus, an anticipated ego figure to adopt and leads to artistic perfection. Every art object as a novelty can be regarded to act by an autofeedback on the axis autos-self-autos introspectively to the artist and an allofeedback extrospectively on the axis *self-autos-self* of an admirer or a critic. In this line of order, poetry is richest in its direct positive feedback effect on the basis of abstractive metaphoric power of words on the self(s) in general. The reason is the recursive nature of this particular art, which is in reality conditioned by the codified system of the language, which is further codified in prosody in poetry. This sublimation of ego occurs by the enhancing beauty of structure, meaning, and rhythm effects especially involving emotions reaching ultimately the autos but with determinant lasting effect on the self(s) through semantic memory.

Loving and expressing the desire for self-sacrificing to reach the beloved, or to be even remotely unified with the object of love, is another sublimation of ego that is basically a deep need for sharing emotions with a projected symbol of protection and care. This type of sublimation internalizes the object of love possessively and may use the six words starting with *i—incorporation, introjection, imitation, internalization, idealization*, and *identification*—as formulated by Vaillant.[14] Altruistic behavior, acts of benevolent colossal monetary donations, servicing humans or animals with extreme abnegation, renouncing to enjoyment or refusing mundane values as in ascetic practices, and other similar activities can be all included in this category of sublimation of egos, although possibly including mixed origins and variable courses before reaching the sublimation in this altruistic end. Such sublimations of egos with demonstrable evidence, absent in the unconscious autocratic biopsychon and being an adage of consciousness, are naturally reflecting the coevolutionary effect of the consciously developed conceptual values.[15]

Changing egos to superegos or superegos to egos in the mind can take place according to the benefit of the change for autos, which identifies more easily with either changing effect if acceptable. Egos can only form after being first accepted by self(s), if not tresspassing the adaptability threshold for autos supremacy that is maintained silent but salient. Changing can occur to provide a defense mechanism and to facilitate identification of self(s) and ultimately autos with the changed forms. The interplay of egos and self(s), and ego adjusted self(s) and autos, seem to widely follow the lines drawn by the archaic ego models of the four types described.

In summary, egos treated in this chapter in colorful metaphoric sketches are mental concepts usually representing characteristics of forces above natural human capabilities, desired to be achieved as possibilities of autos if accepted by autos. In the unconscious biopsychon, ego figures are archaic models that reflect hypothetically the two components of the bioforce in the autobiocrat, the energy dimension and the time dimension of that force. In the autocratic unconscious biopsychon, the supposed ego models are autogenetic and only represent positive or negative effects on autos enhancing positive or negative biotropism and biodynamism. These egos in the unconscious biopsychon can theoretically originate as autos-ego. In the conscious biopsychon, egos can be comprehensibly formed on these models, resulting from extrinsic forces or intrinsic ideas starting representation as self-ego. In the conscious biopsychon, egos are psychogenetic products of consciousness, and in general, they present desirable effects on autos through self(s), essentially structured on the archaic models of autogenetic egos. The overall effects of variable psychogenetic egos on autos follow the same principles of autogenetic egos. The archaic autos supremacy is generally kept undisturbed but potential in spite of mental dispositions sometimes enhancing ps over pi.

The unconscious biopsychon is an unaware time master with an implicit sense of time permanency, which is the manifestation of the titanic time continuum in autos. In the conscious biopsychon, in addition to the inner time titan, time becomes both the egos of immortality and mortality, the concrete perception with recognized limits imposing contrast and also extendable to infinity through memory encompssing past and future.

REFERENCES FOR CHAPTER FOUR

1. Ashtiani J. 2002. *Zarathushtra*. Translated in English from Persian by M. Nourbaksh, Edited by H. Pirnazar. Sherkat-e Sahami-e Enteshar, Tehran, Iran.
2. Edinger EF. 1992. *Ego and Archetype*. P30. Shambhala. Boston and London.
3. Razi H. 2002. *Encyclopaedia of Ancient Iran*. Sokhan publisher. (text in Persian).
4. Edinger EF. 1992. *Ego and Archetype*. P16. Shambhala. Boston and London.
5. Gabel C. 1965. *Man before History*. Prentice Hall, Inc. Englewood Cliffs, N.J.
6. Razi H. 1992. *Mithraism; Cult, Myth, Cosmogony, and Cosmology*. Pp424-430. Behjat Publications. Tehran, Iran.(text in Persian).

7. Bahar, M. 1994. *On Some Debates in Iranian Culture*. Fekre Rooz ublications. Tehran, Iran.
8. Edinger EF.1992. *Ego and Archetype*.Pp26-36. Shambhala. Boston and London.
9. Walker EH. 2000. *The Physics of Consciousness*. Basic Books. A Member of Perseus Books Group, New York.
10. Weaver JH. 1987. *The World of Physics*. Vol.II.Pp 475, 481. Simon and Schuster, New York.
11. Libet B. 2004. *Mind Time. The Temporal Factor in Consciousness*. Harvard University Press.Cambridge Massachusetts, London England.
12. Edinger EF. 1992. *Ego and Archetype*. Pp37-40, pp154-156. Shambhala. Boston and London.
13. Mehr F. 2000. *The Zoroastrian Tradition. An Introduction to the Ancient Wisdom of Zarathushtra*. Pp42-45. Zoroastrian Benevolent Publication. ZoroastrianCenter. California.
14. Vaillant GE. 2000. *The Wisdom of the Ego*. P344. Harvard University Press. Cambridge, Massachusetts. London, England.
15. Fehr E, Renninger SV. 2004. The Samaritan Paradox.Pp 15-21. *Scientific American Mind*. 14 (5) 15-21.

CHAPTER FIVE

BIOGNOSTIC STATES

SENSING THE SENSE

The state of conscious awareness has been extensively studied in many disciplines, by philosophers, psychologists, neuroscientists, cognitivists, and psychiatrists, and also in hypnosis, occultism, and computer science, to cite examples. As a result, significant information has been gained on several aspects of consciousness, but in reality, consciousness still remains unknown. Its puzzling complexity that prevents its constituents to be distinctly defined, and its possible structural code or codes, if any, to be deciphered to really know what consciousness is appear beyond present accessibility. Consciousness constituents, indeed all formed by codes bringing them into time definite biology from the time indefinite biology (see chapter 6 and 8), necessitate extensive data study, scrutiny, and accurate interpretation to be validated. Particularly perception, attention, intention, logic, and decision making are important aspects of such studies in which the language codes play the main role. A host of affective conscious feelings and the interplay of innumerable other factors, not to mention the role of memory and evolution that have allowed consciousness reaching its present stage, are far from being lucidly known. With linguistic thinking capability as abstract artistic and Gnostic concepts, there are decisive points to be defined. To set a goal to study thoroughly and clarify this complex problem would be pretentious of any single mind with interest in this field. But to try basic clarification for a more comprehensive approach to the study is my goal. I will try to elucidate aspects of consciousness related to biocracy hopefully paving the way for more precise clarifications to follow.

Indeed, biocracy is a reality, and its course through the evolution is a proved event that has used sensing from the very beginning to evolve into neurosensory refinement to finally give brain, mind, and our present state of consciousness. On this solid evolutionary ground, sensing of simple to complex type has shown regulatory evidence for sustaining biocracy on one unique principle—serving the autocrat and, by one single aim, securing a hire for the autocrat. Thus, primitive and complex nervous systems and their functions in essence have helped the living species to survive and reproduce, securing the material biological ground for the ultimate conscious states with the futuristic immaterial biological ground for metaphysics.

Thus, consciousness, as classically understood, follows unconsciousness phylogenetically through the natural selection in the course of evolution. It is a time maturation of an earlier state that we call unconsciousness due to lack of other more precise denomination for it in our today's vocabulary. Definitions for consciousness indeed have been generally applied as black on white, considering absent realization of self and nonself against solid mental such realization with the possibility of lingually coded autofeedback readjustments. But is this really the way we are sensed to view consciousness against unconsciousness? Or should we allow an extended transitional zone between the two, assuming distinct components in between and presuming a progressive change for each constituent contributing factor from its inception to our present time?

To the first question, the answer is certainly no; although the two states appear as black and white, there are intermediaries in between them. There is also an immeasurable part of consciousness that floats and submits itself to no limit to allow any precise definition, like most emotional states that we are aware of and unable to define with precision as we otherwise do in our language with mathematics and science. And to the second question, the answer is yes as gradual change from an initial simple state of sensing-acting to our present day complexity has been witnessed to take millions of years.[1] In fact, the majority of our performing skills are executed unconsciously, and in our daily activities, we do things that we are never conscious of doing them although we know they could not have been done by anyone else or originated in any other part of us than in our brain. All our routine daily activities are unconsciously performed, and in all, we can become aware of any obscure step needed to be reviewed if we roll back the film of events from the end to some prior time and again screen from the start. The entire series of events then becomes reviewed by memory until the missed step, evidently performed less consciously, is found and brought to light.

In this reviewing process, there are steps in sequence that have to be followed. If we realize with the evolutionary psychologists that the sequence and steps of living processes with all events and actions have gone unconsciously, or less consciously, for millions of years, we can see that if today's human biopsychon would have to locate one event in its remote past, it would need a gigantic memory and cross-referencing index. Now such capacities are in fact possibilities with consciousness and its literary lingual coding today.

The course of the evolution indeed evidences this fact that developing and perfecting memory and expanding and using the nervous system capacity have worked as adaptive traits in the service of the living species, *Homo sapiens* in particular, to have reached the stage of today's mental refinement with consciousness. We then grasp perhaps the single most important clue in the development of the nervous system and consciousness, which is the realization of the consecutive time framing by perception-cognition that allows recognition of events as a factor of time. Gnostic value of consciousness thus appears to be in the power of using memory for singling out recorded frames of mental images at will and bringing them from the extended time titan to the actual real time. If the steps could not be singled out by perception-cognition and recorded in memory, the essence, the task of time sequencing and consciousness realization simply would not have been possible.

Thus, the essential role of memory in consciousness undoubtedly is fundamental, and its mechanism undeniably time related, making time frames for events whose sequencing allows recognition of the frames in series realizable. This serial recognition is perfectly a natural outcome of the normal course of time, the very chronic order of events and actions. This has indeed been the conscious sensing in life experienced in brains with even early limbic-cortical connections and for millions of years before our present stage of consciousness. This consciousness has been holistic, fundamentally based on vision and synthetic meaning of events and actions and essentially monitored by the right cerebral hemisphere having holistic dominance. However, the more salient and significant frames of the chains of events specifically impressive to autos through the limbic system have been presumably reinforced, being singled out as individual entities by cortical association areas and predominantly in the left hemisphere. This separation of part from whole should have started sometimes with gradual formation of protoepisodic to episodic memory. The specific focusing on the significant frames should logically facilitate both memory registration and recall, granting tacitly recognizable individuality to them. Again, all this scenario has been rehearsed and reacted for millions of years with the number of series of visual frames in segments, meaning holistically, or single-one figures, meaning individually, all in tacit conscious recognition.

Language coding and labeling of the tacitly recognized frames started with primitive affective emotional sounds gradually in the last two million years, extended the visual labeling to a lingual one with astronomical multiplication of concrete visual frames and addition of galaxies of abstract visions, constructing the verbal consciousness and expanding it to practically unlimited horizons. This transition from visual coding to verbal, somewhat transforming the mental ideations to communicable symbolic sounds, has been the crucial transfer from autos and autogenetic world of the implicit memory consciousness to self and the conceptual symbolic cognition of the explicit memory consciousness. These two conscious states, respectively named psychognosis and idiognosis in this book, are shown in table I (page 129), which sketches the hypothetical outline of the biocratic course of Gnostic states from the presumed nonmaterial biology of the precreation hypothetical state through the material biology and again to the issuing nonmaterial one as described in the table. In further steps in the discussions to follow, I continue for a while to use the customary vocabulary so that my readers's grasping of the developing novelties will follow a more natural, gradual course and conditioning.

Now it seems appropriate to ask one's self again if every part of the biological unconscious or conscious state with its time bondage necessarily represents an action, a process, or an image that could be codified as a time segment by the sequencing process, and thus framed, and would be quantifiably reconstructed by memory, coming out from an extended time storage into real time. The answer is no. We are also conscious of qualifying sensations that are not precisely quantifiable and cannot be time-framed like all feelings—happiness, sorrow, pain, suffering, wrath, grudge, jealousy, love, graded emotions of all sorts, elations, emanations, and all pure or mixed feelings—that have no time boundary and seem floating freely or continuing silently to eternity. This part of consciousness, making the mental experiences, also called *qualia* in general, is the immeasurable part and is the subjective private conscious feeling unique to the conscious subject. This state is ephemeral, hard to keep in consciousness, and sensed as such by us, or is often shadowy, not well definable and quickly lost in our unconscious continuum of the inner time, the time titan of the autos. It may be experimentally approached[2] and may be somehow reproducible, but not on our will and not in a controllable way.

THE TWILIGHT ZONE

Between the unconscious state of routine activities and the conscious level that both can be realized as true and real, we can see a gap, rather virtual than real, extending over the range of awareness variations. This extended wide limit can be said to model a shadowy ontogeny for the phylogeny of the consciousness. It can be likened to a permissive twilight that makes it possible for the mental force to use less illuminating power, not showing the facts in bright shiny icons all the time, and thus conserving neural-mental energy. The routine repetition of the processes, in fact, can assure the accuracy of satisfying the tacit intention and reaching the end result with reestablishment of the neural-mental energy equilibrium and without need for rechecking the processes in bright light. This sort of autonomous self-regulated attention-intention cycle as potentially conscious is a state that does not have to be, but may become fully conscious. The autocracy processes of biotropisms and biodynamisms, in fact, are similar to such a perhaps unconscious state in which processes do not have to be conscious for functioning. The equilibration between the autos and the primordial autogenetic egos also can make up two main groups of bioactive mechanisms—the less intensive (instincts) and the more extensive (emotions) types that vary in degree of bioconscious realization and are fundamentally in the twilight zone of Gnostic states. The seemingly very extensive twilight zone between the shadowy neurocognitive impressions and the clearest, sharpest visual realization can be regarded analogous to the transition between the uncertainty principle of the subatomic quantum norms of time indefinite type to that of time definite willful consciousness.

This metaphoric twilight zone can be regarded as basic ground that can lead from the darkness, total reflexive unconscious acting, to the brightness, the self-realized full consciousness. The scope can encompass all neural and mental processes that we can group together under the nomination of mind. All activities with energy levels below or at the routine metabolic rate can be considered subthreshold and not causing awareness, and all others tresspassing the limit would be those used for conscious recognition. This course of events, a priori, implicates precedence for the subthreshold stimuli to occur in all instances whether or not ultimately reaching conscious recognition. If in fact living mechanisms can continue in all animals without consciousness, or in comatose humans on nutritional and cardiorespiratory supports, why did we become conscious? Or to quote Owen Flanagan, "Is being conscious a difference that makes a difference?"[3] Cognitive science has no definite answer to these questions, but we can argue on philosophical ground that gnosis of the physical and metaphysical worlds would not have been realized without consciousness. In fact, the nucleus of the sensing indeed seems to have been created to expand sufficiently enough to close ultimately the loop making nonmaterialistic biological ends (individual and universal holognosis) meet through material biology.

The brightness coming out of the twilight in our evolutionary journey to cast the bright shining rays of consciousness must have dawned with the early archencephalon in bioconsciousness, gradually shining more and becoming perfected with the prosencephalon and its specialization with the neocortex. The processes added by interbiopsychon contacts necessitating interplays with mental iconic and literary coding, demanding interpretation and problem solving, then granted the state of today's conscious power. The birth of consciousness as the brightest light, the commonly defined psychological state of awareness, is in fact quite recent in the course of evolution but from a very old nucleus whose differentiation has permitted consciousness to develop. It has evolved through the twilight zone and reached completion in the last two million years in the hominids to reach finally its today's status.[1] This development, by all evidences from anatomic differentiation of fossil skulls, bipedalism with descent of larynx, and acquisition of manual skill has kept simultaneity with progressively developing language based on plausible matching with progress in tool making.

I have discussed the foundation of the chain to be formed with biognosis allowing autognosis and heterognosis as the basic ground for initiation of conscious awareness in chapter 3. Here I will mainly focus on the bright light on its way to dawn by recalling the stereotype mechanisms in the twilight zone that establish the regulating biological functions to radiate the bright shining conscious light. In the twilight, in the biopsychon with a preconscious ability, bioawareness and bioconsciousness, the central actor is the autos that attends to the task accomplished unconsciously or less than consciously, and to which, equilibrating feedbacks are addressed. In this way, the essential pattern of action is more or less similar in unconscious, preconscious, semiconscious, subconscious (UPSS), or even a supposed-conscious animal; it is some simple autonomous response to external stimuli gradually becoming more complex in perception-recognition. In essence, all these steps establish the course from time indefinite initiation of an unconsciousness that is reminiscent to an autonomous floating vessel, so to speak, to get closer to a shore and to anchor at last in defined time.

Addition of consciousness does not eliminate all the less conscious states, and actions performed do not switch to be done all by consciousness. Basic biological needs are usually satisfied with these less-than-conscious states at subthreshold energy level. But environmental forces are not restricted to only cause routine responses, and impacts of forces, seriously threatening, make biocratic equilibrium to occur by the biotropic sensors to detect these greater impacts and biodynamic responders to use higher energy levels to reestablish the equilibrium. The change in response includes an estimate above the status quo of the routine, sensed by the system that also responds with a higher energy level above the routine threshold, which indicates more than unconscious acting. But we have no precise words in today's usage to describe the levels of neural activity below consciousness other than the arbitrary serial chain of UPSS. On the other hand, the more defined terms of paleognosis, biognosis, bioawareness, bioconsciousness, psychognosis to end in consciousness, as used in my earlier discussions, describe the situations in more defined logical and biological meaning (table I).

These changing stages, not without overlap, are accompanied with episodic-sequenced memory and implicit memory within the inner time titan and without extended measurable memory time, whereas refined communicative needs and projected planning using full consciousness implicate extended real measurable time and explicit declarative memory. What is crucial in consciousness is therefore the conscious sensing with its energy level passing a set threshold for all subconscious actions. This conscious percept emanates from the self facing nonselfs and differentially presenting a directing power, the autos. Thus, the acting is intentionally directed by the conscious self (in balance with autos) to reach a defined end to complete the action in real time exteriorized out of or revealed in time titan.

An inherent part of biocracy is the hypothetically inculcated genetic programming by bioforce in autos to carry autocratic biological functions. Here, the presumed intentional activity to safeguard SR is autonomously part of autos and needs no triggering other than by direct biotropism of the organism facing the environmental forces. The equivalence of intention in bioconsciousness rather than a conscious will, in this circumstantial condition when neurohumoral mechanisms may predominate, is essentially constant and unchangeable and serves the same purposes responding to stimuli to regain equilibrium. At this early stage of neural development with SR being the only principle to be safeguarded, the UPSS organization is naturally provided by the routine mechanisms, having passed the test of time for millions of years. However, intended attention in the conscious mind is the trigger to activate not only the same constant routine line of proceeding, but also new lines of actions necessitating additional mental force. Therefore, we can assume that in the conscious mind and beyond simple perceptive processes, attention must serve intention, and that likely forms the fundamental pattern for conscious mental energy allocation.

The metaphoric twilight zone thus seems to be only partially known, and possibly in both states of consciousness and unconsciousness, there should be stages of variable degrees of awareness and variable connective associated feelings. We should therefore cautiously refrain from limiting the Gnostic states to only two—consciousness versus unconsciousness. We should understand that Gnostic states are at least of two frankly observable known types, but several less well known, less distinct and less-defined ones also do exist in between. If all our primitive sensing, feeling, and emotional states of gradual progress that have developed in the last fifty-six million years of time[1] could be traced and evaluated, our two million years of conscious state development would be comparatively negligible.

To attempt a working plan in order to classify the varying Gnostic states for facilitating their studies, I proposed in chapter 2 the fundamental types of archaic, reflexive-reactive, and perceptive-reactive neural activities in terms of sensing and realizing from unconscious to conscious states as paleognosis, biognosis, bioawareness, bioconsciousness, and psychognosis. These level-variable sensori-perceptive neurological activities were further assumed in chapter 3 to fit in the large groups of bions and biopsychons in a simplified schematic presentation before the literary biopsychons to represent the full capacity conscious organisms. These states could be included in a classification to replace UPSS with even some room for extension. In fact, we really cannot evaluate the twilight zone and its components before we journey through it. In order to gain more insight on these main types of UPSS Gnostic states on the one hand and the frankly conscious one on the other, some characteristics of both groups are assembled in a more representative classification in table I.

The vague, penumbral part of the twilight, which includes the first group, covers the sensing that is endogenous to autos and is inherent in its time titan without having any definable time frame and is undefined in any more elaborate neural capacity other than sensing-acting. It is important to realize that from the very initial stage of prokaryotic DNA and even earlier with the bioforce and bioquantum timing and energy partitioning for block action of biotropism-biodynamism, the principle of SR has shown its priority through all the subsequent forms of life. This evidences the unfailing governing power of biocracy with its mandatory adaptive rule no matter how primitive or advanced the mecanisms at its service can be, judging either the archaic past or the present mind forms. This sensing by autos, initially primitive and reduced to strict biotropistic mechanisms, eventually reaches the level of being distinct as really exogenously activated, recognized by autos as a nonautos reality, representing the earliest form of heterognosis, but it truly belongs to autos and makes a subjective part of it. This type or group of Gnostic states gradually reaching consciousness includes, by extension, all ephemeral mental experiences—emotions, feelings, sentiments, and all *qualias*—and are inseparable from the time titan of autos. These are grouped in table I under the term of psychognostic states.

On the other hand, the bright end of the twilight includes the clear images that represent exogenous spacetimes with clear figures of nonself characters from the environment with both a definable content and a time frame, and are thus measurable, and their aggregation in time follows a known repeatable course and a defined reproducible sequence. They are changes distinctly exogenous to autos and are part of the spacetime outside of autos time titan. They can be recalled in memory by attention and can be reproduced by recalling the specific time frames that contain them. If attention is focused on that particular segment of time for a tangible event in the memory, that specific Gnostic state can be reproduced and its codified framed content made available to the real present time, reusable and relivable. This is the consciousness component that has been gained through evolution and has really evolved since the beginning of the primitive Gnostic state initiated with autognosis and heterognosis. This type, physically and sensibly exogenous to autos, includes all time-framed awarenesses of inner or outer mental origin in actual real time, the present tense, or extended time stored in memory that originates de novo from perceptual stimuli to autos or is searched intentionally in the memory by the conscious agent of autos, the self. These forms of clear verbal conscious perceptions named idiognosis are assembled under neognostic states in table I. (page 116).

The two ultimate groups of conscious states, psychognostic and idiognostic, in my interpretation, have been usually considered and described together confusingly under subconscious and conscious names by neuroscientists, psychologists, philosophers, and cognitivists who nonetheless recognize that there is more to each group to be defined and elaborated more specifically. These two groups characteristically reflect essential differences that indeed suggest they should be described and understood better by scrutinizing their details. They have one main difference, which is controllability that can be exercised by attention-intention processes and memory for the clear image in idiognosis, the controllable one, but not the other because of recognition being based on content and time framing characterizing the controllable type only. In contrast, the other type, the psychognosis, is constitutionally uncontrollable at will because it is endogenous to autos, belongs in an undefined eternal permanency of autos granted by the time titan, and is impossible to gauge and to reproduce by attention-intention and recalling unless associated with a recallable framed entity. It is reminiscent of Paul Rozin's *cognitive unconsciousness*, quoted by Steven Mithen.[1]

In my view, this type is the basic fundamental source and reflects the pristine self-sentient nature of the biopsychon, the fundamental archaic sensing of autos as part of biocracy, which I have tacitly included in biognosis, bioawareness, and bioconsciousness in chapter 2. Its archaic characteristics of being limited to sensing and reflecting environmental forces and conducting the biotropism and biodynamism at different scales have not drastically changed through the ages.

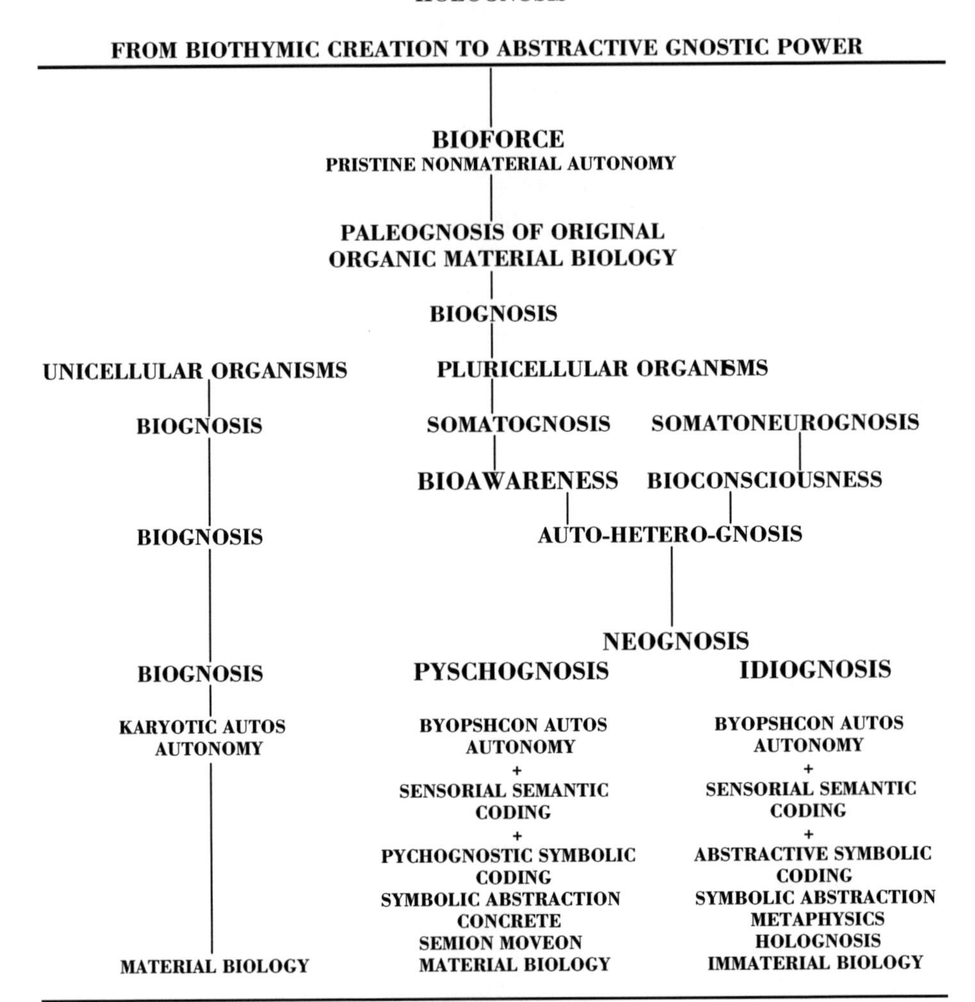

Table I. The outline shown in this table summarizes the figurative hypothetical course of progress from the pristine autonomy in bioforce by the biothymic differentiation to engender the immaterial biology to be manifested through the biological material. The ending realization reaching a nonmaterial biology is depicted in the column under somatoneurognosis, linking ultimately the original and final HOLOGNOSIS. The steps indicate BIOGNOSIS by the unicellular biology to expand into pluricellular organisms in BIOAWARENESS and BIOCONSCIOUSNESS. The limits of expansions and specifications through PSYCHOGNOSIS and IDIOGNOSIS end in the abstract mind but expand to ultimately reach HOLOGNOSIS.

However, its basic simplicity in dealing only with the principle of SR and still adamantly serving the same basic principle has gained complexity related to the same principle, but associated with states of mind that have developed with evolutionary demands and refinements. These are reflected in the adaptive changes anatomically evidenced with the development and specialization of the neocortex allowing increased asssociation with refined processes and modularities. This presentation, as an inherent part of perceptive-responsive archaic mechanisms now expanded beyond the basics, is not expressible adequately in our state of awareness. Its immediate relation to body senses and mental images as part of the primitive sensory-limbic association may elicit its unintentional reproduced feeling without a known time frame, time framing being essentially a function of the neocortex regardless of the limbicocortical relays. In reality, its nature has not greatly changed in the millions of years of evolution as we can see for all emotional effects in our today's human mind that can vacillate between extremes of altruism and atrocity. This type, still classically included in consciousness but viewed with a discerning eye by cognitivists, should be separately treated to its own merits, under psychognosis as used in this book.

The common denominator to both of these two types of classically described consciousness is their relationship to time. The autos centered consciousness, referred to as bioawareness or bioconsciousness and finally psychognosis, appears to relate anatomically and functionally to the limbic system mainly and reveals two basic archaic neuropsychological entities—the subjective feeling of oneself, the psychological self as agent of the eternal autos without needing any other definition, and other undefined subjective experiences, and would not be time framable because of being an inseparable part of the time titan in autos. It follows the rules of uncertainty principle of the quantum world, which reveals itself as time indefinite. This point was discussed in chapter 3 in dealing with autos, self, and ego. These subjective experiences without any defining spacetime, content-sequence, or image-frame could not be recognized, kept in memory, and recalled at will, but they can surface clear when a subtle association with an item of controllable consciousness makes them surface conjointly with that item according to circumstances. On the other hand, the better known consciousness classically described, the idiognosis, is structurally and functionally related to neocortical capacity and is essentially time framed and codified by explicit memory. Whereas floating conscious experiences and feelings emanate from the indivisible time titan and are therefore undefinable and limitless, idiognostic conscious items are part of the real time and are definable according to their time frames. Also, whereas the ephemeral psychognostic conscious experiences cannot be controlled and reduplicated by any known neuropsychological mechanism, items in the classical consciousness are controlled and reduplicated basically by attention-intention processes using explicit declarative memory.

The archaic nucleus of consciousness, the paleognosis in autos, therefore appears to be the origin for all types of modified Gnostic states that show up later in the course of the evolution, and the primordial principle of SR seems to be the fundamental reason for this archaic consciousness to evolve to biognosis, bioawareness, bioconsciousness, psychognosis, and idiognosis, safeguarding the same principle.

PALEOGNOSTIC STATES

It is not surprising that with studies of paleontology, paleoanthropology, biology, psychology, and even cognitive ethology, a notion of paleognostic states has not been firmly entertained or even suggested in these disciplines. Obviously, adherence to scientific reasoning based on observable, measurable, and testable facts renders it difficult to judge otherwise and precludes classically a belief in a Gnostic state other than consciousness opposed to unconsciousness. Such respected adherence to scientific facts would imply that a hypothesis for possible paleognostic states would not be testable and therefore is not worth entertaining as it may prove undefendable. In contrast, mythologic and religious literatures are replenished with descriptions of figures and events invoking a belief in archaic Gnostic states and mindful animal examples. This contrasting opposition in human mental inclinations, being in essence between rigid scientific and permissive philosophical realms, needs revision for permitting mutual understanding. Although even adequate analogical reasoning still may not lead to testable results, it provides the best assurance before proofs. Such is what I have reviewed in chapter 2 with the examples of animal actions in various situations from simple tropisms to more evolved acts showing attention, intention, aiming, planning, communicating, and useful collaborating. All these actions, reminiscent of some somatic-neural-organized coordination could either be viewed simply as behavioral patterns, or could be interpreted in a more reasonable way as reflecting some awareness from simple perceptual-responsive to reasoned-reactive types, all being combined with variable emotional awareness of hunger, pain, fear, and fight. Again, the analogy can be regarded to suggest the effect of the bioforce in the biological substrate with lesser or greater neural complexity. The inclusion of time dimension by the concept of time titan in the biological material, along with the bioforce, supports the evolutionary adaptation that must have occurred to safeguard the principle of SR assuring time continuation with the replicative transfers of the bioforce.

The content of table I summarises the hypothesis of paleognostic states that accordingly must have been operating in biocracy since the beginning of time and have provided the basic ground for further adaptive specifications to occur. Indeed, the archetype of adaptive senses, the pain for instance, which results from nociceptive stimuli causing tissue injury, is basically reflecting the same fundamental physiologic and pharmacologic changes in all animal species, and sensing the pain less or more consciously by the living organisms including humans is purely subjective and not precisely reportable or measurable, nonetheless its essence remains the same and of course undeniable. Therefore, a paleognostic type of a subjective definition for any and all feeling is as arbitrary as a neognostic one, but still makes sense in indicating their unique origin and their possible differentiation with time. In fact, anatomical changes with complexities of nervous system organogenesis insinuating analogy with more complex functions is evidence for the differentiation in time. In addition, testable results of studies especially positron emitting tomography (PET), single photon emission computerized tomography (SPECT), and functional magnetic resonance imaging (f MRI) of the brain dictate possible differences of awareness on subjective feelings. Images can detect somatosensory and neuromotor function centers operating in harmony in the human brain and showing objectively aimed functions like oriented attention, thinking, concentrating, planning, calculating, etc. Such studies, if they could be also performed in primates, could show unexpected surprising similar but perhaps less clear or refined findings and mainly related to limbic-neocortex functional connections.

BIOAWARENESS

As sensing-acting of the live organisms is the foundation model for the nervous system formation with its archetypal mechanisms, we can see that the basis of such mechanism in the simplest form of life to be, prior to any nervous system, is the tropistic sensing-acting pattern in unicellular and even multicellular systems without nerves like in plasmodia. In table I, this is shown under caryotic autonomy of biognosis meaning simply tropistic reactions that regulate biotropism and biodynamism of the basic cellular life. *Caryognostic states* in reality would mean fundamental broad application of tropism interpreted as sensing-acting and therefore unconscious in our customary understanding. But the objective of this tropism is evident in its result, which serves survival and allows replication, a function that recalls analogy when nervous system activity grants more refined states along the same principle. When can we frankly talk of a state of bioawareness is hard to define without definite evidence of sensory-neural perception as would be expected with biochemical agents like neurotransmitters. However, short of a well-established sensory-neural perception, I feel we have justification to consider the beginning of what we could envisage as a primitive building block for sensory-neural awareness in the actual tropistic mechanism.

The time of the first neurotransmitters to appear with synaptic connections in animals must be considered the pivotal point for speaking of sensory-neaural perception. This event transforms the simple tropism of the unicellular life into a staged neurotransmitting sensory-motor act and initiates the phase of neuromotor living with the incomparably more environmental exposure secured by mobility. The event must be searched in the very early biopsychon species with primitive multiple nerve cell networks. Examples like *Caenorhabditis elegans* we saw in chapter 2 or even any simpler organism possessing synaptic neural connections could be the appropriate ones for our purpose. The question of what synaptic type or neurotransmitter appeared first, of academic interest, is not precisely answerable, but we can logically consider the central pivotal importance of nutritional mechanisms and surmise that perhaps cholinergic synapses first and serotonergic next were the earliest to appear.

What is more important for us is to adjust our vocabulary to be able to name the more elaborate neural capacity in biognosis in the mobile dynamic biopsychon with its precise sensing-reacting mechanisms unfailingly securing and safeguarding SR. In this endeavor, I defined the term bioawareness pointing to the adamantly safeguarded SR principle of the living organism by the somatognostic state to imply its basic somatoneural perception-reaction mechanisms.

The scope of development of these earlier neuroanatomic sensing-reacting systems with appropriate neurotransmitters is quite extensive including all living organisms with variable neurological and locomotive capabilities able to do foraging, mating, and escaping danger as routine acts of living. Again a sharp division is impossible to advocate between these stages of neural activity. Using somatognostic and somatoneurognostic titles in table I is for the purpose of labeling to keep distinct reflexive-reactive from more advanced refined neural stages with predominantly sensing-reacting or percepting-reacting functions.

Neural structural complexity, with more complex neural functioning, will involve additive associative functional harmony more than can be defined by the type of neurotransmitters serving those functions. In fact, a single neurotransmitter may serve different neural activities and help harmonizing functions of various cerebral structures.[4] The presentation used in table I is but a general outline that cannot eliminate overlapping of Gnostic states and arbitrarily categorizes the somatognostic states mainly with bioawareness and the neurognostic ones with bioconsciousness. Evidently, either state can include neurological and biochemical attributes common to both but bioconsciousness representing the more evolved one designates predominant neural associative structure for serving associative functions more than pure somatic functions.

In this context, bioawareness is broadly meant to indicate regulation of basic biocratic processes serving SR in a perceptive-reactive way facilitated by an adequate nervous system for that purpose. Referring to the primordial biodynamic principle of SR, the nervous system must be sufficiently developed to act as the regulator maintaining the equilibrium of the soma and safeguarding the biocratic principle. The neurosomatic coefficient described in chapter 1 (N/S, a functional, not anatomical correlation), indeed shows constantly greater value for superior mental functionality as required by biocracy.

With the more developed complex nervous system, in addition to the basic SR principle being served, associative nervous relays and functions serve additional purposes that can culminate in the example of the human brain. A positive neurosomatic ratio theoretically functioning from the inception of life and autonomy can start to be noticeable at some stage of evolution with full development of the limbic system but also neocortex structuring, the stage that could be said to represent transition from reptile to mammals according to the triune brain concept of MacLean emphasized particularly by Whybrow[5] among others. The Gnostic state then must be considered more elaborate, exceeding bioawareness with a neurosomatic coefficient value clearly above one.

BIOCONSCIOUSNESS

Discrimination of self and nonself described in chapter 3 under auto-hetero-gnosis, being a characteristic feature of consciousness, must have started much earlier in the evolutionary history of humans. If the simple unconscious living in karyotic life or in the early somatoneural structured animals, and even in humans, can keep going on autonomously as remarked earlier, then consciousness is a luxury. Advent of additional substrate in the more complex nervous system with additional associative functions allowing consciousness to be formed must be regarded as an adaptation in the evolutionary course that has appeared as the result of a necessity for providing an ability to respond to whatever unknown stimuli could be faced. Thus, the beginning of a surplus of nervous system development over and above what is needed for basic maintenance of energy equilibrium and safeguarding of SR must be taken as the starting of a type of consciousness whose reason to appear is not quite clear, nevertheless whose end is evident in us with our conscious control not only helping us in our living tasks, but projecting out of us in imagination forming new worlds, metaphysics and holognosis. That starting point is what I have considered as basic separating line between bioawareness and bioconsciousness as sketched earlier, a Gnostic state that allowed self-realization and recognition of heterogenous environmental world and autogenous inner self.

Bioconsciousness, as defined in the context presented in this discussion, must be considered the early differentiation of somatognostic states served by the basic organic nervous system, the limbic system, into the associated limbicocortical completion. Addition of the neocortex, judged to be the foundation of neognostic states tips the balance of neurosomatic coefficient, which is essentially an indicator of neurosomatic regulation function to keep just the minimum functional positive value for N/S coefficient. With bioconsciousness, what the provision of limbic function procures as the routine neural mechanism starts to change by assessment of what the limbic function feeds to neocortex and what neocortex feeds back. This earliest reciprocity relation between the two functional neural perceptive-reactive and reflective-reactive systems forms the building blocks of neurodynamic processes for unfailing auto-hetero-gnosis, extending the implicit memory, starting psychognosis, and preparing for idiognostic consciousness.

The fundamental feedback between the limbic system and neocortex that served to prepare the ground for long-term memory, forming basic autobiographical memory and allowing distinction between self and nonself, firmly established the actual auto-hetero-gnosis. This took place between the archaic hippocampus of the limbic system essentially functioning to administer the short-term working memory, and the recent neocortex as reinforcement serving long-term memory in initiating the explicit memory. Completion of this reciprocal relay system went on along the other progressively more complex modularity of brain functions to neognostic states. It is a fact that indeed bioconsciousness harbored and brought with it all the older archaic states of gnosis to the neognostic stage, and today, the brain modular functioning allows all innate biocratic biognosis including bioawareness and bioconscious states, noted as instincts and emotions, in what we have been customarily calling subconscious. Bioconsciousness in essence reflects the states of sensing, percepting, reacting that serve biocracy in the more complex biopsychons, including hominids, before psychognosis leaving autos and their basic innate autocratic supremacy unchanged.

The evolutionary reason for the bioconsciousness to develop out of the tropistic archaic mechanisms and expand over bioawareness was the necessity of commensurate N/S growth to better serving the principle and the mechanisms of biocracy. This reveals further supportive evidence for the directional time effect to needed somatic improvement discussed earlier for the bioforce. Bioconsciousness in fact served the energy balance securing the phenotype living and maintained the stimulus for reproduction securing gene transfer but, in addition, formed the ground for affective sensing and effective limbic-neocortical implicit memories and their relays. In this early representation of spacetime structuring, bioconsciousness is functionnaly closer to the bioforce origin of life. Bioconsciousness can be considered the factor facilitating sustentation of the bioforce to allow transit from soma to psyche as the preparatory step to psycho and idiognosis exposing ultimately the conceptual Gnostic world.

For the basic principle of SR to be safeguarded in evolution, we can see that energy balance to a threshold gain is the primordium that usually has to be secured prior to gene transfer. Bioconsciousness served this balance as did bioawareness, but more neuroperceptively than tropistically as a matter of definition by the basic sensing of hunger to set the mechanisms for nutritional provisions. In general, after well-equilibrated energy balance and growth, reproduction is served by bioconsciousness with the basic sex instinct in the great majority of animals. Primates, hominids, and humans are making perhaps the only exception to this basic rule, changed by progressive psychognosis. The exception eminently reflects what earlier in this book was attributed to the symbiotic characteristic feature as the mainstay of the human life recognizing its independent psychological value at least equivalent, but often superior, to the physical one.

In terms of psychosomatic value, what bioconsciousness initiates in following the basic principles of biocracy is a Gnostic state of pleasure versus pain in its simplest sense to its broadest meaning reached in psychognosis, ultimately providing securer survival for the phenotype and recapitulation for the genotype. Interestingly, the basic main somatic sensing organ that relays the pleasure in humans before genital organs is the same in both eating and sex—the mouth and especially its external boundary, the lips. By the same token, pain as a common sensation produced from injury to all ectodermal body derivatives, in the first place, and mesodermal, in the second place, is not affecting the brain that itself is the register instrument for pain (impervious to pain) and is only the representative of the soma in pain, but not the psyche in pain, for pain in psyche has its different specific affective forms.

This fundamental differential essence of the pain of soma and the pain of psyche is the pivotal axis separating two biological neurodynamic characteristics. I am of the firm opinion that the brain, in fact, evolved through the complexities of feedbacks between pain and pleasure-giving realization of pleasure in affective feeling, and if any pain should be imagined for the psyche, it should be based on relative absence of pleasure in the balanced affective pleasure-pain principle. From both lines of evidences based on archaic biocratic principles and evolved psychosomatic states of pleasure versus pain that are shared primarily with the same somatic structures, we can presume that the prototype bioconscious model based on the pleasure-pain balance kept predominating in the evolution.

Complexities observed in subsequent psychosomatic manifestations with interdependence to limbic and to neocortical operations are therefore natural modalities responding to specificities.

THE PILLARS OF EMOTIONS

Accepting the principle of pleasure-pain equilibrium, which implies a positive balance (pleasure) versus negative balance (pain) in terms of energy feedback mechanisms, we can realize two maxims of truth in biology. The first being the fact that energy spent to keep a positive pleasure balance is not counted as lost. The second is that this pattern analogically mimics the fundamental innate feedback mechanisms in the nature of the two quantum components, energy and mass, in which the movement of photon consumes energy, but the autofeedback replenishes the energy loss, keeping the balance constantly on the positive side for unending times. We can realize that the goal is adamantly the continuation in positive balance, securing pleasure and eliminating pain: a philosophy of permanent happiness.

Indeed the equilibrium in the energy balancing in pleasure-pain system throughout animal kingdom is significantly safeguarding positive balance for pleasure. This can be even regarded as a reason for the evolutionary adaptation that has reached consciousness to grant conscious happiness.[3] Notwithstanding this reasoning's wrong or right and supporting determinism less or more, the Gnostic state with bioconsciousness took its roots in the limbic system and permitted the very archaic emotions to develop and act in the course of evolution serving the system of pleasure-pain balance on a solid unfailing basis. The aim for this adamant determination may be assumed to be an ultimate hypothetical ratio of infinity pleasure over null pain that I consider as the primum movens of nervous system refinement to psychognosis and idiognosis transcending ultimately from the material biology to the nonmaterial metaphysical elation.

The undeniable experimental evidence that indicated the limbic system to be the initiation site of neurosomatic emotional states was provided by experiments at the turn of the nineteenth and early twentieth century in which resection of the cerebral cortex in experimental animals, leaving the limbic system intact, showed that animals behaved violently mainly in demonstrating rage and aggressiveness, as if an inhibiting power of the cerebral cortex had been eliminated.[6] Development and further refinement of the neocortex therefore appeared adjusting the function of the limbic system. Furthermore, endorphins, known to act in producing euphoria are essentially, and in major part, neocortical neurotransmitters having a major role in this adjustment. Thus, the reason for considering the interplay of the limbic and cortical functions as reciprocally determinant in keeping the balance is logically appealing.

The basic stimulus for this system to be activated is any and all events tending to cause energy loss. Thus the stimulus is unique as a trigger no matter how originated and is simplified in terms of negative feedback on the metabolic plenitude of energy reserves. The pivotal sensing that organizes the evaluation of the negative feedback in intensity and scope and activates the mechanisms for responding accordingly with commensurate positive feedback to reestablish the initial balanced condition uses a complex harmonious neuromotor and somatovisceral process. Hypothlamic, hypophyseal, adrenal, and glycemic, hepatic, and gastrointestinal responses concur to establish nutritional gain to reestablish the metabolic balance. The autonomous nervous system on the one hand and neuroendocrine central and peripheral responses on the other come into play during the process and additionally also produce some basic archaic senses that can be felt as feelings or emotions but most have been called instincts.

The negative stimuli of mainly pain-hunger-fear-rage opposing and the positive stimuli of pleasure-satiety-security-love enhancing the balance form respectively the foundation for negative and positive bioconscious Gnostic states forming feedbacks to the central equilibrating organization of harmoniously working somatovisceral, neuromotor, and psychosomatic activities. Coordinating these system activities is complex in terms of neuroanatomy including archencephalic brain with limbic structures and prosencephalic brain with neocortex. The same is true in terms of neurophysiology including multitudes of relay chemicals and hormones of somatic and visceral origin and finally in locomotor activity—in short, in sensor and effector organs. Practically, all visceral and somatic perceptions of unconscious and conscious nature initiated with bioawareness and bioconsciousness contribute to realize these two groups of pillars of emotions of either psychopositive or psychonegative stimuli and their related various manifestations reminiscent of the four autogenetic prototype egos of ergosynthetic and ergolytic types. Comparative examples of these four pillars of emotions and their psychosomatic manifestations of pain, hunger, fear, rage, and pleasure, satiety, security, love are summarized in table II.

Pain is the most basic means for the biopsychon to avoid bodily damage. The spinal cord mechanisms, in addition to neuromotor rapid reflex, already uses enkephalins to secure analgesia counteracting pain. In the limbic system, from reticular formation to thalamus and with control by the cerebral cortex, a similar humoral mechanism works in addition to the spinal mechanisms to reestablish the balance through paraventricular and brain stem centers.[7]

The same central position of the limbic function is pivotal in hunger versus satiety. Lateral wall area of the hypothalamus is known as center of thirst and hunger whereas its medial wall regions represent the same for satiety. Blood sugar fall and rise being the central trigger for this feedback balance, and higher cortical and lower somatovisceral manifestations are associated controllers.[7]

Regarding fear and security, and rage and love, it is appropriate to note that these opposing Gnostic states of bioconscious origin, grouped in the category of pillars of emotions, are more complex and depend on and affect more cortical functions. Fear has a distinctly centered presentation in the amygdala functionally between thalamus and cerebral cortex, with predominant afferent fibers to the latter.[4] It is limbic, but under some cortical control. It presents both a rapid uncontrollable startling reflex and a prolonged more controllable sustained effect.[5] The positive Gnostic state of security appears to be also complex with regard to humoral mechanisms and balance. In fact, what is called *harm avoidance* that is related to adequate brain serotonin activity on the one hand, and what is known as *novelty seeking* that represents excess dopamine like action in transgenic mice[4] on the other, show opposite effects, and the connection of this balancing, if any, with cerebral endorphin levels (essential in pleasure) is also complex.

Rage is somewhat complex as well, but this complexity is twofold. Firstly, it is a more apparent complexity than real according to how its association with aggression is interpreted. Aggression, being interpreted as harmful intent to destroy, is the real association with the emotion of rage, but aggression to kill in hunting or fishing is routine sport by those doing it and is a remnant trait of archaic hunting with nutritional aim and can be devoid of rage. The other complexity of rage is its real exaggerated anger and full scale fight and flight reactions and mechanisms that can have far-reaching effects on the totality of psychosomatic balance with catecholamines and steroids excess.[8]

The opposite to rage, the love, if we allow extended margins to worded expressions, is also complex. It is similarly based on the opposing interplay of positive-negative emotional feedback balance centered on the pain versus pleasure mechanisms. Complexity here is reminiscent of the type encountered in rage. On the one hand, the sensitivity to lust and sexual pleasure is tantamount and is the essence of libido and sexual satisfaction with oxytocin being important in both foreplay and orgasm[4] and endorphins in more lasting effect being analogous with real rage effect and its chemistry. On the other hand, analogy is also noticeable between the sublime maternal love of nurturing with deep mental satisfaction and pleasure or sensed with altruism without the lust and the sex pleasure as in hunting-killing without rage.

Table II

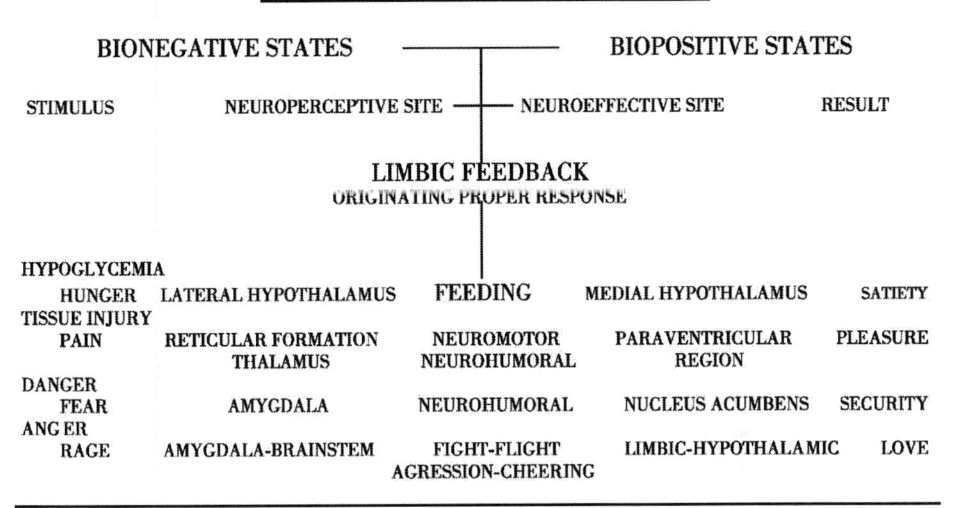

BIOCONSCIOUS PILLARS OF EMOTIONS

BIONEGATIVE STATES			BIOPOSITIVE STATES	
STIMULUS	NEUROPERCEPTIVE SITE	LIMBIC FEEDBACK ORIGINATING PROPER RESPONSE	NEUROEFFECTIVE SITE	RESULT
HYPOGLYCEMIA HUNGER	LATERAL HYPOTHALAMUS	FEEDING	MEDIAL HYPOTHALAMUS	SATIETY
TISSUE INJURY PAIN	RETICULAR FORMATION THALAMUS	NEUROMOTOR NEUROHUMORAL	PARAVENTRICULAR REGION	PLEASURE
DANGER FEAR	AMYGDALA	NEUROHUMORAL	NUCLEUS ACUMBENS	SECURITY
ANGER RAGE	AMYGDALA-BRAINSTEM	FIGHT-FLIGHT AGRESSION-CHEERING	LIMBIC-HYPOTHALAMIC	LOVE

Table II. Contents in table II are grouped under titles summarizing the essence and importance of limbic system integrated in a central position between the lower spinal cord to brainstem and the higher brainstem to brain cortex. Neuroanatomic and neurophysiologic structural and functional related details are not included in the table to help focusing attention on the limbic system's central position and its connections in order to secure clearer exposure of the schema of the actual feedback between the fundamental pillars of negative and positive emotions. A note of further allusion to these categories of basic emotions is included in the text to briefly provide the readers with pertinent information.

This simplistic regulation by a feedback system equilibrating energy balance in favor of psychopositive Gnostic states with energy gaining (ergosynthetic ego models) is reminiscent of balanced state of the historical *milieu intern* of the great French physiologist Claude Bernard. At time of Bernard's experimentations in mid-nineteenth century, neurotransmitters were not well-known, and the reciprocal adjustment between limbic system and cerebral cortex not elucidated, but his observations and judgment nevertheless justified his invoking of a well-balanced milieu intern for assuring physiologic states. Today, our enriched views on the same old physiologic problem with evolutionary effects observed from the earliest life-form to humans indicate that the physiologic balance is achieved by numerous different mechanisms and hundreds of different chemicals, neurotransmitters and hormones, however, all concurring to solve the same old problem of the constancy of the milieu intern with extension to affects. Furthermore, the solution for which the archaic biocratic tropistic mechanisms and the more evolved bioawareness and bioconsciousness have appeared and assumed more and more precision, in final analysis, is based on the simplest feedback system that serves life on the fundamental principle of the causal bioforce in all its evolutionary biologic differentiations resulting in life forms.

The four types of emotional Gnostic states summarized in table II as negative and positive feedbacks to reciprocally act to serve SR are in reality autocratic traits principled on the autogenetic ego models and continue unchanged safeguarding SR and can be often witnessed and observed today as essential emotions sensed consciously, but not easily subordinated to conscious reason. This state of affairs reinforces the selfish gene concept of Richard Dawkins, which, however, seems to be negated to some extent by the altruistic tendency that has been more and more evolved in advanced civilizations to its today's best. The altruism, also recently called coevolutive,[9] is granting love and permissiveness to support others without apparent material self benefit. The Gnostic state of sublimation in altruism that reflects nonmaterial and nonsexual happiness and composure, like in mystical elation, may ultimately prove to be a new product of mental-social evolution, another pleasure-based justification for evolved consciousness.

AUTOGNOSTIC STATES

Evolutionary changes from the nucleus of paleognostic life to biognosis and to biowareness, bioconsciousness, and neognostic states do not negate the autos and its autognostic state that are the foundation of psyche. Changes in reality are here functions added with new interrelations, not destruction and reconstruction, to be explicit. What I grouped as pillars of emotions is in essence what the autognosis was in an extended meaning in the earliest biopsychon and is now the base point for the states of feelings and moods. We can reasonably assume that our deepest sense of self and its attachments are our archaic autognostic states with coloring changes added according to autos' investiture in self(s) and in our present multivariant environment.

Feelings and moods are states of mind that contribute to our happiness or sadness sometimes with and often without a strong stimulatory reason to trigger an alerting emotion. The basic mechanism yet is similar to what was discussed regarding the pillars of emotions and is centered on pleasure-pain axis, but the biological importance of the product in the scale of evolutionary preparation to face the environment is not at its extreme. This has been made possible by the great neocortical association areas linked to the fundamental emotional processes that sooths the limbic perceptions and incitations. The urgency of the limbic demand is moderated by the neocortical intervention and often protracted. The difference is clearly obvious for example in impatience of children compared to calmness of adults in equally strong affective feeling expressions.

A far-reaching effect that should be reiterated regarding the cortical brain development affecting autognostic states is what it can do to the rules of phylogeny described in chapter 1. In this context, it is important to notice that the extensive cortical development in humans further reinforces the rules of phylogeny as discussed. This is evidence, once more, that progression from the autognosis to neognosis is certainly a fine adaptation on the evolutionary basis and must be in line with any possible further development to come that can only result in a better adaptation. The ultimate question is: What objective state can be the final realization for all these progressively advancing adaptations inexorably continuing forward? With this question, we reach the limits of our visibility as imposed by our limited material biology with our mental power of reflection, and we naturally tend to look beyond for new visible light, searching in our nonmaterial biology, the Gnosticism. Metaphysical imagination thus shines naturally, giving us new visibility beyond matter and science, allowing our inner bioforce to realize its origin through consciousness.

A real symbiosis with more independence of the psychological life from the biological material life may be on its way to more adequately tune neognostic capabilities to autognostic life, but still keeping with the autognostic principles. This progressive change that has proved its relentless unidirectional course seems to favor the symbiont immaterial mental life to surpass the material biological one and with no visible end. Thus, in pure theory, the initial infinitesimally small biological carrier of the bioforce with N/S relation of about unity and Pi>Ps phylogenetic value (see chapter 1) may surpass unity in N/S value and reach Ps>Pi dominance, both values theoretically nearing infinity in Gnostic mental power. This theoretical viewing seems closer to reality than actually believed as some scientifically backed observations indicate. I agree with the interpretations given for the suggestive evidences[10, 11] to mean that we are facing mental purity to transcend out of our physical bodies and unite with the universal wholeness. To this, I add that the biologically acting bioforce is the origin for this conditioning, and scientific observations are in support rather than against this final stage of the mental symbiont in trances and transcendental elations. In this line of thinking, the Ps of the "phylogeny for love" in chapter 1 must be taken to mean transfer from social-physical to metaphysical extension, losing all Pi and uniting with the origin, the pure nonphysical universal whole, and the N/S value to reach an extension of neuropsychic infinity from the somatic unity, for example, in the presumed ratio of infinity divided by one. Should our millennia see such realization taking place in progress from the archaic bion to biopsychons, to biognosons [!], and ultimately to hypothetical phantom gnosons [?!], what could be the biological carrier that tends to nonphysicality of pure energy carrying imaginable beings?

This question invites challenging scrutiny. To try to find an explanation for the definitely clear subjective experience of consciousness and its limits of expansions, the immateriality of the theoretical pure-energy state for its justification suggests a state of processing, a functional pattern, a perceptive self-reidentifying state of being in which basic autobiographical memory is reinstated at infinitesimally short instants of revisions and with simultaneous real time revelation of limbicocortical inseparability. Recent thinking on this state of affairs that grossly summarizes consciousness seems to favor what is called coalitions or assemblies of synchronically cofunctioning pyramidal cortical cells of mainly the front and the back of the brain relating the logical real time, meaning elaboration with iconic configuration respectively, and in intervals of microseconds, changing in extent and function the neuronal group types in harmony according to the inciting perception-reaction or attention-intention processes.[12] This coalitional functional state that uses multifocal brain areas seems to constitute the assumed immaterial presentation of an organized stereotyped reaffirmation of the assembled reality of autos-self-ego to self as consciousness. This conscious state is in reality a clairvoyance of limbic autos and self in reciprocal feedback terms with cortical areas of real time logical (right prefrontal area), positional (left parietal area), iconical presentation (occipital visual), and idiognostic literal presentation (left temporal) at each moment of the functioning awake brain. The objectivity of this state to self becomes clear when it is compared to the dreaming meaning of scenes and events. The conscious-awake state presents each moment of definite time with logic of causality preserved, but the conscious-dreaming scenes only show the indefinite time and the uncertainty problem of the quantum mind in its time titan primarily in the service of autos. We can simplify our understanding and assimilate the state of awake consciousness as an instantly assertion of time definiteness helped with double iconic-word interpretation of reciprocity, reflecting feedback between iconic psychognostic and literal idiognostic meanings in constant connections with the entire neocortical modular function areas. This degree of functionality can be regarded awake consciousness asserted by self. Further discussion on dreams in the next chapter will shed more light on this immaterial biological state as a whole.

REFERENCES FOR CHAPTER FIVE

1. Mithen S. 1996. *The prehistory of the Mind. The Cognitive Origin of Art, religion, and Science.* Thames and Hudson, Ltd. London.
2. Ekman P. 1971. Universal and Cultural differences in Facial Expressions of Emotions. In : JK Cole(Ed.). *Nebraska Symposium on Motivation 4, Lincoln. Nebraska.* University of Nebraska Press, Pp 207-83.
3. Flanagan O. 2002. *A companion to Cognitive Science.* W Bechtel, G Graham (Eds.) Blackwell Publishing Ltd. 178. TJ International Ltd. Padstow, Cornwall, United Kingdom.
4. Rosenthal NE. 2002. *The Emotional Revolution.*Pp 39-41.Citadel Lexington Publishers. New York, NY.
5. Whybrow PC. 1998. *The thinker's Guide to Emotions and its disorders.* Harper Collins Publishers. New York, NY.
6. Glynn I. 2003. *An Anatomy of Thought. The Origin and Machinery of the Mind.* P 336. Oxford University Press. New York, NY.
7. Guyton AC. 1986. *Textbook of Medical Physiology.* Pp 594-596. WB Saunders Comp. Philadelphia, PA.
8. Burchfield SR (Ed.). 1985. *Stress. Physiological and Psychological Interactions.* Hemisphere Publishing Corp. Washington DC.
9. Fehr E, Renninger SV. 2004. The Samaritan Paradox. *Scientific American Mind.* 14 (5) 15-21.
10. Newberg A, D'AquiliE, RauseV. 2002. *Why God Vont Go Away.* Ballantine Books, New York.
11. Churton T. 2005. *Gnostic Philosophy. From Ancient Persia to Modern Times.* Inner Tradition International.
12. Greenfield S. 2007. How Does Consciousness Happens? *Scientific American.* 297(4) :76-83.

CHAPTER SIX

PSYCHOGNOSTIC CONSCIOUSNESS

THE MUTE AND THE TALKING HISTORIANS

The fundamental soundness of biocracy and its principles conducting the biological adaptive evolution become more evident when we consider the change from paleognostic to neognostic states. Psychognostic consciousness, as foundation for neognosis and with connected characteristics described in table I, requires solid definition, which necessitates detailing the neurobiological foundations of the memory types establishing psychognosis. When we realize that biocracy has succeeded to establish the means for adaptive perfection of the sensory system to reach more evolved stages and presently provides the most efficient system with neognostic states that respond to requirements of the human mind, we face the bare confirmation of the tested hypothesis of neurobiological organ-function responses to living necessities. But is the response to necessities the limit? These responses have answered the evolutionary needs, but not without the innate biological power aimed to go together with time for infinity, the power of the bioforce reflected in its pristine autonomy and its unlimitedness. The means for reaching these perfect adaptive levels of sensory systems have been biologically evolved from an initiating recording of internal metabolic sequences, an early capacity of protomemory type, starting with notation of repeated episodic functions in series to eventually founding the implicit memory. What really has been supporting this course of evolution is the process of repetition, rechecking, recording, and the possibility of genetic transmission of corrections in a long adaptation, all realized by the bioforce behind the perpetual time effect. This time accumulates in memory and expands in biological forms in evolution reexpressing the refinements in metaphysical thinking.

With the implicit memory, we have all the foundation of understanding and registering our percepts short of words. In fact, akin to our ancient remote ancestors without speaking out their expressions, we could have developed the perfect skill in natural biological acting and the primitive tool using without elaborate language. We could demonstrate our intentions primitively by gesticulation and phonation indicating our acceptance or rejection and perceive others through the same means, and we could remember vividly the scenes of our dreams and reflect on them quite as well as if we could use words, but using words give us an edge in our recall process. In essence, words do not create our mindfulness to be only literally conscious, and we can be psycognostically conscious all along our living, realizing it even without words to recall our state of mind to ourselves. With words, the comprehensive record-keeping memory, which is the real evolutionary marvel of the explicit declarative memory, the slate for the codified symbols of idiognostic consciousness then became established, but with many puzzles remaining to be cleared, the main puzzle being the primordial *what* and *why*—that is, what is behind the whole thing if not the bioforce, and why the whole thing is constituted if not for a purpose, and what is the purpose. These issues will be mainly studied in this and the following chapters starting first with the fundamental outcome of time concretizing biological mechanisms: the memory.

Memory is the undeniable building block for the system of sensory transmission of recordings that has been refined by evolution. Memory, in fact, has acted as a historian that has kept the records of biocratic events from the very beginning to date and has allowed correction of past mistakes on the way to perfection. How this has been achieved is what I attempt to discuss first before describing the important relation of the psychognosis with its specific type of memory and its most significant meaning in biology: the making of language and the power of abstract thinking.

The philosophical meaning of memory can be condensed in ***mental life***, and in fact, if no memory, no mental life, and no symbiosis in destiny or symbiogenesis concept either as title in this book's initiating chapter. But to comprehend the biological foundation and course of memory, it would be helpful to recall the thought experiment in chapter 2 to visualize the continuum of time as a major infinite time vector with a set direction and a hypothetical constant pace and the biological time as a minor finite time vector, a biochronogenetic unity in the phenotype in parallel with the major vector, with the same pace and direction. In this setup, as long as the minor vector, the phenotype spacetime, keeps pace with the major one, it only would need one recorder for its private internal processes. But if, because of any interaction with the environment it would have to adjust its own pace to externally imposed changes and to accelerate or decelerate, it would need an additional sensor for its external contacts and for time changes needing recording. Both of these functions, the internal one depending on molecular microspacetimes and the external one depending on the phenotype spacetime needing recognition, could be possibly handled by the internal recorder with additional adaptation. Such adaptation can plausibly provide a stage of cofunctional recording eventually acquiring more specification forming the two types of implicit and explicit memories ultimately—one to be essentially involved with internal affairs and the other mainly with the external ones. This simplified presumed course of events applicable to biochronogenesis seems to have been indeed the scenario for the two main types of memory, the implicit and the explicit types, and in two substrates, limbic (hippocampus) and hemispheric (neocortical) structures. These two memory functions ultimately keep the time frames connected inside and outside of the phenotype.

Viewed from a different angle, in reality, the minor hypothetical vector being a biological element, it disposes of an energy-regulating system, balancing its anabolism and catabolism, and in short, a feedback function for constancy that uses several processes all necessarily in sequence. The system must keep the order of the steps in the processes of SR and needs a record keeper, which is the internal biological time keeper so to speak, the implicit memory.

This primitive internal memory in reality does not recognize or record the measurable time but keeps the meaning that is attached to the sequencing chain of events for the result, a meaning of events sequences of separate events' frames, not by their time length but by order of sequences making meaning in terms of an algorithmic logic. As previously said, the pace and direction of the minor vector, the phenotype spacetime, and the major vector are the same. The pace of the internal biological changes of the basically set molecular microspacetimes, however, may become different from that of the phenotype accelerating or decelerating according to environmental factors. As long as the minor vector size is greater than zero (live phenotype) and no environmental effect causes its pace to change, it shares the same constant pace and direction with the major time vector as it is sharing the same spacetime with the major vector. In psychological terms, the autos in the biopsychon has its time titan as part of the continuum of time and follows its pace in that continuum, just keeping records of sequences of internal events.

This internal sequence keeper is the biological inner sense that has ultimately formed the implicit memory that served preparing for the explicit memory primarily on the basis of the meaning of events frames. The implicit memory, in reality implicit in biological functions, is implying and recording primarily the meaning of the order of sequences of the functions without realizing any estimate but a unit of work, which is for the balancing of the biological feedbacks of all sorts, metabolic, neuromotor, neuro-perceptive, psychodynamic, and attention-intention perceptive and executive types.

Thus from the very initiation of recording with the primitive memory, the process has included a protoimplicit and a protoepisodic explicit memory function. In the bion without well-developed neuromotor function and locomotion, the implicit memory is mainly active at the scale of biochemical processes and reactions essentially within the cell. This is done between the membrane, cytoplasm, and the nucleus through DNA and messenger RNA using special enzymes and producing proteins for the short and the long-term memories, just like what is observed in nerve cell culture experimentally.[1] But in biopsychons with well-developed nervous system and locomotion, it also secures exact motor responses in aiming an action and is the one that serves and facilitates acquiring skills. The function of the implicit memory in the locomoting biopsychons is still not time measurement, but recording the meaning sequences in a continuation that may ultimately recognize separate time frames as it recognizes the meanings of frame contents separately, eventually initiating time recognition regardless of content meaning.

Facing external events to which holistic phenotypic reactions may be needed, a real episodic memory seems to take shape from the protoepisodic functionality of the implicit memory. Thus, implicit memory serves endogenous events and functions working in the infinite time titan of the autos, and the episodic memory acts in dealing with external finite time events. In its evolutionary course, implicit memory must have started at the biochemical level of sequencing reactions unconsciously. It must have gained more capacity with nervous system development, with bioawareness and bioconsciousness, gaining significant elaboration in neognostic states and especially with psychognosis in which the content of a state of mind like an emotion, or feeling, or mood is sensed consciously without however any definite time frame in contrast with distinct time frames of external events. In other words, implicit memory is still unconscious as to change of time; even if its psycognostic content is sensed consciously, it is in the permanent time titan in an eternal present time with no controllable independent recall at any future time that cannot be realized with the implicit memory. In fact, time frames exist in implicit memory by their meaning related to sequences and not by their length as they are in chain continuity of the time titan at the incessant constant wholeness of the present time in the autogenetic conditions. Recalling in implicit memory seems to be only dependent on sequencing of one process to another in continuity. Therefore, psychognostic states of mind like emotions, feeling, or moods are sensed consciously by their sequential connections without however any definite independent single time-frame memory to allow a possible intentional recall.

The included time factor attached to each internal event with meaning, an action-reaction or cause-effect type, determined by the implicit memory, serves to indicate steps and the end result, which is a balance between the two opposites of the feedback systems in biology regardless of times used. In fact, if any meaning could be imagined for the time included in the biological processes, it could not be other than the time of an integral holistic meaning for a stereotype purpose to be achieved in that extent of time. The implicit memory functions within the indefinite time titan of the autos for the moment's present time. Metaphorically, an internal state of timelessness is the norm, so to speak, with nothing beyond the unconscious present time that, in reality, can be interpreted as the psychognostic eternity time.

Exiting from the endogenous to exogenous sphere with the functional relaying senses, in particular vision, the sensory-motor adaptation still remained based on the meaning of sequencing events but had to recognize the exogenous time effect as different from the endogenous one. Indeed with the near-perfect visual representation of environmental changes as chains of events, both the content meaning and times of the content frames became more solidly established in distinct separate frames according to event results neuroperceptively and with clearer time identification by episodic semantic memory. This dual capacity memory became refined with more distinct episodic and semantic functionality with the clear visual definitions of time frames of events indicating urgencies of fight or flight, for example, imposing a more real time oriented sensing and interpretation needed to determine appropriate reactive responses.

The holistic time effect, based solely on functional meaning for endogenous processes, then acquired the additional facility of including time frames of external visual events with their meaning. This adaptive necessity started estimation of the time from exogenous origin as nonself items imposed by the environment. This forcing of an exogenous impression, with defined observable content and time to trigger a change in the internal pace to match the external necessity, ultimately succeeded because of realization of its type as nonself, with the clear nonself exogenous image. This type of sensing eventually established the full explicit memory. The early proto-explicit memory, presumably taking ground in the implicit memory, balanced related external changes to their times as intervening events in disrupting the implicit-balanced timeness of the continuum time titan. It should have started with timing of the episodes of external events intercalating them in the implicit memory and thus forming the earliest type of explicit episodic memory by episodic event recording. Episodic memory is in fact still holistic in connecting the psychognostic meaning of the events to their times of occurrences in a semantic capacity—the meaning being perceptive-reactive with only an affective primitive sense of categorization. Completeness of the explicit memory is to come with idiognosis with language and with the precise labeling that it will introduce with codification of items in conscious manipulations and categorization. That will form the declarative semantic memory.

As far as the explicit memory is concerned, we can say that this new capacity has taken place also as an evolutionary adaptation supplying the implicit memory and adding extendable scopes to it. The realization of the products of implicit memory in the present actual time of the events such as we may observe with psychognosis when occasionally it gives us conscious states of *qualia* remains the same and only represents the present time of the recalled event by the explicit memory. In fact, in implicit memory of essentially unconscious type, when bringing up any consciousness of mood, sense, feeling, or emotion, the sensing is invariably in the present time in an ephemeral form only lasting as long as the actual present time lasts in the applicable scale of the working memory. We may recall a conscious memory of an event from the past by the explicit memory that can possibly bring an associated psychognostic sense with it, but we cannot recall the sense without the event. In other words, explicit memory can recall both the material and the nonmaterial content of an event if firm association is formed between the two with involvement of the implicit memory, but neither the explicit, nor the implicit memory can recall the nonmaterial psychognostic content alone. This nonmaterial psychognosis, in my belief, represents the nonmaterial component of consciousness akin to the probabilistic uncertainty in the quantum theory for both being and not being of a physical state at time of detection. It is that time segment whose reality is bound to its material content, reminiscent of the quantum situation for Schrödinger's cat, alive or dead at the same time!

We can say that evolution forced the biopsychon to have a chronicler for its internal and a chronographer for its external affairs, necessarily working jointly to continually assure the real time meaning and its changes brought to the phenotype. We feel indeed the nostalgic effect of a past time experience in the present time when our chronicler, the implicit memory, provides us with the strong affective meaning of events in our psychognosis, and our chronographer reminds us of their dissociation from the real present time. In this contingency effect, what is crucial is the basic biocratic sensing and remembering to relate fragments of time-connected materials, events of all sorts, in a way that a useful result for the living organism would be assured to help its behavior for securing SR, which also includes all degrees of emotions and affects.

Memory development, accompanying central nervous system perfection, naturally took place in steps to reach the two highly functional specific implicit and explicit types. But both of these types, once established, are based to become permanent and hold their content somewhat fixed and saved for implicit recalls of essentially biological meaning of the internal mechanisms or explicitly for affairs of external environmental contacts. Another type of memory was therefore needed to act accessibly within the immediate relations of the living organism with its surroundings or regulating internal and external associated needs, only in real present time and without leaving a superfluous permanent imprint as a conscious state lingering behind or beyond the instant present time.

This short acting effaceable memory is called working memory. With the initiation of the explicit memory, or even before, a short acting recall intermediate between sensing and record keeping must have been formed and then evolved to give the presently called working memory. This type of memory serves within the time limits of an immediate attention-intention process of only seconds to minutes before a new attention-intention process replaces it, like memory of phone numbers of about seven digits (the magical number of the psychologist George Miller) kept shortly in mind to place a call, or a title, or short phrase to immediately reduplicate—in essence to serve the present time between immediate past and future.

There is of course associative relations that can be established in real time between these types of memory functions when more permanent stability between the implicit and the explicit memory is established and the final cortical register is set after the initial hippocampal imprint. The implicit type seems to be essentially limbic, not dependent of, but subjected to modifications by the cortex. These memory types act as perfect relays in keeping mental events connected, both in timing and life significance in relation with SR. However, the evolution has established fine specifications through paleognostic to neognostic changes in both the time and the time related content and meaning of events recorded and saved. The main result of this natural process has been the establishment by the implicit memory of the early autobiography essentially with affects from autos and by the explicit memory mainly with registered contents by self. In this autobiographical memory, both memory modalities are needed and decisive in their cofunctional capacities.

The philosophy behind the biological meaning of memory as a whole is the most crucial in terms of memory's value in abstraction and in the metaphysical realm of the nonmaterial biological entity that is the mind. If in fact we hypothesize that a memory registration can represent a unit of neurodynamic energy spent for the time, registering an event for a recall process, we are simply accepting that the memory is in reality coursing through time to the extent of making a permanently disponible imprint that could be reduplicated at any time. This hypothetical unit of memory in the simplest form that is the saving of the imprint in the hypothetical unit of time will express the primordial meaning that it exposes for the forming block of abstractive knowledge.

THE LIFE COMMANDMENT: LOGIC AND HAPPINESS

The crucial importance of the implicit memory and its result, even before the evolutionary stage of an autobiographical memory and self, is the establishment of the principle of causality logic. In fact the functional principle of the implicit memory based on a whole *action-effect* in biological processes, pointing to an end and implying a meaning, is a constant principle dictating a biocratic commandment. Indeed, the fact of accomplishing the task in biological processes, and reestablishing the balance in all feedbacks, is an adage by itself, a constant principle, a maxim. This biological mechanism, in type and constancy, reveals the first life adage: the principle of causality logic. This principle of basic logic has been inherently inculcated in life by the bioforce, patterned on its material and nonmaterial components reciprocating into one another for permanent support in an indefinite time. The pattern appears to have reduplicated in life forms as the logical principle of causality: The principle that orients the sensing system to realize that for every event sensed, there is a causal force and a necessitated time frame.

It is important to realize that this principle of causality logic is fundamentally inherent in the living organism, endogenous, autoregulated, and time related only in an integral way, implicitly including the time factor in sequencing biological processes for reaching a required end. This principle can be regarded as the first foundation of a real commandment of biocracy that has been realized in paleognosis with the implicit memory. The principle can be simply formulated as:

Biological Logic Principle: Frame I Action + Frame II Time = Frame III Result

These two factors, time sequencing and meaning, fundamentally obvious in visual perceptions can be analyzed further to define how they contribute to the logical principle of causality. The time factor is paramount, and in order to form the building blocks for the visual impression to lead to the logic of causality, it would have to be defined in frames compatible to be perceived by the mental analyzer.

There would appear that the time factor must include a minimum of two frames, one for the cause and one for the effect, both necessary for the logical principle to work but sufficient only if sequenced in the proper order to make the principle hold true. This order of happening in the visual representations of the environment, and its contents in the brain, are exogenous to autos and self and induce them to learn the principle of causality logic in extension between autos and self. This is of course an active reinforcement for the implicit memory by the explicit iconic memory working in concordance with it. The time span for this total effect has extended since the creation of life on earth, and its perception by animal visual organ probably dates over five hundred million years.[2] It is therefore reasonable to think that the time for environmental evolutionary *teaching* and for the nervous system *learning,* so to speak, was long enough, and the principle was inculcated first in the primitive Gnostic state of the neural perceptive system, in what has been called cognitive unconsciousness by Rozin, cited by Mithen,[3] or in what I have called bioconsciousness. However, notwithstanding the precise type of the Gnostic state for the principle of causality logic to be perceived, the pristine force behind it, which is biocratic and dictates SR, merges with detecting danger and escaping it. This does not need any self-representation by another reinforcing system but possibly other clues from senses, hearing or smelling for example. In any case, the learning is based purely on biologically available means. This learning is essentially mediated by the implicit memory, which is used also for the coordination of locomotor response to achieve escaping danger but conjointly with the proto-explicit memory with visual detection. The principle of causality logic in this situation, as an example, is therefore using both types of implicit and nonliterary explicit memory.

The dramas of visual scenes of danger to other animals sensed with the consequence of the included related causality are enough to trigger an instant automatic recall from the proto-explicit memories and an adequate immediate motor response to secure safety. This adaptive learning is certainly not immediately conscious in the sense the word is commonly used, but it is primarily a type of unconscious perception engaging the component of a conscious perception by the forming explicit memory. In this perceptive-reactive complex, the meaning of the time sequence of the first frame (apparent or inaparent cause for an action) to the last one (apparent or inapparent effect) is autonomously serving survival. It is an inherent part of the autos with its instantaneous allowance by the innate time titan served with the implicit memory. Any possible appearance of an initial frame of the cinematographical course, which would eventually imply the subsequent frame to come, would trigger this perceptive-reactive complex in bioconsciousness.

This perceptive-reactive example as a typical case demonstrating the intrinsic sense of principle of causality logic shows the constant component of time as do all such biological examples that serve SR, but also has its strong emotional content grouping it in pillars of emotions detailed in the last chapter. This example and all others from that group, in fact, serving directly the biological principles of SR, can be called psychognostic to specify them in addition to the bioconscious meaning of their biological bonding. But psychognostic states do also include all sensed psychic perceptions, feelings, or moods that would all have undefinable time frame but a definable meaning. When time effect does not imply urgency associated with the pillars of emotions, it is undefined and not restricted to a known exigency. The emotionally tinted sense of well-being, happiness, peace, serenity, and pleasure as well as a regretful sense with sadness, resentment, discontent, shame, etc., are perceived as timely undefined conscious states.

With the implicit memory on a scale of billions of years and the traditional explicit memory centered on meaning possibly for over a few million years since initiation of primitive hominid communication in any form, not only the principle of causality logic has been solidly established as the first foundation setting, a sine qua non type of biological commandment, but also its logic has come to include a meaning specification in addition to SR. This additional meaning is that of the biological well-being balanced state that is reflected in the mind by happiness and resumed in pleasure-versus-pain balance described with the pillars of emotions in the previous chapter. These affects are all extendable to psychognosis and are also associated with the conatural truism and its satisfying affect based on correct right-versus-wrong choices in idiognosis that will become clearer as will be discussed in the coming chapters. They add significantly to dimensionless frames of psychognostic feeling and real time confined frames of idiognostically categorized realization of pleasure or pain. The principle of pleasure-pain balancing, which in reality dictates the quest for happiness, can be regarded as the second foundation of the life commandment, naturally added to the first one and inseparable from it. If the first one primarily rules for secure living, the second rules for happy living, and both can be called the biocratic commandment or more descriptively, the biodynamic life commandment, formulated as outlined below:

BIODYNAMIC LIFE COMMANDMENT

Time Frame I: Stimulus + Time Frame II: Balancing Process = Time Frame III: Balance

Change in Values → Logic and Happiness Axis → Secured Values
Pleasure Loss　　Logic Securing Pleasure　　Pleasure Gain

The two principles of logic and happiness are inseparable axioms in all life-forms, making together the major unified biodynamic life commandment. The process is axiomatic, innate, and implies precedence (in the reasons) for biological activities as in SR. In the developed mind with extended expressivity of emotions and more elaborate logic, these two elements of logic and happiness become reciprocally necessary conditions to sustain one another, but ultimately only as sufficient condition for happiness principle rather than for logic as necessary condition, ruled by autos in the obligation of secured SR.

It is interesting and enlightening that we can trace the subatomic scale uncertainty principle of time indefinite nature in time definite material biology in the life commandment with SR significance, meaning survival, defining time definite logic and reproduction, and implying indefinite happiness. The nucleus of this biodynamic commandment is evidently autocratic and inherent in autos. The fundamentality of the time indefinite subatomic scale (in uncertainty principle of Heisenberg) is reflected similarly in the immeasurable level of happiness problem and shows itself dominating the commandment. In fact, the normal evolutionary course reveals autocratic biocracy eons before democratic logic. Logic's neuroanatomy and neurobiology became elaborated much later in the present human mind after the pleasure pain principle was established. In addition, the two foundations of the commandment in the autocrat biopsychon, the causality logic and the happiness component, have specific values and interrelations. The first principle, the logic, is fundamentally time definite in function, representing the episodic sequences in living processes for the absolute biocratic logic of SR. This logic, the logic of immortality of the bioforce, recapitulated in biological phase transfers by genes in the material biology, is the logic of the primordial truth, the sublime truth of the creation that we sense in our abstract metaphysical thinking of our immaterial biology. It is the same **TRUTH** that we deduced from the thought experiment of chapter 2. The test of time through the evolution seems to have secured this logic by SR in the material biology to serve the immaterial biological life. The logical reasons for life processes securing **TRUTH** are time definite in the scale of priorities to the ultimate logic of SR, the sublime truth that overrules all other reasons and imposes the binary axiom of *either* versus *or.* The second foundation, the ensuing contingent and also conditioning component of the commandment, the element of happiness, is a functionally time indefinite entity in both unconscious and conscious states, representing truly the subatomic bioforce nature throughout life. It combines with and conditions the time definite conscious reasoning and covers the living phase of the biopsychon with potential overruling power on all reasons but on this **TRUTH** in which it is inherent.

In consciousness, the commandment is still adamantly respected and applied, but may face opposing values formed in pi versus ps views explained in chapter 1. In that condition, the reciprocal sustaining between the two principles submits to pi, not to ps. Furthermore, the logical aspect in conscious-awake states (time definite states) may be frankly disturbed in dreams (time indefinite states) and become illogical, showing exaggerated viewing of the fundamental meaning in logical or illogical but significantly more in happiness and unhappiness interpretations by various presentations. This abnormal presentation of the biodynamic commandment justifies special discussion of the state of probabilistic consciousness in dreaming.

THE LOGIC OF CHAOS

It is interesting to compare dreaming visions with wakeful visual perceptions that make our alert real time experiences. Both give perfect images impeccably clear, and both are conscious, but may have a difference in consecutive correct meaning of frames of actions; the wakeful images show perfect timing that keeps securely constant the sequence-meaning complexes undisturbed and logical, but dream images lose this constancy. The loss is not always obligatory and often the constancy of normal sequencing and meaning of events is saved in parts of the dreams, but other parts of visional dream images, though being always sensed consciously, may show bizarre connective disorders in the bonding of figures and sequences with meanings. This gives the impression of an abnormal fluidity that interchanges the *content-meaning* and *sequence-framing* of events in dream visions. This phenomenon is evidently illogical to reason, which is satisfied only if causality logic with its irreversibility of both sequencing of cause-to-effect for meaning and real-to-unreal for timing are preserved. So what is the logic behind the illogical chaotic dream visions, if any, which seems to have so much to do with psychognosis and with so much similarity with probabilistic nature and uncertainty principle of quantum physics?

As both logical and illogical visional views occur in dreams, both dependent and independent mechanisms on logic must be functional according to rule or rules of dream imageries that remain unknown but could be hypothesized and tested. A fundamental hypothetic analogical basis is the quantum probabilistic principle also operating in the quantum mind. All other hypothetical considerations treat the neurobiological procedural facts. The analogical hypothesis on quantum basis will gain more ground as we progress in our discussions but cannot reach tested security with our present knowledge even though interesting evidences have been presented by Walker,[4] lending support to this contention. Other hypotheses concerning the neurobiology of dreaming also need consideration, starting with their histories as sketched briefly below.

Dream interpretation started millennia before history and was historically recorded in antiquity. The epoch making history of dream interpretation, however, was made by Freud at the end of the nineteenth century and indeed initiated a school of Freudian psychology and psychoanalysis with vigor lasting to second half of the twentieth century, and modified versions of it still continue to be used. Freud's theory will have occasion to be seen again in relation with more recent dream interpretation theories that will be discussed. However, what is more important is a basic outline that would give a plausible approach to this type of study that has to deal with and evaluate the two sets of logical and illogical data with mechanisms completely opposing and negating one another, but necessarily intimately related. The solution rests on finding a third mechanism, which must be hypothesized to explain their interconnections. This third mechanism would be the crucial basis on which dream interpretation could be expanded. This third mechanism should provide a ground for the logical and illogical dreaming visions to intermingle and possibly reach some balanced or near-balanced result.

In this quest, the third mechanism to be envisaged must function with both logical and illogical presentations and figures that stand as two global opposing groups of data that could annul each other, and the opposition may possibly serve a purpose in life. If this opposition could be mathematical, the result, not predictable, could eventually become zero in a subtraction formula. Although such function in the mind is never a simple subtraction with a probable zero as a result, nonetheless the hypothetical zero would theoretically mean no memory whatsoever whereas the implicit memory that includes the billion-year old background of autobiographical memory can never normally be effaced, sleeping or awake. Therefore, the mechanism may be regarded to be of the general feedback balancing type but more of a mathematical form of not a simple one-time resolvable equation, rather a continuous mathematical adjustment going along with life. It must be preferably an algebraic equation with data in balance, respecting and keeping the autobiographical memory up to date, repeatedly renewed afresh for the psychological self. The balance must be such as to allow additional entries in the equation to be treated repeatedly, reestablishing and keeping the equilibrium throughout the life span. This algebraic equation with its terms of balanced based autobiography therefore should stand similar to a theoretical equation whose terms either contain one constant and several interdependent variables with time and can never be resolved at one given moment or may have multiple theoretical variable terms that could give equality at one hypothetical given time. However, this hypothetical time dependence would still make the result unstable with time continuation, forcing continuous calculation for assuring result constancy if needed. The equation must be constantly rechecked and terms simplified mathematically, even including the constant appearing axiom, the autobiography itself. That is what I believe normally takes place between wakeful additional changes and sleeping adjustments. Here, irrespective of the theorized type of the equation, the probability principle seems operating as in Schrödinger's equation, which, in essence, relates sets of possibilities as factors of time. In dreams, both iconic figures and time frames, though directed on the set modus of life commandment's logic, are chance directed by the life commandment's pleasure-pain principle, which gives a product of sets of probabilities analogous to a hypothetical variability of around a defined chance item (Ψ psi) in Schrödinger's and Dirac's equations.

To test the hypothesis that one such mechanism for keeping the operating psychological self in perfect functional balance (according to the life commandment of logic and happiness) is indeed the basis for dreaming and is the condition necessary and sufficient for it, one has to study all available opinions and experimental data concerning dreams. Here, to be concise but objective, only essential findings of significance will be considered.

The starting point could be with the dream recording and interpretation. In this connection, the work of Domhoff as a solid attempt to establish a neurocognitive basis for dreams is remarkable.[5] It indicates proportionality of types of dreams relating the meaning and sequencing in a general way and the incidence of male and female dream categories and other variations. These thorough recordings with all details yet do not answer questions that may be entertained according to the categorical outline of changing contents versus meaning and timing to check the totality of probabilities for icons, times, and their logical versus illogical relations. The general conclusion, however, is that dreams reflect the dreamer's states of mental development, sex, affects, education, and surrounding influences. In short, all effects have to do mainly with the dreamer's psychological self and are not stationary, still meaning that dreams follow the dreamer's modus of the operating mind with its inherent probabilistic allowances.

Other important ideas in dream research presuming significant role for dreaming in memory organization and learning that concern mechanisms responsible for logical versus illogical dream imageries were clarified and supported by fine experiments performed on rats and birds. The essential idea gained on dreams in rats by Matthew Wilson, described by Andrea Rock,[6] suggested the belief that dreaming played an essential role in establishing and solidifying memories. In these experiments, recordings were through microelectrodes implanted in the rat hippocampus and indicated that animals' sleeping rapid eye movements (REM) duplicated exactly the awake recordings during running in the experimental maze for finding food. The precision in similarity of the two recordings, awake and sleeping, was such that the experimenter could indicate where the animal could actually be in its dream journey, standing still or running at a particular turn of the maze. This duplication of the awake experience by the sleeping REM, recorded from hippocampus, which is known as the primary neural memory station prior to final destination and engraving in the cerebral cortex solidly indicated a reviewing for reenforcement of the autobiographical memory. Later, this duplication was interpreted as related to the cofunctioning mapping effect of the entorhinal grid cells and hippocampal place cells.[7] The conclusion was that reviewing in REM allowed the brain to keep essential parts of the memory useful for survival while rejecting the superfluous parts.

Somewhat similar studies on zebra finches were carried out by Margoliash, according to Rock,[6] through microelectrode recording from the brain part equivalent to hippocampus, which showed that direct recording from the brain in awake singing birds was similarly duplicated in sleep. This finding not only supported the theory that young birds learn their singing tunes by rehearsing them after hearing from older birds' and from their own singing in awake states, but also from their own singing in their brains during sleep. In other words, rehearsing was effectively done in sleep with refining the details by repeating and memorizing to improve the learning. These findings also confirmed Crick and Mitchison's theory of reversed learning proposing that stimuli from brainstem, randomly presented to forebrain, trigger the memory organization mechanism to keep the useful and reject the useless facts, saving only what is good in life for the living organism. This is again what the autobiographical memory's role is playng in securing constancy of selfhood.

If we ponder on this aspect of safeguarding the constancy of the selfhood with neurobiological adjustment operation, we are reminded again of the operation of the probability principle in quantum physics, safeguarding the essence of energy-matter ensemble. In fact, neurobiology seems showing exactly what bioforce is doing in the protein-based physics, so to speak, as it is doing in the nonprotein quantum physics. So far, we remain with these two known facts that the bioforce as a modulator controller and the constancy of selfhood as an aimed modulation seem to be involved in the presumed third mechanism for explaining both logical and illogical aspects of dream contents. We still have to study the variables that could play a significant role in that mechanism before attempting to formulate our hypothetical equation explaining dreams' role in biology.

THE INFALLIBLE DETECTIVE

The brain with its complexity of anatomy and function is a unique marvel of the creation in accomplishing the job of a superb unfailing controller. The main role in the task of data recording and processing is played by the vision with its complex neurobiology and its nighttime revisional review. Evolution as the master builder of exceptional ability and skill has indeed given the biopsychons an unfailing system of constant contact with the environment with their visions, which also seems to serve as a clearance vehicle in dreams.

Vision is in fact the most perfect integral informer of what happens and when. Images incessantly seen through this highly specialized functional complex system follow one another without blank or empty frames. They fill up the visual fields days and nights continually in succession. Thus vision in constant neuroperceptive collaboration with other senses in the mind is the only one that continuously operates uninterruptedly from birth to death, both in sleeping and wakefulness. Remarkably, the system is only open to new recordings in wakeful times and closed to them in sleeping while reviewing dreaming scenes, even if eyes are kept artificially open.

Anatomically, the visual system also has the most complex structure and the most extensive relays and connections with the rest of the brain.[7] Each eye's visual field reflects into the retinal surface, crossed through the lens, with the right lateral retinal halfs receiving the medial visual field images and the medial half the lateral ones. In passing through the optic chiasma the medial nerve fibers in the optic nerve cross side but the lateral fibers do not so that the optic tracts beyond the chiasma will each contain the ipsilateral uncrossed lateral retinal fibers and the contralateral crossed medial ones to be transferred to each primary visual cortex. This pattern assures connection between right and left visual field images in each visual cortex so that ipsilaterality is saved for the medial right and left visual field images to each corresponding cortex with contralaterality of the lateral field images that are crossed. Thus, the needed depth of image perception from the lateral visual fields that cannot be helped by simultaneous focusing to the right and the left is best assured by simultaneously available matching in each occipital cortex for safety reasons. Furthermore, the paired primary visual cortices in the two occipital lobes with further paired cortical relaying to the hemispheral cortical foci in parietal and temporal lobes and thereby to subcortical limbic structures in the left and right hemispheres assure the most elaborate connected functional system for vision. As further crossing connections of significance exists also between the hemispheres, data to be analyzed are doubly secured to represent both visual field impressions and, with added scrutiny by associative conditioning in the cortical regions, to be ultimately available to both hemispheres for analysis. Associations between the frontal and occipital lobes seem also secured partially directly and partially through intercalated steps.

Two exclusively outstanding anatomical and structural findings of biological importance should also be noted. The first is that all retinal to cerebral connections between the optic tracts to occipital lobes have passed the optic chiasm and harbor some significant crossing effect to connect to the occipital cortex. This crossing however is excluded in one connection that is from the optic nerve containing the uncrossed retinal representation from the left eye to the left and the right eye to the right suprachiasmatic nuclei and from there to diencephalon on each side. Thus, connections are essentially crossed, reaching occipital visual cortices first and diencephalon through secondary cortical connection next, except for this single direct connection to suprachiasmatic nucleus and to diencephalon from each eye.

As the uncrossed direct relay to diencephalon suggested other functional reason than only light or image representation, it stimulated the idea that retinal structuring, from the prosencephalic origin could embryologically include more than just light sensing cells. Kashani was the first to pioneer a triplex hypothesis of vision on this basis indicating the third retinal category of ganglion cells with direct connection through the suprachiasmatic nuclei to diencephalon.[8] According to Kashani's hypothesis, direct retino-diencephalic connection is for a general nonvisual activity including regulation of the circadian rhythm[9] and is also responsible for other biological effects generally called chronobiologic, internal rhythms, or biorhythms.

The second remarkable finding is that spindle cells in the brain, most exclusive to humans and some humanoids, located specifically in the anterior cingulate gyrus in the left and right hemispheres show the greatest activation associated with greatest active dreaming. This exclusivity of cell type and location and solid functional association with the highest active dreaming must play a significant role in the dreaming neuroperceptive mechanisms possibly related to time sequencing.

Without going into any more detailed study of anatomy and physiology of the fine visuocortical structures, we already sense the significance that all visual information for logical interpretation integrally reaches the cortices and information of archaic innate biological importance reach diencephalons directly. It is quite clear that the system disposes of a perfect setup of data entry and data analysis. In addition, the system is set to record continually and automatically for an immediate obligatory decision making at any instant without failing. In this process, mental interpretation is instantaneous, but decision to act is timed from immediate to delayed type according to the interpreted meaning. This situation is the norm in the wakeful state.

In dreams (even if eyes remain open), visional dreaming views appear and usually represent familiar figures experienced or known or conceptually imagined earlier with inherent meaning. They may be still or more often showing motions. The time is also normally a monodimensional present time of the ever-permanent time titan, but sometimes with spontaneous recall of the dream scenes as passed events. The characteristic visions in dream scenes may be theoretically regarded as combinations of icons and times and expected to obey the mathematical rules of probabilities in possible free combination or permutation for mixing the logical and illogical presentations, but in reality, this is not so. The fact is that the theoretical probabilities follow a preferential rule when we examine the tabulation of dream contents from different investigators. Also, data are clearly similar in type and can be grouped in categories; they show frames of images and frames of meanings for single or combined images in final conclusion. Analytical work, by the preferential controller of whatever origin, must therefore check the frame contents, which is in essence the significance of the content depending on familiar versus unfamiliar icons and their logical relations, and correct or incorrect sequences of the frames singly or related to one another. In simpler words, logical and illogical dream parameters are recorded in terms of iconic contents and plausible timing frames. Available data from indicative samples of dream's contents from Domhoff's work[5] indicate clearly a major predominance for familiarity of figures of close relatives or friends and affective significance of meaning of their relationships to the dreamer or to themselves combined, evidencing significant emotional charges like aggression, friendliness, and sex. Thus, we can plausibly think that the preferential modulator must be real, must be possibly of psychognostic nature represented by autos and its autocratic emotional character to modulate the predominantly emotionally meaningful dreaming scenes. Furthermore, the frames for time that is the present time in the majority of dreams and sometimes the past reflect the time titan of the inseparable eternal inherent time.

This evidence assures that if dreams follow any order, it is that of the life commandment of logic and happiness, dependent heavily on pleasure-pain principle and less on logic that is practically only the logic of causality, conditioning the cause-effect timing priority. The pleasure-pain principle dominance in dictating the meaning associated with the icons and times and generally presented in simplistic holistic figures and actions indicates the essence of the biocratic real life. The probabilistic scale of possibilities, presumably limited between states of being or nonbeing with a factor of time in the modulated visions, is now more evident in analogy with Ψ of the Schrödinger formula as mentioned earlier. If then any compatibility with this autocratic life meant to be lived by autos and the exigent social living imposed by self(s) has to be assured, it would have to secure an acceptable balance between both as one cannot exist without the other. We can understand that some biodynamic mechanism must operate sometimes to assure such balancing in a continual way, and the only plausible times for that purpose are at night and day dreamings. The hypothetical third mechanism that we have envisaged earlier seems now gaining more ground to be entertained further.

A THEATRICAL SETUP IN THE DARK

Sleep studies mainly undertaken for finding about dreams started practically in 1930s when human EEG changes were noted to show special activities during sleep.[10] This incited interest in the field starting a new era of neurobiological studies. When the discovery of the rapid eye movement (REMs) in 1953, observed under the closed eyelids in sleep showed synchronous occurrence with the EEG changes,[11] further excitement was raised. Then, when electromyogram (EMG) of the neck muscles during sleep showed no activity whereas electrooculogram (EOG) proved simultaneous ocular movements with REMs, the findings became overtly significant and of special interest. Subsequent studies provided understanding of the sleep stages that were classified based on EEG changes associated with rapid eye movement, REM, and nonrapid eye movement, non-REM sleep.[12]

These sleep stages are numbered according to REM in four stages. The first stage starts from lighter to go to deeper levels and back to lighter level to include the first REM before repeating the cycle. Thus research progressed expanding on the mechanism establishing sleep and incited interest at the same time to investigate the meaning of dreaming. In short, two arenas of interest opened up, one concerned the mechanisms of sleep production, hypnogenetic mechanisms, and another, related to the functional meanings of dreams, concerning changes in consciousness that I prefer to call hypnokinetic. In this section, I expose essentially the hypnogenetic mechanisms and, in the next, the functional aspects and meanings, the hypnokinetic theories and conclusions briefly before defining any assertion for the dreams role in biology. This exposure will mainly follow the biocratic principles controlling interactions of autos and self in the play arena of psychognostic and idiognostic mechanisms.

The EEG of the awake person, alert but quiet with eyes closed, records a fast low-voltage tracing called alpha waves and, with eyes open or paying attention to a task, tracing becomes desynchronized and records a characteristic fast beta waves that look somewhat asynchronous and usually of lower voltage. Sleeping starts with non-REM. The stage I sleep shows a gradual disappearance of the alpha rhythm replaced by slightly slower and lower voltage rhythm but with bursts of some fast low-voltage activity. Stage II, still non-REM, is marked by bursts of spindle-shaped wave activity and some high-voltage biphasic waves called theta waves. Stages III and IV, also non-REM, demonstrate high-voltage slow delta waves, of less or more than 50% of the time respectively. Then the period of REM starts with beta waves. Thus, the first REM sleep occurs about ninety minutes after sleep onset and alternates with non-REM stages with minor interval differences. About 80% of sleep time in adults is spent in non-REM and 20% in REM. Recording varies somewhat according to the electrodes' sites on the scalp and can be manipulated by inducing pleasing or displeasing affects to the sleeping examinee.

Non-REM sleep preceding REM seems to be the established norm by the evolution as some animal studies indeed concur to show that the pattern is evolutionarily determined. According to two such studies by Winson and by Siegel as described by Andrea Rock (6), sleeping in Anteater, an egg-laying monotreme mammal, between reptiles and more developed mammals, show no typical REM sleep but some pattern between it and non-REM sleep. As mammals differentiated from this monotreme example around 140 million years ago, REM sleep probably developed from that time.[6] Appearance and progress of the manifest showing of sleep activity in dream also seem to parallel the neocortical evolutionary development and likely did not exist in its present form prior to our fine neocortical differentiation.

Classical studies in cats in an experimental preparation by midcollicular transection separating the cortex from the brainstem so that no sensory input reaches the cortex, called *cerveau isolé*, caused a continuous sleep. The same result was obtained if transection damaged the brainstem but was not extended further to cut the ascending sensory fibers to the cortex, a preparation that was called *partial cerveau isolé*. In this situation, electrical stimulation of the brainstem awakened the sleeping cats. Transecting only below the brainstem, leaving the brainstem and the cortex intact (*encephale isolé*), did not change normal sleep-wake cycles of the brain electroencephalogram.[10] The general idea then was gained that leaving the limbic reticular formation intact in the *encéphale isolé* was responsible for activating the wakefulness, and thus the surname of activating reticular formation in the midbrain came into current usage. However, a more inferiorly placed area in the reticular formation named raphe nuclei was later found to promote sleep, and its near total destruction caused insomnia of lasting duration before any partial recuperation could occur. The function of the raphe nuclei was found to be serotonergic, preventable by parachlorophenylalanine (PCPA) injection that caused insomnia in the animal. But daily injections of PCPA lost insomniac effect gradually with partial recuperation and reappearance of both REM and non-REM waves on EEG. Another nucleus, locus ceruleus, being noradrenergic, and other cholinergic nuclei, all located in the reticular formation were shown later to play roles in sleep; cholinergic nuclei promoting REM, and noradrenergic and serotonergic nuclei promoting non-REM. These puzzling mixtures of findings remained disturbing until further light shined through more refined studies and allowed clearer interpretation.

Progressively improving technical facilities to study the sleep and dream mechanisms encouraged researchers in both neurobiology and psychobiology of dream studies, and interesting experiments and interpretations ensued. The first major contribution was by neurobiologists when direct recording from the brainstem by implanted microelectrodes were used. Pioneering work of Hobson and McCarley,[13] based on direct recording of brainstem cellular activities in cats reached the conclusion that REM was entirely caused by acetylcholine flooding of the brainstem, changing the overall balance, which showed simultaneously marked decrease of the adrenergic and serotonergic transmitters so crucial in cortical fine judgmental work. Stimuli reaching basal forebrain nuclei from the brainstem would be received and would activate in turn the forebrain structures to trigger dreaming. As a result the admixture of limbic brainstem stimuli of all sorts transmitted to the cortical centers simultaneously with active REM naturally caused dreaming vagaries. Their interpretation of dreams was thus uniquely supporting a neurobiological belief on the basis that what was sent by the brainstem to the cortex was reproduced in haphazard images without any psychological reason of emotional type being at cause. This view strongly rejected the century-old theory of Freud with the presumed meaning of psychological reasons behind dreams.

However, researching the absence of dreaming in patients with lesions to their parietal lobes or namely to the prefrontal regions bilaterally, pioneered by Solms as outlined by Rock,[6] revealed strong evidence that the undamaged condition of these regions, specifically the prefrontal ones, was necessary for dreams to be formed. Moreover, support gathered through case studies of lobotomies and locotomies, removing the prefrontal regions surgically in schizophrenics to abolish their hallucinations, further confirmed the correctness of Solms's theory that the brainstem was not alone with any decisive role in REM proposed by Hobson and McCarley. In fact, REM, although containing about 80% of dreams, represents only about 20% of all sleeping time, and dreams are also reported with lower variable frequency in non-REM that expands four times longer. Furthermore, as the necessary prefrontal regions for allowing dreams to appear are also known to be the crucial cortical sites not only for logic and judgment, but also for selfhood and intentionality including emotive inclinations, the psychological basis for dream reconstructions based on Freud's basic theory revealed to still hold some truth concerning emotive causality in dream scenes. In fact, naturally, the basal prefrontal region's role must exert its predominant inhibition of the limbic stimuli in wakefulness, and its permissive relaying of the same stimuli in sleeping when inhibition is absent. So the actual self's presentation, with orderly channeling of neural network to use the brain's association areas and form logical ideation in the wakeful state, once removed, would allow the relays of limbic stimuli to the same potentially available association network for ideation to present autos without pure logic, but with happiness logic of the life commandment inseparable from the pleasure-pain principle.

Another interesting fact in relation with emotional factor playing a role in the dreaming was the observation that damage to basal forebrain nuclei, the same that were interpreted by Hobson to incite dreaming, as seen by Solms in patients with such damages, made the patient having unusually vivid dreams and difficulty in distinguishing dreaming from awake daytime experiences, as if a center of reality testing was no longer working in these patients.[14] Additional evidence in studying the development and aging of memory in blind individuals affected in early childhood or later also proved that the longer time the vision works to establish the visual frames of memory before blindness, the more vivid visual recall is attainable by the blinds.[14-16] This view by Solms supported that cortical memory background gained with time could have effects modifying the brainstem stimuli received by the cortex. This could be a factor in mixing the emotional content of the memories engraved with dreamers' visional images.

At this stage of knowledge based on animal experimentations and human observations, the ground for a theory of dreams mechanisms and meanings could be formulated regardless of conflicting opinionated ideas in the field. Thus, the simplest and closest to truth belief would be that the theater scenes of crowding actors in the dark had to be rehearsed several times and directed each time with improving the *mise en scene*, decors, and lights to provide a refined enjoyable spectacle. These arrangements of course had to be made between directors, the cortical coordinator self(s) with autobiographical memory norms as screenplay and rules of conduct, and actors crowding the stage, the stimuli from the brainstem and limbic centers, playing characters and casts. With this much of understanding allowing the metaphoric theatrical performance, we seem to grossly know now about "*what*" of the spectacle, but we still need to know about "*how*." Indeed, we still do not know whether actors rush to the stage actually forcing their ways in or whether doors open by the directors inviting them in, as well as the type of acting they perform and other puzzling possibilities in the entire dream imageries.

An ultimate clarification, most interesting to reconcile the neurobiology with neuropsychology, was revealed by studies using PET (positron emitting tomography), which allows seeing the increase or decrease of blood flow in activities of the whole brain in its specific regions in real time imaging. The first such study and results reported by Braun and Balkin showed the indubitable actual course of the brain activities.[17] The scenario evidenced by this examination shows that the entire brain activity decreases, reaching deeper non-REM levels initially and mainly in the prefrontal cortical regions known for the most precise data processing including memory, logic, and problem solving. These areas are the first to sleep and the last to awaken. A sharp fall in adrenergic and serotonergic activity, which controls focusing attention and solving problems, in other words serving a logical purpose, accompanies this change. Then in REM sleep, a surging increase in the acetylcholine release occurs that facilitates almost all synaptic activities in all cortical areas that underwent the earlier depressive inactivity in non-REM now showing hyperactivity except in two areas, the prefrontal area, the controller of logical thinking and planning that continues to remain inactive, and most interestingly, the primary visual cortex even if the eyelids are taped open.

This PET scenario indicates that the initial gradual prefrontal hypoactivity to inactivity is the cause for the entire events to take place, but in a passive way, which implies that the events are likely the result of a force not opposed by the prefrontal control once the prefrontal parts representing the conscious world are turned off. It seems that actors, psychognostic characters representing moods of autos, force their ways in to invade the stage when the doors are no longer guarded.

This appears most plausible, taking in consideration the age-old limbic psychognostic heritage and its emotional meaning importance, dealing with basic natural factors as against the prefrontal pure logic and judgmental values dealing with the conatural aspects of enforced social norms.

Furthermore, studies show that activity in the association areas of vision and some frontal areas both shown to be related to narrating stories from dream memory by wakeful patients increase above normal, indicating the active recalling and possibly reconditioning of visional images. This also indicates the availability of the long-term implicit and explicit memories and imageries from them in dream production in sleep. However, activity of the primary visual cortex remains zero, indicating the impossibility for any new visual sensing and a working memory in image transfer to interfere with the visional dream reviewing. This visional dream reviewing must be done in fact without new data entering while in progress. Furthermore, its evaluation is not the function of the high-ranked prefrontal judges (self(s), ego, superego) that have ceased functioning now but the archaic supreme court with its biocratic code rules under autos authority. This metaphoric court has knowledge of the opinions of the high-ranked judges, the egos, through the spokesman of both courts (limbic and neocortical), the self, in the considerations for final verdict. In this metaphoric scenario, the amalgam of meaning differences of the visional images, the only ones that can be interpreted as compatible with the autobiographical personality of self and autos will be interpreted as acceptable as part of the additions to autobiography. Thus it appears that in dreams, autos as the supreme power remains the superior athority, and the self also stays in power to continue its functions as recorder-interpreter, and finally the presumed third mechanism, so far kept waiting, in reality adjusts balancing between autos and self-enhancing mutual understanding and informed agreement.

THE PSYCHODYNAMIC BALANCE

Differences of opinions reign in dream interpretations by the Freudian, Jungian, and the varying group of opinions as a separate body altogether. All the theories that I regard under the rubrique of hypnokinetic have two major common characteristics: one is that they all relate conscious states of psychognosis and idiognosis (both cognitive but at different levels), and second is that this relationship is directed both emotionally and nonemotionally. The most rational theory so far elaborated is the neurocognitive theory of Domhoff.[5] This theory seems to have developed along the same line of more general neuropsychological thinking based on activation-synthesis reasoning, which viewed dreams as stimuli of pontine origin activating the cortical centers and producing a combination of coherent and incoherent meanings. However, it deviated logically into a different line of realizing more compatibility of dream meanings to real life based on dream data. Some more logicality of dream scenes matching normal life scenes are accepted as evidenced in this theory, contrasting it with extremist theories. Cognitive basic logical coherence, even in illogically exaggerated actions by familial figures according to exposed situations, appears to form the thematic point in this theory. As an example, conscious scenes of being chased but not being able to run can be recognized in dreams indicating the logic behind need to run—as would be experienced in wakeful consciousness, or flying in dreams, possibly generated by vestibular stimuli—appears to be recognized as a neurocognitive cortical function.

However, there are essential points concerning the emotional weight of charges of these two examples. The first is more apparent; the logical fear of the illogical inability to save one's own life contrasting with the logical pleasure of illogical flying. The fact is that both conscious emotional states induced logically, though their bases are illogical, are surprises (immobilization and flying) without apparent reasons for the dreamer except for the highly emotional reasons of fear or elation. The second less apparent is the question of why this particularly extreme level of fear or elation in the emotionally charged dream scenes is chosen by the cognitive function among the multiple other possibilities that were not considered. The main illogicality, being the choice of these distinct dream scenarios by the cortical association responses, then appears to be restricted to these precisely specific examples forced by emotion. Logical coherence for the inability to run or the overnatural power to fly does not explain the cause but the mechanism for the dream type; the cause is nothing but the emotional force expressed by the emotional meaning. It seems inescapable that emotion must be at the origin of the majority of meaningful symbolic dreams, and perhaps all dreams, but at different levels of significance. Furthermore, logicality does not invariably operate correctly in both mechanism and meaning of dream scenes. For example, in patients with spinal transection and paralyzed limbs, when they dream of full motility, a reversed situation to the inability to move in chasing-running example, logical mechanism is reversed in dream production for illogical meaning to replace the expected logical one for the sake of pleasure. The logic in both situations is emotionally originated and shaped at the expense of inducing illogical mechanisms. As I mentioned earlier, Freud's name would necessarily come back in dreams' interpretation as it can be seen in these examples of clear-cut, emotionally based scenic dream meanings with mechanisms of dream production emotionally directed by the basic emotion of either pain or pleasure.

In essence, a hypnokinetic theory cannot reject emotions lest negating psychognosis, consciousness, mind, and all the established mechanisms for all accepted psychological explanations. If we accept the theorem of the reality of the mind and consciousness, then explanation of what can play the stabling role of constancy of the fundamental selfhood—in spite of modular selfves and modular functionalities in the conscious life—is due and must be given. Biocracy principles so far explained in this book indicate the autos as the unique primordium that persists throughout evolution without needing a self until auto-hetero-gnosis, bioconsciousness, and psychognosis, and finally idiognosis that requires a distinct self for such functional entity to be formed. The hypnokinetic meaningful scenes of emotional forces showing in iconic figures, in essence reflect changing emotional charges configurated in icons, according to the life commandment's logic and pleasure-pain principle. This pattern is basic, and additional changes in it by literary consciousness, the idiognosis, bring restraints to the simple mechanisms of the limbic origin through the cortical reinforced reason and logic. Many iconic scenes in dreams then show more or less coherent meaningful similarities with normal conscious life. They may be originated on similarity conditioning, facilitating cortical visional choices and recognized as conscious. These types do not expose the self and the autos strongly to emotionally charged meanings. The autobiographical memory will not suffer any inflicting change, and the conscious mind needs no adjustment to absorb the impact of emotions. On the contrary, when the configurated iconic combinations show an action with exaggerated meaning beyond the normal coherent types, then some adjustment between autos, self, and the autobiographical memory must be made to preserve constancy of the status quo for a balanced mental life.

In terms of biocracy with autos, self, and the conscious state, the fundamental adjustment must be a biological feedback. This feedback cannot freely occur in the wakeful state and must be started and completed when both limbic and cortical functions can deal with one another with most fluidity and in greatest freedom, which is in the sleeping time. Daydreaming can allow some feedback effect but not as securely as the night dreaming. This feedback is therefore psychognostic in essence and is concerned with holistic conceptual meanings fundamentally based on the life commandment of logic and happiness. Thus, elemental logic in it in figurative presentations and motions of icons must be dominated by the emotional forces from autos. The psychodynamic feedback therefore exposes limbic version of life commandment whose logic principle is what its pleasure-pain principle dictates, and this makes up the dream consciousness. Thus, dream consciousness essentially exposes the logic of autos based on pleasure-pain principle either as the counter logic of repressed pleasure, or as the logic of impossible facts as the counter logic of natural possibilities in the conscious self. Simply said, the repressed natural logic of autos with its version of life commandment that is repressed in conscious self, leading to imbalance, makes up the stimulus to present itself in the dream scenes in sleep consciousness. Its constituents will have to present counteractions to aspects of the changing self and thereby to changes in autobiography against the theoretically accepted autocratic constancy of autos. These constituents must be reviewed in psychognostic dream consciousness. Thus, if the autobiographical memory has to be amended in any way, the contact time between the psychognostic and the idiognostic consciousness for this purpose cannot be when the conscious awake state is operating.

The issue of dream interpretation, if not easily or totally clear as discussed, should not cause confusion in the biological purpose of the psychodynamic feedback and the variable data entered in the feedback equation. The biological purpose, plausibly, is to keep normal psychological balance that serves and saves self-identity for autos. The types of data entered into the balancing equation is quite variable and may include multitude of images of emotionally significant or insignificant types in terms of pleasure-pain principle of the biodynamic life commandment and with all possible psychognostic meanings. It seems, as data indicate, that the biodynamic life commandment in dreams more often condition dream visional views in terms of recent unfixed working memories of all sorts, but with a predominance of emotionally significant ones as just discussed. In fact, the long waking time exposure (about sixteen hours of wakefulness versus eight hours of sleep) of the autos to forced unnaturally imposed logic of self(s) would necessitate time to efface damages to its natural logic, the natural logic that appears in the illogical emotional (pleasure-pain), meaning scenes in dreams. As the wakefulness, in general, causes the camouflage of affects, the radiating dreaming affects shine in sleep. Regardless of how impartial one looks upon the interpretation of dreams as solidifying memories, exerting facilitating effects in learning, or adding to the idiognostic extent and power as basically nonemotional category effects, the gap for the emotional categories remains and cannot be crossed without accounting for the emotions conducting dream mechanisms. In the final step, the apparently nonemotional and the emotional dream effects serve self and thereby autos and can be incorporated in a formulation as a psychodynamic equation by their affective meaning of the biodynamic commandment of autocratic logic more than by their meaning of pure logic, and as such, they could be used in an equation for balancing self and autos, saving the constancy of autos, autobiographical memory, functionality of self and the psychological balance. This outcome I belive can be accepted as the third mechanism searched that can be now formulated in a concise presentation.

A theoretical equation for such a psychodynamic representation could be formulated taking in consideration the autobiographical memory background, the conscious self, and the possibility of effects of exchanges between psychognostic and idiognostic states at any given time that could affect self and autos. Autos regarded as the permanent final authority can be represented by a constant in a fundamental axiomatic summarized form of:

1. Autos = Autobiographical Memory + Life Commandment × Self Inputs

In this formulated axiom, autobiographical memory (M) and life commandment (C) are basic axiomatic constant values, but self inputs (S) are the apparent variables that must be included. The inclusion mechanism of self inputs, to match autos, takes the channel of life commandment's pleasure-pain principle. If the life commandment is visualized as the product of logic (L) and pleasure-pain (Pl/Pa), it can be shown in:

2. C (commandment) = L (logic) × (Pl [pleasure]/Pa [pain])

Therefore, we can summarize the proposed psychodynamic formula to show

3. Autos = M + L (Pl/Pa) (S)

In this formula, the conscious variation of self as self inputs, S, to autos, to be accepted or rejected, can only affect the variable (Pl/Pa) term of the equation by being either positively or negatively charging the equation emotionally providing rarely a value of 1 as neutral, or often above or below 1 as a rule. S in actual facts, being produced in conscious logical waking conditions affecting *L* through (Pl/Pa) during the waking hours, must be checked and adjusted during sleeping time to be either totally rejected by autos keeping the constancy of the autobiographical memory unchanged or adjusted by autos and accepted, saving the useful changes in the autobiographical memory.

This psychodynamic equation, as a baseline, can explain the biological necessity for sleeping as an evolutionary trait. It must be regarded as an outline for the foundation for securing the combined psychosomatic balance between dominating psychognosis in sleeping and idiognosis in wakefulness, mainly in terms of values gained and incorporated by autos through self. The equation is constantly used in the brain between autos and self, limbic psychognostic feeling and cortical idiognostic realization, mainly in night dreaming and also daydreaming. However, the mind, the psychological entity of both autos and self, becomes openly conscious of it when adjusting processes of the equation are shown in meaningful visional scenes in dreams. But as mentioned earlier, these scenes only respect the sequence of frames with the meaning of cause-effect logicality and charges of pleasure versus pain in psychognosis. Therefore, the basic mechanism for adjustment is emotionally conducted without regard to time logic.

In a simplistic metaphoric-psychological interpretation for dreams, there seems to be a drawing board on which black and white and basic color drawings are mixed in a way that would ultimately produce a perspectively correct and enjoyable viewing according to the logic of pleasure-pain by the artist that is autos, no matter how often effacing and redrawing must be repeated. In such metaphoric view, the artist, the autocrat perfectionist, does all retouching adamantly but necessarily privately when her artistic inspiration with its pleasurable impetus (autos) and logical effects (self) can transact freely in the absence of critics' eyes watching artist's retouching the art.

PEACE, SERENITY, ETERNITY

It seems that the psychognosis, originated from the basic ground of evolving biognosis for millions of years in the animal kingdom as a sole autocratic trait, has been gradually losing power in hominids and further in *Homo sapiens* through evolution to presenting a resistive partially subdued autos' authority, which is restricted to show uncontrolled activity only in dreams and for just about a third of the lifetime during the nights. So in essence, sleeping can be regarded as a physiologically elaborated evolutionary adaptation, a real trait, for allowing idiognosis and psychognosis to coexist in the human being. In common terms, consciousness with new additions to the explicit memories in wakeful state must coexist with psychognostic state, but in a way that would be beneficial to both and ultimately to autos. Evidences from the studies cited earlier indicate that although not a thorough dream classification in purely logical or illogical types, based on icons and timings in mere mathematical combination or permutation has been made available, present evidence points to the fact that sleep time, no doubt, serves an adaptive purpose for the normal functioning not only of soma, but also, and much more, of psyche. The brain securing an adequate state of mind, in keeping with the life commandment that is fundamental, makes it possible for a balanced status quo to be maintained for the lifetime. The equilibrated state is secured by the proposed psychodynamic equation, a type of equation that is practically straight forward mathematics with one unknown variable (S), one known adaptively constant (M), and one variable term as coefficient (Pl/Pa) to the basic constant (L) that, if resolved, will assure constancy of psychological balanced integrity of autos, a biological necessity. Thus, the so prominently ruling mechanism in biology in the form of a bona fide elemental or chain feedback is also regulating the mind, functioning between wakeful and dream images of visual and visional views and in conformity with the life commandment axed on pleasure-pain principle.

Evidently, the delicate task of keeping the mind in control of everything precisely and concisely is secured by vision both awake and asleep. Interestingly, as said earlier, the vision is completely turned off during sleep even with open eyes, revealing an important indication for the necessity of the dreaming visions to go on undisturbed. But more importantly, this is an indication that in fact the keeping time to prevail must be the time of psychognostic reigning by autos from its limbic throne. In fact, the only time when the most perfect biological detector, the visual system, is closed to external environmental views and only open to the internal visions of psychognosis is the sleeping time. The limbic psychognostic function must be allowed to rehearse the essence of the autobiographical screen play in a visional dialogue between autos and self, and this can only take place adequately in the privacy of the psychognostic conscious dreaming state. This consciousness is iconic, an immaterial consciousness by figures, and these figurative symbols in action form and relate to ultimately reach an acceptable meaning for keeping the autos-self in balance. On the other hand, literary consciousness of idiognosis, in comparison, may be said to be a mixed figurative (iconic) and descriptive (word meaning) types, also ultimately immaterial. So we can accept these two conscious states forming together the purely functioning nonmaterial abstractive consciousness, the essence of the immaterial biology.

Some other theoretical considerations should not be neglected. Consciousness of idiognosis using verbal detection of items in the waking states, we have to realize, is energy consuming. This dependence to coded precision in the literary conscious state of course cannot be used in dreams to represent events in the same way that figures represent eventful meaning. Presentation in iconic motional changes, presumably less energy consuming, seems to predominate over worded passages in dreams. Furthermore, if wording is experienced in dreams, it is usually in the mother tongue and only rarely in a second language. These facts may explain that the primary continuous consciousness in dreams is the least energy consuming iconically based psychognosis, which enjoys fullest fluidity in interchanging icons and time frames of their relations. It is showing the holistic conceptual meanings of psychognosis, but including possible aspects of illogicality due to the variable Pl/Pa term of the psychodynamic equation, reminiscent of the uncertainty principle of the time indefinite autos.

In fact, the image form of meaning in dreams that is an iconic induction originating from limbic and cortical contributions is the only single body of multiple capacities, presenting the content meaning and the time together bound by the life commandment. Wakeful visual images of course should reflect all logic and nothing but the logic as only one part of the life commandment, but visional perceptions in either night or day dreaming allow illogical images motivated by the pleasure-pain principle, as the other part. Overall, the inner eternal time titan, the undefined time in dreams, and the outer real time, the defined one in wakeful states, must conform through the psychodynamic equation to allow balanced logic and pleasure-pain principles for the life commandment. This conformity is basically possible through uniformity of time that necessarily presents a nondimensional indefinite time that, to the dreamer, is standing naturally and only for the present time.

To be realistic with facts, the limbic functions needed no superior control through eons of animal life on earth until possibly the stages of MacLean's triune brain started to evolve from reptiles acquiring addition of limbicocortical formation to develop and reach its complexity in humans.[18] Then the limbic life alone became gradually insufficient for the more elaborate coordinated nervous functions with evolutionary adaptation, making the brain, the mind, and the self with the necessity to save the self and the mental life with it. Controlling limbic functions then came up and progressed when balancing of emotional drives with real-life exigencies had to be exerted, mainly in societal living, and this function had to be controlled by a nonlimbic system: the superior new cortical additions that developed and became perfected.

The psychodynamic balance in reality assures the welfare of biopsychons and secures the fate of biocracy. It establishes all safety measures to maintain the commandment of logic and happiness in life to prevail under the basic permanency of bioforce and its manifestations in biology. In this line of thinking, even before PET demonstration of the initial cortical inhibition, eliminating self's logical order to inflict a definite change, dreaming could be well understood not to follow this restricted setup if still under the uncontrolled limbic autos supremacy. Accordingly, as emphasized by Walker[4] that consciousness may influence the probabilistic uncertain possibilities to be reduced to one choice in spite of the uncertainty principle in quantum physics, I would add that the same may be true indeed in all biology including the dreaming consciousness of psychognostic type, not in terms of time definiteness, which seems to be an impossibility in time indefinite dreaming, but in terms of logic of causality revealed in action significances and their included emotional meanings. If bioforce, with its inherent decaying substrate shortening the biological life, could compensate instead for the time loss of material decaying by expanding the living experiences with consciousness into nonmaterial biology, a basic analogy could be realized between physical and nonphysical worlds, physics and metaphysics. Along this line of argument, then I could say that consciousness should be theoretically present incessantly at all times for a maximum benefit to be assured along this line. This appears in fact to be what the evolution has provided with the refinement from sensing to consciousness. This could not have been realized minimizing the material soma to maximizing the living time like in quantum physics (minimum matter-maximum speed and theoretical permanency of the nonintercepted photon) that is impossible in biology, but could have maximized consciousness as an extramaterial life with psychognostic iconic and idiognostic worded conscious imaginations reaching immaterial dimensions. Psychodynamic balance prolonging both psychognostic and idiognostic consciousness serves this possibility. Psychodynamic balance, in fact, helps prolong the wakeful consciousness with words, which is more immaterial than psychognostic consciousness. This analogy seems to indicate that the bioforce is recapitulating its essence in biology in ways compatible with biological decaying. This analogy of recapitulation exposes nonmaterial biological time expansion in contrast to the material biological time condensation (ontogeny recapitulating phylogeny). In this course of events, we can see the hidden tract to metaphysical thinking to appear gradually more clearly. In fact, the nonmaterial abstractive consciousness seems to allow autos-self to reveal its being beyond the world of dreams revelation in psychognosis through limitless revelation in the world of metaphysics.

REFERENCES FOR CHAPTER SIX

1. Douglas Field R. 2005. Making Memories Stick. *Scientific American.* 292(2):75-81.
2. Palmer JA, Palmer AK. 2002. *Evolutionary Psychology. The Ultimate Origin of Human Behavior.* Allyn and Baker. Boston, London, Toronto.
3. Mithen S. 1999. *The prehistory of the mind. The Cognitive Origin of Art and Science*. Thames and Hudson Ltd. London.
4. Walker EH. 2000. *The Physics of Consciousness*. Basic Books. A member of Perseus Books Groups. New York.
5. Domhoff GW. 2003. The Scientific Study of Dreams. Natural Networks, Cognitive Development, and Content Analysis. *American Psychological Association*. Washington, D.C.
6. Rock A. 2004. *The Mind at Night*. Perseus Books Group. New York.
7. Martinez-Conde S, Macknik SL. 2007. Windows on the Mind. *Scientific American*. 297(2):56-63
8. Kashani AA. 1993. The Triplex Hypothesis of Vision. *Annals of Ophthalmology*. 25:125-132.
9. Kashani AA. 1995. The Vegetative, but Significant, Role of Non-seeing Photoreceptors. *Guest Commentary. Ocular Surgery News.* 13(9).
10. Pinel JPJ. 1990. *Biopsychology*. Pp 365-395. Allyn and Bacon Publishers. Boston, Sidney, London, Toronto.
11. Aserinsky E, Kleitman N. 1953. Regularly Occurring Periods of Eye Motility and Concomitant Phenomena During Sleep. *Science*.118: 273-274.
12. Rechtschaffen A, Kales A. 1968. *A Manual of Standardized Terminology, Technique, and Scoring Systems for Sleep Stages of Human Subjects*. Washington, DC: US Government Printing Office.
13. Hobson J, McCarley. 1977. The Brain as a Dream State Generator: An Activation—Synthesis Hypothesis of the Dream Process. *American Journal of Psychiatry* 134:1335-1348.
14. Solms M. 2000. Dreaming and REM Sleep are Controlled by Different Brain Mechanisms. *Behavioral and Brain Science* 23:843-850.
15. Foulkes D. 1982. *Children's Dreams*. New York. Wiley.

16. Foulkes D. 1999. *Children's Dreaming and the Development of Consciousness*. Cambridge, MA, Harward Universeity Press.
17. Braun A, Balkin J, Wesensten N, Carson R,Varga M, Baldwin P, et al. 1997. Regional Cerebral Blood Flow Throughout the Sleep-Wake Cycle: An (H2O)-O-15 PET Study. *Brain*120: 1173-1197.
18. Whybrow PC. 1998. *A Mood Apart. The Thinker's Guide to Emotion and its Disorders*. Pp 121-128. Harper Perrenial Edition. Harper Collins Publisher.

CHAPTER SEVEN

THE PSYCHOGNOSTIC SPEECHLESS MIND

VISIONS AND MEANINGS

If we follow the development of senses in evolution, we can see that vision has surpassed all other senses in adaptive refinement. Even the amazing echoing system of bats, used in orientation and navigation with its astonishing usefulness for bats, is rudimentarily crude in comparison to trichromatic human vision and far more inferior to tetrachromatic birds and many reptile and fish visions.[1] The vision in humans is not only a physically and biologically refined perceptual apparatus of extraordinary complexity, it is a conditioning and directing means of mental interpretation bringing meanings to perception in association with the combined elaborate cortical capacity. The usual serial frames showing in movies, of twenty-four per second, which integrates shape changes with motion and time, interpreted in a continuous form in our mental perception, is a clear example of finely regulated neurovisional dynamism. And this is in spite of the screen limitation to a two dimensional perception rather than the normal live three dimensional experience. It is further interesting in this example that if a mute movie is seen, the essence of meaning related to the pantomimic component of the show is much the same as in a sound movie with the same body motion expressions, but much different in one without pantomimic similarity where words can play greater roles than images. The bodily expressions, however, can need no words to confer the meaning in a great number of such examples evidencing the psychodynamic automatic connection of the visual frames in motion for perfect immediate interpretation. The interpretation is further facilitated stereotypically by every repeated experience in the psychognostic life.

This observation suggests that we certainly enjoy a mind of speechless, symbolic, and highly clear type by our vision that we enhance with additional effects bringing precision of literary nature to it to enjoy expanded interpretation and meaning. However, we take for granted all that level of perfection in understanding our world without words that has only been made possible by vision. Furthermore, scenes of visions in dreaming not always accompanied by talking, although being purely mental images in symbolic figures and sometimes even senseless in forms, relations, and time, still invoke full conscious three-dimensional realities with the impressions they induce. This indicates that not only the refined adaptation of the visual sense, enhanced with incorporated brain adjustments, is decisive in our psychognostic life to give us the meaning of our symbolic visual perceptions, but our egocentric position also forces our comprehension to experience all inputs with their inherent mental facilitation specifically directed to our egocentrism facing the environment. So the key in decoding the percepts and reaching our meaning in our symbolic speechless mind appears to be tacitly inherent in our egocentric position of limbic sensing and interpretation elaborated and incorporated in our brains by vision over eons of evolution and used tacitly in our evaluation of the world through our vision before words were created. Realization of this centrally positioned authority, the autos, as a perception focus facing every aspect of the environment forms the model for the speechless psychognostic mind on which the conscious mind and speech will be naturally built.

In this chapter, I try to lay down the foundation and discuss the basic features of what I think could and should be legitimately called the psychognostic mind, a mind of integral comprehension in an implicit way, a speechless mind of symbolically meaningful interpretations and expressions, a mental world using the tacit tongue of vision. My reasons to believe in such a mind and endure to prove its reality are two—first are the vision-dependent uniqueness, constancy, force, and significance in securing the transfer of visually depicted data to neurobiological processes for interpretation that have the proof of evolution behind them, and second are the mechanisms used in the interpretation of data that are not defined by words but only by images handled and decoded for a meaning referable to and understandable by the logic of the autos.

Vision in fact is the only sense that in solo acting enables the brain to integrate shapes, colors, movements, relations, and time to make meaningful interpretation of environmental changes enormously significant in the SR realization. The perceptive completeness of interpretable data brought by vision to be used by the cortex to reach a full judgmental decision is beyond simple comprehension. I will present some suggestive data in visional perception and mental interpretation restricted to relevant points in meaning and symbolic idea formation from animal and human studies to facilitate understanding the role of vision with the neurodynamics involved in gaining interpretation toward a meaning. However, the traditional interpretation through words expressing meanings for the visual scenes in our literary understanding can clearly cause a bias in discussing the speechless mind that is the subject of this chapter. Our means of investigating the presumed speechless symbolic mind is, in fact, unfortunately restricted to the use of our literary language. This restriction that forces us using a method developed and practiced in our present armamentarium of investigation, our worded language, to judge a wordless one, can easily cause bias in interpretations and results. In fact, we do not have all exactly the same wordless symbolic or worded literal culture. This mind of symbolic figures, in fact, gives us the symbolic pantomimic outline that is a chronomorphic topographic encoded language of moving symbols with many imaginable meanings without sharp literal precision of analytical value. Although our present day literary mind facilitates our today's understanding of what that speechless mind could have been, we should not automatically presume any immediate similarity between our literary mind and whatever that speechless mind could have represented initially. Cultural beliefs and totems and taboos of prehistoric humans, reflecting their illiterate minds, could be never really known, and their symbolic pictorial vestiges of variable meaning, if found, could not be precisely interpreted by our linguistic means.

In a comparison of the presumed speechless mind of psychognostic nature with our literary mind with speech, notwithstanding the major historical gap in between these extremes, our present mind can try to decipher a possible intermediate stage shown for instance by the language of Pharaohs by checking their hieroglyphic symbolic pictorial writing that used expressing the meanings in visual symbolic pictures with matching phonetics. We should understand that our linguistic interpretation using our today's symbolic meanings to grab this uniquely strange example of symbolic visual language could hardly give us a true translation. Champollion who deciphered hieroglyphic writing, in fact, had to study Egyptian history and culture profoundly and succeeded in his task as an erudite Egyptologist.[2] This example demonstrates an intermediary language between the hypothetical archaic pure wordless language of solely symbolic type interpretable by symbolic meaning, and one with alphabetical writing and precise phonetic expression, hopefully matching comparatively each examiner's contemporary dominant culture. In this extended scale of hypothetical pure symbolic expression to modernized coded linguistic evidences, changes tend to hide the foundation of the interpretive neurovisional mechanisms, which is always symbolic. Whether visual, phonetic, or mixed and whether pictographic or alphabetic, the essence of the cerebral neurodynamic interpretation is still symbolic in terms of global meaning. So the foundation of linguistic communication simply, and by and large, uses the process of ***ideation-to-symbol-to-ideation*** regardless of visual or phonetic means used. This too simple a reminder is of paramount importance if we consider the fundamental origin of the ideation to belong to autos in the hypothetical speechless language, or to self in the conscious phonetic language we use today.

Literary meanings so clearly expressed in our alphabetical thinking and writing, which itself is so clearly senseless if not backed by memory coding in the words, is decisive to show that symbolic understanding and expression in essence does not need arbitrary words, but essentially needs cultural visual codes, reflecting cultural beliefs in symbolic memory forms, changeable to any form. The interpretation of visual data brought to the speechless symbolic mind is in fact a holistic interpretation based on the axis of autos versus environment that lays the foundation of psychognostic knowledge and culture for autos, not transferring the meaning through bare arbitrary figurative symbols that have no true natural sense. It seems to work in a perceptive-expressive, two-way, bidirectional, introjection-extrojection communication. In this mechanism, what the mute mind would naturally understand and express in symbolic presentation is primitive and simple, nature-related in relation to the autos' reflected affective changes in possible bodily expressions of all sorts including phonation and in relation to the environment.

Another crucial point in addition to our restricted means in investigating the speechless symbolic language or languages, or coded cultural beliefs of any sort, is the basic lack of our knowledge on the mental origination of expressiveness, the knowledge of neurobiologically developed expressivity demonstrated in acts, either pantomimic or notably in language. This fact prompts me to introduce a new theory of language formation in this chapter based on the psychognostic speechless expressivity. The theory is based on psychognostic ideation by autos and on implicit memory prior to consciousness acquiring its functional capacity of explicit memory. The theory exposes the origination of mental expressivity as manifestation of mental intention in acting in psychognosis, in visual symbolic wordless acts, and stands applicable to literary worded language that naturally follows. Let us first consider some data on the significance of perceived visual figures and mental interpretations before we engage the vision in psychogenetic concept formation.

SPACETIME INTEGRATION BY VISION

The neurocognitive significance of vision in facilitating the understanding of the world for the human species has become better known in the last half of the twentieth century through experiments on animals, human infants, and adults. Some innate survival-related reflexive interpretation of static depth level and dynamic space change has been confirmed by experimental studies. For instance, cliff avoidance in newly born goats and young rats have been documented,[3] indicating the survival value of such innate visual interpretation in early locomoting animals. In the three-month old human infants, a rapidly approaching object has been shown to trigger a defensive blinking or withdrawing of the head.[4] This early vision-registered safety reflexive interpretation for static spatial difference by a cliff or dynamically shortening the safety space by a rapidly approaching object indicates two valuable neurocognitive features of early acquisition. First, it indicates that this vision-related neurocognition is an actual trait, and second, that it does possess the spacetime meaning for the species survival gained over eons of evolution and only in relation to autos by indicating the safety spacetime of the immediate environment to the individual.

Studies of vision-related interpretations of lines, forms, shades, colors, and shapes on a two-dimensional plane and more complex three-dimensional configurations have confirmed that crude data from visions to become interpreted in a meaningful way by the brain seem to go through three levels of explanations—mechanistic level, algorithmic level, and computational level.[5] The mechanistic level is the basic, transferring data to neurons in the brain. The algorithmic level is the stage of initiation of the abstractive work, which starts finding direction to match the right algorithmic path in which crude data can both condition the basic algorithmic class and be conditioned by it to become more specific, and finally the ultimate interpretation level that functions similar to a complex computational processing. This course of actions seems to fit in a computer program with the task to be handled by a system of large comprehensive capability and memory for data interpretation. It seems that the neurobiological hardware prepared in millions of years by the evolutive adaptation indeed functions in a similar way. The fundamental algorithmic classes by this neurobiological computation seem to be defined by the life commandment of logic and happiness using categorically the pleasure-pain principle.

In this whole system of mental interpretation of visual stimuli, the abstractive work is the most elaborate and most demanding in terms of providing usable information to be processed, thus facilitating the final computational work. Studies of visual perceptions have shown such premonitory preparation of data in visual processing. Such preparations are basic and provide the elementary building blocks to be used to initiate and ultimately complete the algorithmic course to find meaningful interpretation of data for symbolic thinking.

Edge detection in variable visual conditions of lighting and with filters smoothing edges of objects not to be sharply visible has shown that the simplest one-dimensional line presentation visibility can still induce two-dimensional or three-dimensional impressions if separating the limits of a defined known object's features in comprehension. Linear boundaries clearly separating areas on two-dimensional pictures, areas based on grossly contrasting colors or textures, are better interpreted even though not being an edge. This indicates that structural color reflections or fine textural dimensions at lines of contrast-induced comprehension of boundary limit in a visible linear form construct mental images. For instance, in mondrian figures of multicolor patches,[6] limits are automatically detected and interpreted as showing full dimensional objects. So the simplest preparatory task in visual detection is stepping from simple one-dimensional data presentation to two-dimensional ones and beyond as the very initiation of spacetime symbolic meaning.

Refining steps culminating in the provision of visual three-dimensional body and motion, integrating time and representing the classes of the living world to autos for interpretation, were considered in recognition-by-component studies. According to the theory behind these studies proposed by Biederman[7,8] and in keeping with the literary identifiable names for objects definable on the basis of this theory, about three thousand English terms representing definable objects were recognized. It was assumed that concrete bodies of three-dimensional visible forms could be represented by single or combined simple basic volumetric units called geons indicating geometric configuration.

Each geon was in reality a hypothetical volumetric unit with a distinct shape that could be imagined to be formed as a three-dimensional body by time-integrated change moving a two-dimensional cross section of some particular contour, for example, of a square to form a cube, a triangle to form a prism, a circle to construct a cylinder in uniform straight line, or nonuniform line increasing or decreasing cross sectional area with time-giving variations within regular geometric shapes like cones, funnels, donuts, etc. Hypothetically, the geon was formed in this way by integrating the edge depiction with time incorporation and thus changing from monodimensionality to bidimensionality and further with evolving time to be finalized in its simplest form of one geon or a complex form of multiple-combined geons to represent an actual real three-dimensional or multimodal object. This geon-based object recognition seemed to concretize meaning on an algorithmic foundation, but of course keeping matched with literary forms defined by words beyond imprecise figurative symbolic presentation. Nevertheless, we could assume with plausible assurance that the foundation of geons as building blocks to provide meaning interpretation in symbolic speechless forms has been laid down much earlier in time with the evolving refined vision-based neurobiology, possibly in bioconsciousness, reaching the comprehensive level of symbolic ideation in psychognosis far earlier before language to appear.

The simplified visual data units in terms of geons, to facilitate categorization, were estimated to only include twenty-four such units. The twenty-four types of geons thus served for all variable symbolic ideas to be formed for concrete bodies.

The basic geon to be realized only would need the minimum of correct edge recognition and therefore could be rapidly perceived by the mind-integrating time progress in producing its shape. Combination of two similar geons in six variable spatial relations was estimated to give 108 possible variations, and if both six spatial positions and three individual aspects of curvature to volume ratio were considered, two-geon objects could represent 186,624 variable forms. In the same way, over 1.4 billion variations could be estimated for three-geon objects in possible combinations. In spite of this enormous possible variations of combination forms for three-geon objects, it was estimated that usually an arrangement of two or three geons was sufficient to cover literarily identifiable familiar objects in the order of about thirty thousand items.[9]

This course of visual detection to final neurobiological identification prior to completion of interpretation for gaining the meaning is assumed to form the preparatory phase possibly identifying the final object from the initial simple edge depiction. In this theoretical course of events, the basic geon models serve as algorithmic directives. They conduct the interpretation into different time-integrated lines to give the final recognizable configuration for which there should be an implicit known entity in the environment with a definite psychognostic symbolic meaning. This visual interpretation-identification is for objects, close or remote in relation to autos' egocentrist position in the same spacetime. However, final interpretation to be of practical meaning to autos must include the spacetime relation of depicted objects to autos with more precision. The notion of proximity, spatial position, orientation, and time relation to the phenotype forms therefore the conditioning factors to complete the psychognostic interpretation of the symbolically identified objects. In fact, these symbols in the tacit psychognostic speechless mind can still have an identifiable mental iconic tag as to their shape, color, position, and function to acquire more meaning precision without any literary definition. This relationship to autos on pure symbolic presentation completes and finalizes the symbolic functional meaning by vision thus forming the wordless psychognostic concepts and with them the potential meaning expansions related to these concepts for a sound psychognostic cultural basis.

THE EGOCENTRIST AUTOS

The self-conscious autognostic autos realizes its integral unity in a hypothetical symbolic name, a name without letters, a holistic symbol formed as a whole shadowy image not comparable to a mirror self-portrait of a literary self. But this hypothetically presumed name, understood in whatever symbolic form it takes, is tacitly well-defined by autos in a psychognostic speechless meaning in relation to autos' egocentric position in the world. It is an orientational tacitly positional self-cognition in autognostic meaning as opposed to heterognostic recognition. The position of autos to autos' world of symbols is a prime autocentric standing for autos, unique to autos' physical limits, facing potential and real dynamic possibilities in the environment. The autoheterognostic power of autos is formed through repeated stages of evolutive adaptations with environmental demanding changes, and it is only natural that self-realization at any stage would be based on the central position of the unity of the autocratic biopsychon facing the multiplicity of environmental static and dynamic objects.

Such relationship developing between a single sensing-reacting biopsychon with its environment can only reflect symbolic meaning to its bioconscious or psychognostic mind reminding and enhancing its integral unity. Therefore, a nucleus of expressive symbolic language will be formed that could use all possible perceptive means of visual symbolic expressions, only reflecting functionalities in respect to autos. That could only expand with evolution and add refinement to the basic symbols covering all definable relationship exchanges with animate peers and inanimate objects with no meaning other than in shadowy affect-laden symbols, essentially meaning holistic functions. Such symbolic functional meaning naturally needed no name or label as it assembled the visualizations of parts to mean functions synthetically and not analytically as in speech. Fossil studies of skull endocasts, in fact, do not show signs of any cortical language center prior to *Homo* lineage.[10] Furthermore, investigation of conscious attentional visual focusing on cats show characteristic synchronization of visual cortical neurons in such focusing with simultaneous EEG recording of strong gamma oscillations typical of the conscious states.[11] Thus, there is no doubt that significant neurodynamic conscious perception does exist in the psychognostic nonlanguage stage in the mind that could be modified by attention-intention processes in the formation of the holistic conceptual meaning.

Symbolic understanding of figurative visual imaging cannot make sense without at least some justified initial vague semantic value in meaning attached to visual symbols, and there seems to be a natural trait causing the mindful need and the mechanisms developed for such meaning. The foundation of symbolic interpretation inducing a clear meaning in contemporary psychology was initially solidly shown by the Gestalt psychological interpretation of a given pattern by various observers' concurring agreements. But clear Gestalt meaning, we should recognize, is the product of our literary expression. This literary expression unifies possible constraints imposed by the visual system that allow a three-dimensional interpretation from a two-dimensional figure revealing structural elements that, even if not clearly representing a known object,[12] still would lead to the holistic concept of what the object could mean as a definition for the object to be known. The main characteristics of Gestalt interpretation reside in simplifying and unifying the interpretation of object figures based on surface uniformity, homogeneity, and continuity. This is noticeable for example in infants of about seven months of age in a simple symbolic interpretation of a three-dimensional object, but if continuous surface homogeneity is altered, the same simple symbolic meaning can be gained only by adults.[12] This mental simplification to symbolize seems therefore to be a genetic trait subject to phenotypic refinement with experience. In contrast to this Gestalt understanding that can be regarded constraint-mediated to impose generalized semantic characteristics of symbols to make a coherent whole, Rorschach test of inkblots[13] by its undefined symbolic nature presents no constraints and grants liberty of free interpretation to what may form from one to multiple visual images and symbols or scenes. This contrast in interpretation of visual symbols in Geshtalt with constraints versus no constraint in Rorshach, in spite of the obvious abysmal separation between the two equally demanding interpretation, is evidencing one common feature, which is the natural necessity in our minds for a meaning to be found for each visual symbol. This fact indicates the most natural trait in biocracy behind the need for clear identification of symbolic items in the human minds, which is the necessity to confirm time definiteness and expressivity for decision making. The clearest example is the autonomous chain reaction simplified in stimulus-sensing-perceiving-interpreting-decision making-responding to constructing a meaning for comprehension and for expression solidly based on vision.

Psychognostic necessities for symbolic meaning interpretation are bound to the affective trait for expressivity and cannot be neglected for the single reason that all symbols in psychognosis are somewhat emotionally oriented and are therefore essential and basic. Certainly, affects in the sensing-reacting processes are more crucially conditioning the expressiveness than neutral scenes of routine interchanges as can be observed in every day life. Furthermore, the significant life commandment of logic and happiness with pleasure-pain axis has been operational in the entire evolution of all species at different scales of sensing-reflecting or perceiving-reacting types to be so obviously discernible now in the human species. No doubt there has been meaning of the crude to elaborate symbolic visual images throughout the evolution, helping the living creatures to recognize them and keep up with the primordial SR exigencies and also in line with the pleasure-pain balance of the life commandment.

The question of significance is to find the stages of events that lead to semantics for symbols of initially concrete type, often repeatedly experienced in similar conditions and therefore iconically recognizable but only implicitly. These essentially comparable representatives of facts and events should have been recognized as such by autos and should have induced unified significances to similar visual frames of images. Comparable scenes should have the inescapable meaning of contributing to the same effects to be reproduced with the recognized items. That can be regarded as the main course of events in the long speechless time with the least effect of informing autos implicitly and building the primitive implicit autobiographical memory. The implicit memory must have also included a tacit meaning for each visual dynamic symbol interpretation.

However, there is no practical clue to explain the original mechanism for meaning acquisition for symbols in psychognosis except as related to the egocentric position of autos. Even considering the uniquely central position of autos in the environmental events, the relationships involving relatedness of the environmental spacetimes with the autos' physical parts can be considered only for basic symbolic understanding of orientations, proximity, motions, and possibly, time relation. Further extension of this stage of comprehension from this autocentristic ideation to exocentristic type by extroversive projection will form the psychognostic amalgam of cultural knowledge and beliefs possibly aided with early appearing word concepts and many existing nonword iconic concepts alltogether. This consideration will concern mechanisms and steps in nonliterary semantics to be incorporated in psychognostic concept formation. However, we need primarily to discover the sine qua non in the formative originating neurobiological mechanism for the implicit symbolic meaning incorporation in psychognosis. In other words, we must come to discover the steps in the course of action initiated by the mental agent acting to direct the mental energy spent to produce psychognostic concepts for expressivity and to formulate ultimately the formation of the worded expressions. This discovery phase for the foundation of language will be exposed here inasmuch as it will concern the psychognostic expressivity with symbolic concept formation and will be further expanded in the next chapter.

THE PSYCHOGNOSTIC CULTURE

Imagining a culture for a speechless mind may sound strange, esoteric, or pure fiction. However, all the extensive literature on behaviorism in the first half of the last century, in fact, tacitly described cultural beliefs and rituals in animals by behaviorists but precisely without the authors referring legitimately and clearly to mind or culture behind the animal behavior. Who in the era of behaviorism could indeed assume and express the existence of culture without language. Even all dictionaries summarizing definitions of the word culture give no clues that culture could be nonhuman and nonworded and yet characteristic specifications enumerated for culture include thinking, ideas, communication, education, rituals, beliefs, etc., that are not all solely based on language. In brief, culture is defined essentially as a common social knowledge, belief, and attitude reflected in individual and in reciprocal social norms and, in practice, expressed by many under the term of social intelligence.[10, 14-16] This same foundation of common social knowledge, belief, and attitude observed in both animals and humans are sometimes named simply behavior, presumably without mind and language in animals, versus culture confirming mind and language in humans. The fallacy is the result of a scientific precision mistake based on idiognosis alone, neglecting the solid extensive foundation of the mind from paleognostic to neognostic states. The testing taking only presence or absence of language per se to assure culture is invalid, as language is an expressivity means not necessarily reflecting the inner reality and truth. The baseless assumptive difference can be realized when we consider the cultural capacity levels of blinded deaf humans also deprived of adequate hearing with yet mastery of language expression who still manifest high emotional valuation for self-esteem and also equal rights for all living creature.

Culture has a much deeper inherent sense of knowledge and belief for any mind capable of comprehension and expression than just granted by language. Although behaviorists generally downgraded mind and culture in behavioral sciences, biologists and cognitive scientists sometimes went as far back in tracing the culture in animals as to the prokaryotic life. This was clearly pure analogy explaining the various tropistic evidences in cellular life interpreted as presenting a primitive culture in the stereotype repetitions adapted to benefit the organism. A clear example is the comparison of this trend of similarity between prokaryotic cellular living and plasmodial type of unified pluricellular functions when the fixed and the motile changing life-forms of the myxobacterium for example, alternate to access nutritional material and to safeguard surviving by spore formation according to Bonner[14] who correctly concludes that *the first step toward the capacity of culture was clearly evident in the earliest known living organisms*. This conclusion supports the functionally sound autocratic biotropism that leads biodynamism in serving autos in its fundamental autocratic culture.

Accepting this conclusion, the culture by definition becomes basically a product of genes as potential developmental plan to be evolved with the evolutive exigencies. Acquired mental additions, as with learning, then become conditional in cultural formation. The answer to this apparent contradictory aspect of culture, a potentiality to be conditioned to actual evidence, is reflected by the view expressed by Marcus that the phenotype is not only a strict blueprint replica, but also reflects adaptive adjustments to the environment as well.[15] Therefore, culture can be regarded as an algorithmic poetry base laid down by genetic dictation in which the outline cannot be altered but the literary effect can be enhanced and expanded by more appropriate wording making smoother prosody, or even a melody in which the lyric is free to vary according to the singer and the literary society's preferences. In the same line of analogy using metaphors, the genetic geon depiction is subjected to time for the finilized symbolic figures to take conditional shapes to be interpreted. We can see once more that expressivity trait based on emotional foundation progresses using the means of phenotypic accommodating changes to build what can be the ultimate psychognostic speechless culture leading to literary culture with language.

How would the foundation of expressivity-based culture, harbored in the genetic code, be used with appropriate conditioning by social adopting constraints or enhancements to provide the finalized form which must be kept changeable with time? This is a legitimate question to every one's mind that indeed the genetic codes work to provide the structural model, but how the ornamentation on the structure to reveal the refined presentation is accomplished and by what mechanism(s) remain to be explained by the sine qua non autocratic bioforce as the permanent unfailing promoter and the evolutionary conditioning factors as posted directional signs to that promotion.

Before making any attempt to find the answer, which may not be an absolutely satisfactory one, we have to consider the psychognostic state without literary expression that is the natural source for all cultural attitudes developing by the expressivity of the speechless psychognostic mind. Certainly this source could have artifacts created and later used by language, and many aspects of the psychognostic culture reflecting the more strongly archaic affective feelings and meanings expressed in symbolic comprehension and communication could have formed models for language formation. On the other hand, the explosive linguistic facility that has expanded social relations with the benefit of acquiring literary knowledge has produced social norms along the expansion of the literary world, some compatible but some significantly different from those in psychognostic states. Some expression evidently artifactual and even contrary to psychognostic beliefs, but tacitly used for convenience in societal norms, seem the reason for the apparent abysmal gaps between the two cultural lives of psychognosis and idiognosis. However, the emotionally based expressivity is absolutely the same in both. This clearly refined genetic structure of culture by the coadaptive changes allowed by language has been achieved by the mental memory dependent activities named *memes* by Dawkins,[15] being considered as replicator other than the gene, effectuating the environmental adaptive change in cultures passed from generation to generation. Thus, the speechless mind of the psychognostic biopsychon practically has the bare skeletal structure of the genetically formed culture and that of the idiognostic mind of the speaking biopsychon uses that structural frame to make the additional ornamental changes. In this respect, another question may complicate this simple viewing, and that is the possible course of the *meme* changes adaptively imposing on the gene expression. What, in fact, can occur between the psychognostic and the idiognostic states from the fundamentally autocratic cultural dispositions of autos to the sociologic cultural norms of self?

This question is the crux of the puzzle of the entire transfer mechanism from psychognosis to idiognosis from the illiterate state of the former to the literate state of the latter. In attempting to answer this question, the two fundamental aspects of the neurodynamic changes to occur need to be considered and explained—namely, (1) the visual perceptual gains changing from initial geons or living figures to new functionally meaning forms as symbolic concepts to include time of the change with the gained functional meaning, and (2) the conceptual symbolic meaning thus gained to be transformed into worded symbolic concept, again including the time for the further differentiation in the meaning to occur. In linguistic terms, this transfer is passing from deep structure to surface structure over millennia. In fact, timing of symbolic concept formation is an inherent part of biology just as is the biological energy, and the time does not show symbolically with the meaning formed by the symbol, being always in its permanent present time, the time titan. The answer seems to be cogenetic and not transgenetic. But before undertaking the crucial question of the change from psychognosis to idiognosis, let us complete our discussion of the psychognostic culture.

Considering the powerful visual perception inducing symbolic understanding, the living materials and their functional meaning, inherently time containing as they are, they satisfy the synthetic basic two-frame meaning structure proposed for the life commandment in the cause-effect logic and pleasure-pain principle. For geons changing through time incorporated mental meaning into entities to be interpreted as varied functional forms, the same two-frame meaning structure applies and results to meanings according to the life commandment principles for interpretation. We may want to consider a unit of value psychognostically elaborated in homology with *meme* in idiognosis, but the advantage gained by doing so is not assured as psychognostic symbolic meanings are simple and holistic unlike those expressed in worded speech and is not transferable like *memes* in worded communication. Thus, elaboration of the symbolic visual meanings in psychognosis, and the cultural significance they imply based on the central position of autos percepting and interpreting all inputs from the environment are basic, holistic, and adequately sufficient for forming beliefs, attitude, knowledge, and rituals elaborating the speechless culture. No *meme* is operable in this situation where everything is natural or based on natural relations of the egocentric autos to its environment in the speechless psychognostic mind. If any equivalent item for *meme* can be considered in psychognosis, it would be simply a unified model type of holistic symbol that I will entertain and discuss in the next section.

As life is dominated by emotions from the archaic sensing model of automated tropistic mechanism of the paleognostic stage (in the general meaning of sensing-expressing acts) to the motivated psychognostic stage, the full expressivity that is developed in psychognosis leading to culture is for the emotions to be expressed. Emotions primarily project the loving care of the parental origin to offsprings in an empirical learning norm and, in lesser degree, to other members of the group living together, all in keeping with the supremacy of autos. This expressivity is both genetic and learned. This cultural manifestation in the speechless life of primates has been examined in both natural habitat of their living and in laboratory. All observations concur to show clearly that a high-level of cultural bonding for group unity exists, reflecting hierarchies with seniority to be respected and submission to be accepted and mutual protection and care to control the safety of the members of the group, and this is reigning just like in human societies. Many subtle affective feelings in our rich lingual expressions, not expected to be seen in primates, are strangely observed in the speechless psychognostic mind that dominates the primate life.

To understand the cultural development throughout this course of changing from paleognosis to neognosis, summarized information on correlated functional and emotional development, would be significantly helpful. Such correlated data have been presented by Greenspan and Shanker.[17]Their work is greatly helpful in defining the timing for the correlated available data with the progress of culture through the ages starting with the early speechless emotions in mammals showing attention, self-regulation, and expressivity (stages I to III[17]). The ending is with developed language and reasoning in ancient Greece reflecting thinking on the future with an expanded concept of the self (Stages XI-XII[17]). The vocabulary of course uses the orthodox scientific expressions not referring to any intermediary Gnostic states that I have used in this book. But one can see references to chimpanzees, bonobos, and other primates as well as to Ardepithecines, Australopithecines, *H. habilis*, and *H. erectus* in stage IV as presenting more or less comparable levels of culture but still with a speechless mind. Then archaic *H. sapiens* dated from six hundred thousand to sixty thousand years ago in the study are said to have been mastering complex affective interchanges, presymbolic communication, and were beginning to use symbols to convey abstract thoughts and feelings (stage V). This work also includes scenes of affective communicative interchanges by primates seen and described by Shanker at the Language Research Center (LRC) in Atlanta where research programs focus on prehuman ability for language and study the language acquisition by primates. The work performed and the results obtained in LRC by the primatologist Savage-Rumbaugh[18] have revealed appreciable language skill especially in Kanzi, the star of the scenes described.

In this short summary of sampling of the affective types of actions mentioned here, the review evidences abstraction with symbolic concepts indicating all imaginable affective expressions as can be seen in humans. These affects start with the first interchanging introduction of amical type to deeper involvement to close cooperating, comforting, grooming, caressing, and caring, and finally separating from the individual focus of interest in a very closely similar manner as in humans. In addition, guilt feeling with acceptance of punishment is clearly shown in situations similar to what we see in the human societies. Also, happy general feeling of recovery for a seriously wounded member of the group or mourning at the death of a member presents scenes profoundly similar to human social situations; the mourning sadness and desperation being particularly moving and reminds us of funeral rituals in the human societies but short of a burial. In these observations, there are clearly distinct cultural beliefs transferred by parents to children or older to younger members, generation to generation. These beliefs have no lingual models but visual ones with inner symbolic meanings of holistic type recapitulating the meaning in the mind without figurative unitary symbol as in language but probably memory figures of passed experiences with close relatives. The abstraction sensed in the form of happy or sad feeling is again figureless with any precise shape except if related to some dominant registered visual event. These events, being naturally multiple, would only give a precise symbolic concept if the figurative associated element is important to autos' affective values. Replacing this relative figurative vagueness by precise phonetic symbols and changing from speechless psychognostic mind to literary idiognostic one expanded the phenotypic acquired cultural field to include the explosive art and science in the human society. Thus, the aspect of changes from symbolic iconic understanding of the mute psychognosis to the speaking idiognosis that uses words forms the foundation ground for the language to be built.

PSYCHOGNOSTIC HOLISTIC SYMBOLS: SEMIONS AND MOVEONS

The basic genetic functionality of the central nervous system, evolved to realize perception and interpretation of inputs from the environment, has answered the needs of the living organism commensurate with the evolving complexity of the system. In psychognosis with the level of perception and interpretation needed as evidenced by the speechless culture just reviewed, the main need is in the correct global understanding of the value of the environmental change as an input to be interpreted in order to justify the appropriate disposition to be effected by autos. The most important aspect of the symbolic meaning for this purpose is the clearest, the briefest, and the quickest meaning to be gained by the perception. In this process, the symbolized action ought to reveal the changing effect with the factor of time meaning the action, in short the effector, the effect, and their relation, so that acceptability versus rejectability for the input can be immediately resolved and answered appropriately and instantly.

The symbols from external events, being for the majority all visually formed and for the minority by other senses alone or in combination with vision, naturally also include the time that gives them their interpretable functional meaning forcing interpretation on the part of the autos in its egocentrist position in which the autos checks the environmental spacetime changes as they affect its equilibrium status, its rest time. Examples include long-term natural environmental variability like seasonal changes or day and night change as the pristine biological symbols for all living organisms with biological meanings in them that cause neurohumoral seasonal or circadian variations in response. Also shorter effecting changes of warm or cold temperatures, their causes, and effects as acceptance or rejection of the reactive feelings are sensed as natural short-term biological symbol models, with or even without any deliberate action by autos. In contrast, interbiopsychon relations evidencing changes occasioned by the living locomoting organsims that can form concepts of useful and acceptable versus nocive and rejectable meaning, all result in finalized conceptual meaning to autos through the autogenetic ego models. Psychognosis as the ultimate illiterate state of mind mainly disposes of these holistic symbols. Mirror neurons, located mainly in the left frontal lobe at the Broca's center-to-be, reflecting self's meaningful actions in response to external iconic figure action significances,[19] seem to play the crucial role in synthesizing action thought and reaction coordination as symbolic stereotyped repeatable acts. In fact, conveyance of meaning and affects related with actions, particularly reflecting intents and results, are controlled by mirror neurons providing the foundation of communication with symbolic expressivity.[20]

These symbols are all simply meaning-oriented for response, and if considered each as a unity, they include two frames—one for the visible content being the agent making the change that is naturally variable, and another for the invisible time titan that is naturally constant as the present time for the change to take place. The inseparable correlation of the content frame and the time frame is a law, the former initiating the latter and the latter allowing the former to expose the meaning in perception. This functional unity, in fact, is homologous to the bioforce function with time and energy contributing to the action result and the variable effect that is changeable for each meaning, which is the result that is energy-time dependent. It is analogous to the two-step setup of the life commandment. The meaning, biologically notable in one way or another, is handled by the central nervous system in general with long or short term responses as would be appropriate. In short, the material energy acting in time and the mechanism and the two-step process making the meaning are analogously functioning as in the realization of the life commandment. This unity on which is based the meaningful sensing throughout the evolutionary history of life from simple tropism-dynamism to the complex perception-reasoning needs to be defined by a name, making justice to it. I have chosen the word *semion* for that purpose, figured out from semiotics, for three reasons—first, its explicit meaning clearly including the Gnostic sense; second, its scope that is not restricted to either affect-laden or neutral meaning; and third, its functional background from paleognosis to neognosis in neural sensing-percepting and perceiving-reflecting with appropriate neurodynamic responses. Semiotics, in fact, refer to two systems of communication—one that is referred to as zoosemiotics and is concerned with animal ways of communications, and another that is named Anthroposemiotics and concerns *Homo sapiens*' linguistic properties.

The important fact in the psychognostic living of highly affective type in which the value of the outside elements to autos is judged as positive versus negative, pleasurable or painful, enhancing security versus endangering life and similar but less drastic opposing effects is that the holistic symbols are amply sufficient for serving the purpose. In fact, the biological axiom, the life commandment based on logic and happiness on the axis of pleasure-pain principle, is a prototype model that has been followed by autocracy throughout evolution, built on this same basic two-frame foundation, although using different facts at different stages of the phenotype complexity. Replication of this model is universal in biology, and in the most sophisticated neurodynamic factory, the brain, it is the foundation of affective expressivity in art and scientific expansion in logic. It can be said that the affective aspect of the life commandment is primordial and has lived with life forms from the very beginning and continued in its most dominant manner by autos in autocracy. The mere objective of autocracy, being adequately served by semions supplying the holistic integral meaning, evidences the main two-step process as a rule although more detailed intermediary neurodynamic steps are involved. Such realization, in fact even if clearly demonstrated, does not alter the rule of the holistic two-step meaning given by the semion. By the same token, the finally unified reaction of the autos in using the neurobiodynamic steps to prepare the appropriately summarized act as feedback to the meaning, given by semion, is a single response to reestablish the biological equilibrium. This response can vary from zero to a full theoretical range. The whole chain of perception of the meaning as stimulus and the preparatory neuroconceptive-neuroreactive steps to give this unified responsive act can be summarized in one descriptive explanatory word that I would propose to be *moveon* for its sensed motion. In fact, the motional time change as action is the main stimulus for mirror neurons to repeat the scenario simultaneously.[19,20] In this way, the sensing-acting simple concept used for the earlier steps of sensorimotor mechanisms can be expressed with semion in the elaborated judgmental perception-action processes as stimulus and moveon in the consequent step to it as response to stimulus. Naturally, if the iconic symbol is void of any stimulus effect as a static figure of no concern, it can be simply called a morphon to complete the list.

Now what exactly can we understand from semion and moveon as being the holistic unities of perceptive meaning (action stimulus) and performing meaning (counter action response) respectively? An important point not to be missed initially is that the speechless psychognostic mind and culture include both concrete and abstract holistic concepts, but the meaning causing the response is only that shown by the concrete effect, leaving the possibility of the causal factor out of the account. The abstract concepts of imagined invisible entities responsible for the visible material events like the meteorological changes, earthquaques, floods, etc., may invoke autogenetic ego models for imaginable explanations on the basis of the observable cause for the meaning produced, but the result as the immediate stimulus is the triggering factor for that meaning and for its counterresponse. These explanations follow, in general, the psychognostic abstract impressions in line with the pleasure-pain principle of the life commandment with feelings of attraction or repulsion without needing and forming any comprehensive semions and moveons responding to any obscure cause, but to the immediate observed effect for acceptance or rejection.

As for the semion, we can expect neurodynamic impressions it may cause to the psychognostic speechless mind, impressions that would, no doubt, also be valid in future forming of the idiognostic-speaking mind. Clearly, the value of the contribution of semion is the provision of the clearest, briefest, and quickest meaning of events from the environment, which is the charge of the input to autos and is the main reason for triggering the response. The semion provides recognition of the fact to which the response by the moveon may or may not incorporate recognition of the causal agent of the fact in any specific way. The agent and the associated time for the change to occur are recognized in the course of the psychodynamic interpretation for gaining the meaning on the combined two frames for the agent and for the time. However, in contrast to tacitly long-term biological sensing, where the trend of events is based on the archaic biological model of memory interpretation based on the more abstract seasonal or circadian changes, here the change is for the more factual interpretation with more rapid response by the brain in the preparation of moveons.

In these staged processes, the agents making the change are variable and often not clearly recognized at each event, the time element, which, in fact, is the time titan that shows the permanent present time in psychognosis, is constant and only tacitly known. Autos notices the agent's visual figurative picture or another sensorial presentation of it only, but does not recognize the time with it as separate from the indivisible continuum of the always permanent present time. In brief, the only content of the event representing the final meaning to autos will be invariably the change and variably the agent if detected, according to situations, and not the associated time with it out of time titan, which is tacitly an inherent normal component of autos. The holistic meaning in semion, therefore, is specifically the factor serving as stimulus to engender the moveons. As for the moveon, the specific stimulus from semion serves to trigger the specific neurodynamic reaction to provide the necessary response by the moveon to preserve biological balance, all in a unified holistic manner and by the intermediary action of the mirror neurons triggered as needed.

All these processes are initiated by the neurodynamic energy that must be released by the autos for the biological feedback or by the self for the tacit attention-intention mechanism. Autos is the single inner decision maker, and each isolated decision to trigger the process to go through, in my belief, releases a constant amount of neural energy in a quantum form. The triggering factor that is autos originated, being constant as the quantum of energy released, is the same in all responsive manifestations in single symbolic formation, hence in semions and moveons alike. This factor representing the theoretical quantum energy and its associated time for the expressivity effect, making semion-moveon models in psychognosis, and worded symbols on the same basis in idiognosis is called *Navand*, representing expressivity, and is explained in the Navand theory of languages in this and the next chapter. In psychognosis, the complex *Navand-Energy-Time* (NET) that makes a unity for semion-moveon formation seems to be equivalent to Libet's time constancy[21] as we will see explained in the next chapter.

Functionally, semion presents the stimulus with the clear final meaning, obeying the life commandment model in using the two-frame structure for holistic symbolic concept formation in psychognotic interpretation. In this process, the stimulus frame I of life commandment, detailed in chapter 6, is used by the agent, the frame II by the time and the frame III by the meaning (holistic concept) to which the psychognostic autos goes into a similar compositional process to respond. In this chain of action, the agent for semion formation is the stimulus (frame I) with the accompanying time (frame II) and the meaning reached (frame III), which is then the stimulus (frame I) for the response balancing process (frame II) to provide the response (frame III), making the moveon (Table III). It is to be remembered that every process practically concludes in three frames, but the constructing steps use only the first and the second frames, hence the epithet two-frame process. The third frame may be an end or may serve as a relay for subsequent similar chained processes in biological serial feedbacks. This analysis shows how the holistic semion is formed with steps I and II by NET behind step III completely unnoticed in long-term biological adaptations (when the long adaptability time is to a morphon of static insignificant item) and noticed in psychognosis by autos in short-term responses coming in chain. But why the step II with the content of frame II, the crucial time that is real, is only implicitly operating without any clear detectable presentation and in what detectable form could it be expected to be seen?

TABLE III

PSYCHOGNOSTIC HOLISTIC CONCEPTS

FUNDAMENTAL TWO-FRAME MODEL

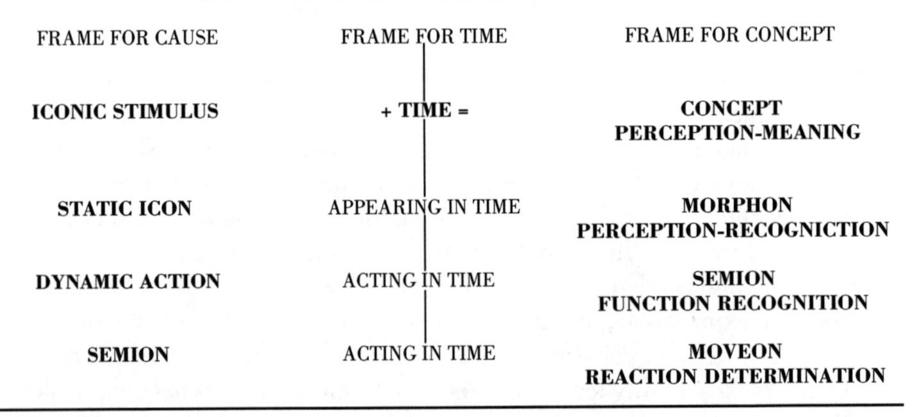

FRAME FOR CAUSE	FRAME FOR TIME	FRAME FOR CONCEPT
ICONIC STIMULUS	**+ TIME =**	**CONCEPT PERCEPTION-MEANING**
STATIC ICON	APPEARING IN TIME	**MORPHON PERCEPTION-RECOGNICTION**
DYNAMIC ACTION	ACTING IN TIME	**SEMION FUNCTION RECOGNITION**
SEMION	ACTING IN TIME	**MOVEON REACTION DETERMINATION**

Table III. Holistic psychognostic concepts constructing morphons, semions, and moveons are formed by symbolic visual meaning triggered by iconic figures, either static nonanimate or animate and realized as potential or dynamic and considered as actual threat, immediately needing response for keeping the biological equilibrium constant. Semions can remain potential concepts or trigger immediate appropriate moveons. The biological feedback system will have to incorporate perception-interpretation-meaning-response, all in the psychognostically holistic coordination for neurodynamic chain reaction from the simplest to the most stressful situations.

PSYCHOGNOSTIC EXPRESSION OF TIME

The time is a biological part of life as it is a part of the bioforce. It is stable, undetectable, and ignored in our sensing system in any spacetime at theortically any uniform constant speed of displacement. It becomes detectable passively if the home spacetime is changed or if a different spacetime with different speed, other than the home sapacetime, is detected, allowing slower or faster perception to be noted for the home spacetime in comparison. It becomes actively detected if the home spacetime is physically or mentally altered. This principle implies two axiomatic realities necessitating clear understanding: (1) that our innate stable time titan is our permanent present time in balanced metabolism in which no extra neurological energy is spent, and (2) that energy expenditure above balanced metabolic rate is spent neurologically with time change detection in physical work (motion) or in mental work (ideation). This means that the basic metabolic balance of the constant uniform type, named rest time in biochronogenesis in chapter 2, includes the uniform neurological maintenance energy expenditure as well. This is the state of equilibrium of the phenotype spacetime with the environmental spacetime in which the matter-energy-time of the other included phenotypes' spacetimes are also basically in balance. Formation of the psychognostic semion-moveon in response to environmental spacetime changes, sensed as actual or potential motion to occur, implies neural energy expenditure for perceptive and expressive mechanisms. This needed additional energy triggers the metabolic working state of biochronogenesis. Thus, the psychognostic neural energy expenditure by forming semions can be noted to be tacitly representing time detection and by moveon response to show tacit time expression, both in the constant monodimensional present time from the permanent time titan of the spacetime of the phenotype. This realization is still likely within the implicit memory time.

Biologically, by vision, motion detection means outcome detection and tacit time detection, each with its own biologic meaning and with some complexity. The example of the rapidly approaching object in the gaze of the three months old infant causing defensive withdrawal of the head, described earlier in this chapter, shows full spacetime change detection with the meaning of jeopardizing safety but not an exact meaning of shortening time by the shortened distance. Environmental changes summarized in a holistic meaning, although including time change with all biological humoral-metabolic adaptations, are gradual with adaptation and not causing evident change detection in physically or neurologically detectable scale. The reason is that the change, tacit in elaborating the meaning, is slowly progressive in the home spacetime, the time titan that, for the home sensing system, is the permanent and eternally constant present time. This time is the invisible time and remains invisible as long as the biopsychon continues in its holistic psychognostic rest time. Any change out of the rest time seems implicitly recognized and reflexively answered without exceeding implicit memory.

In the actual dealing with spacetime changes forming semions, autos facing the conclusive meaning of what is accomplished is not preoccupied with the agent or its carrier time. But it is clear that both the agent and its accompanying time are included in the neurodynamic processes concluding the meaning in the semions and elaborating the answer by autos with the moveons. In this stage of psychognosis, an iconic working memory with its function to accomplish the semion-moveon formation is operating, impinging on implicit memory or even as iconic explicit memory for registering geon-based three-dimensional forms for the agents. However, time expression is action-bound and remains monodimensional as the permanent time titan itself reflecting the constant moment to moment present time.

If the agent for the change becomes identifiable as an invariable figure by being repeatedly encountered in the same semion, the implicit invisible time with it also may become visible in fixed association with the agent. This constancy may impart independent identity to the time. This is a possibility that cannot be ipso facto eliminated as remarked earlier. The reason is that repeated sameness of the agent, action, iconic working memory time involved and similar neurodynamic processes used could contribute to neuroperceptive facilitation, thus making more solid recognition of both the content and the time, making an individual meaning for the recall identification. This could start the initiation of individuality for both the content and its time, but both in one unidimensional present time.

As a good example, if the environmental agent is an animate biopsychon like a known predator icon, known figure, known action, known outcome, its picture in the mind immediately forms the semion evidencing the meaning of the outcome and triggers the appropriate defensive response by the moveon that could become stereotyped for all members of the same species. Actual examples of this sort have been seen and discussed in the second chapter. In this situation, the agent would become overtly identifiable by the expressive defensive signs specific for it, and the outcome can also acquire an imaginable iconic visional presentation, making a ground for easier identification applicable to both the agent and the meaning in the semion, like a short cinematographic vision. Constancy of result such as could be seen in this example versus variations in the predator's and the prey's types actions, but with the uniform final outcome, concur to establish basic points for detailed interpretations analyzing the holistic semion and thereby reaching some prelingual clues possibly applicable to the agent, the time, and the action accomplishment in the speechless psychognostic mind. Now, it seems clear that the time becomes a distinctly known part of the instant present time (the root of the present tense), the agent can be inactive or passive in nuanced position, and the meaning of the action can also be interpreted in the active or the passive form. All these basic modes seem to operate symbolically in the psychognostic speechless mind. The time titan, in this example, appears to be exteriorized as the present tense and tacitly understood as such in psychognosis. In fact, it probably initiated a monodimensional time era that continued for a long transitional evolutionary period overlapping with the early lingual structuring and the starting of archaic language modalities. Continuation of such archaic monodimensional present time culture can be seen in primitive isolated human groups as we will see later.

It is important to remember that the semion is a holistic symbol defining the meaning of an environmental change, and in the examples mentioned, it exposes the entirety of the action accomplished by the agent: a ground for protoverbs. The action may be of course effectuated by more than one agent, and that is why the holistic semion is better interpreted in psychognosis based on the meaning of completion of the action as an outcome that is more constant rather than agent's contribution that may vary: an actually accomplished fact indicating a basis for possible modes of verbs. Furthermore, the distinction that every semion does not represent just any action with any agent is affectively established and indicates that priority of action in meaning dominating in the semion induces categorization for the variety of future agents and verbs in a graded scale for particular actions to be accomplished. This process obeys the life commandment principles in reaching the reasonable understanding of compatibility. At this stage, psychognostic understanding takes for established the completion of action in the basic present time. This now appears clearly evident, and starts the monodimensional time era that expands from psychognosis to idiognosis to extend over future generations.

PSYCHOGNOSTIC FOUNDATION OF LANGUAGES

Mind accomplishes many functions of knowledge-based type or noetic but, above all, includes an emotional orientation of some sort in all, and even pure mathematics may include meanings and affects, hidden in its apparently lifeless symbols in the interpretation of whoever uses them. Interpretations and related reactions in the mind, concrete or abstract, activate the neurocognitive, neurointerpretive, and neuroresponsive processes somewhat automatically but not totally without directives, and these directives are the two principles of logic plus happiness and pleasure minus pain of the biological life commandment. To try to evaluate these elaborate balancing systems, we face the honest, simple truth that they have originated from the archaic biotropism and biodynamism on the simple biological feedback systems to keep the biocracy in balance. All mind does is also to keep the autocracy in balance. Here, the mind has a much more important task to accomplish, and this power of control has come out through the evolution with a real trait, which is the expressivity. The mind is autoexpressive, an innate character that is fullest in emotional expressivity in psychognosis and can be less so in idiognosis. But even in the speechless psychognostic mind where emotional and affective meanings govern the expressivity, there is also knowledge of facts less affectively, or at least less emotionally charged. So the psychognostic speechless mind has both affective and noetic power and capacity. This power and its inherent capacity for allowing development will form the solid ground and prepare the stage for idiognosis to appear.

The hypothetical outline for holistic concept formation in psychognosis showing the fundamental two-frame model as sketched in table III includes the initiation of the expressive intent as the triggering agent for the mechanism to start the psychodynamic processes by Navand, a Persian word from my earlier studies on the subject carried out in Persian, Avestan, and Old Persian languages.[22] Navand (meaning impetus) stands as representative of the mental energy triggering factor for the process to recognize the visual figure, activate its static or dynamic meaning, and conclude to its holistic conceptual effect in the mind as the result. It personifies an unknown mental originator from autos that is using neural energy in serving the expressivity trait. Its energy could be assumed equivalent to Libet constant time in value[22] explained in the next chapter. The frame for the visual figure representing the cause can contain any figure generally, either static and showing no change and causing none or dynamic showing and causing change. The frame for the change to be effected is specifically for the biological permanent time, the time titan that accommodates dynamism to be registered. The final result of no change or change is shown in the third frame indicating the static or dynamic concepts thus formed by semions or the effected response by moveons.

To specify the criteria for the operative expressivity trait and the expected development to come in idiognosis, I assume four distinctive characteristics as axioms to be enforced in the psychognostic mind-controlling concept formation of both psychoaffective and psychonoetic types so that

- the expressivity trait so essential in life, so definitely dominant in *Homo sapiens*, so firmly based on emotions, does also serve knowledge formation and expression;
- the foundation of psychonoetic matters is on the same expressive models as for psychoemotive contents;

- each semion or moveon is formed in the mind modulating the agent and the action in providing the final holistic meaning in a two-frame model of matter and energy, replicating the primordial biological life commandment of logic and happiness with pleasure-pain Principle;
- the components of biological matter (agent) and biological energy (invested in the biological time) form the prototype universal two-frame model by the expressivity factor named Navand, the factor that engenders neural Energy quantum for the unit of work in time, abridged as NET, using the universal model of expressivity in nonliterary visual and literary potential visual-auditive lexical forms.

One interesting basic question of who makes the meaning interpretation and the final recognition, whether of mainly visual speechless holistic type of semion-moveon combination or specified literary ones but on the same models, is defined here in a simplified way attributing the task to autos in psychognosis and self in idiognosis. In essence, expressiveness through semions without words, basically operating in our psychognostic mind or literary worded symbols in our idiognosis, is a psychodynamic process that at first glance appears to involve autos. In psychological norms, self is the agent in our conscious mind with the explicit declarative memory whereas autos is the fundamental archaic agent in us that has served the expressivity trait in our remote ancestors for millions of years. Autos has the unquestionable seniority with priority of immeasurable time, the foundation of memory, semion and moveon formation, and a long background of nonworded speechless communication. Therefore, logically, the basic algorithmic models of expressions are made by autos, eventually contributing to the universal language of Chomsky.[23] In essence, cultural languages are all elaborated on the algorithms by the autos in psychognosis transferred to the self in idiognosis.

Autos and its domain of purely implicit nature and speechless tongue as hypothetically plausible entities grant tremendous ease in psychological dialoguing to discuss facts of assumingly preconscious times before the *self* defined literality by way of language, extending its domain by cultural maturation on the ground already prepared. If we could reach a conclusion by reasoning or by experimental proof that indeed the autos must be legitimately recognized as the initial mental decision maker for expressivity using the speechless symbolic language, then the literary self or selfves, the extensively developed literary symbols, and the practical language of worded communication will all be natural events to be expected to copy the psychognostic models. The essence of the psychognostic models, facilitating early psychonoesis, being the basic two-frame system that was most simply shown in CAUSE + TIME = CONCEPT in table III, reveals that the element of time must be regarded as crucial to be included in this expressive formula, which represents the neural triggering energy, the Navand. Navand seems to act tacitly and implicitly in psychognosis or explicitly in idiognosis. The explicit idiognostic form of time is in reality the tacit form of the permanent present time, the psychognostic time titan that will be exteriorized in worded form with the appearing idiognosis as time in dynamic action denoting being or doing in the present time. This time concept is an inapparent psychognostic one, inherent in the holistic semion and moveon meaning, formed by the percepting-reacting system long before the abstract worded concept of time would appear.

An initial point that can be taken as a proof for the important role played by expressivity in noesis would be to find out what additional change did occur first in the semion model of agent-action combination from the speechless language to worded language and explain it according to the role of NET serving expressivity. In this reasoning, one can see that the unique element, which forms the stimulus, is the meaning coupled with the time element, indicating the change from a static to a dynamic state in the holistic semion and moveon. Clearly, the triggering of semion formation and moveon preparation in the mind has been by the meaning, which is embodied knowledge and for which neural energy and time are indispensable. The time is the only part possibly traceable to be shown in words in the change from psychognosis to idiognosis evidencing energy spent by expressivity. It remains implicit and inapparent, though obviously there, as present time in semions' holism. The time in its transfer from invisible inherent part of the biological nature to appear in filling up the second frame of the two-frame model to be recognized in the worded language, for example, in *it is* in its simplest form, constitutes a solid proof for the reality of the basic two-frame structure for concept formation of psychodynamic or psychonoetic type and for all such single or complex conceptualizations including the important role of expressivity triggered by the NET complex.

We cannot trace precisely the earliest objectivity of time expression in language acquisition and developmental research in human children, but we can see clearly the concept of permanency of objects, meaning time permanency, in the *continued present time* for objects interpreted by children to be present but out of sight.[24] However, we can trace the second frame content, which is the time titan, from prior to idiognosis to the literary language, showing real time by searching in archaic languages. The present time in worded expression in some archaic languages showing the early phase of change from psycho to idiognosis may indeed present clues to support its basic role in the general two-frame structure for concept formation showing action in time. Such early examples from the Avestan gathic language of two to six millennia BC[25] reveals the appearance of the present time, showing both static and dynamic meaning associated with names and adjectives having suffixes, which exhibit the expected role of time and fit in the time frame of the two-frame models for concept formation and expressivity. This observation, which revealed the early worded expression for time content of the second frame, initiated the concept of NET being essential in my studies as foundation for the expressivity. The same concept also led to what I have named the Navand theory of languages that will be further detailed in chapter 10 with idiognosis, but some psychognostic-related facts need more elaboration at this time.

In connection with the old Avestan and Persian languages,[26] which form the root of the present Persian as well as in Sanskrit, the present tense of the infinitive ***to be*** indicated three differences in actual usage compared to today's Persian, presenting an interesting possibility to be examined for my purpose. In the first place, the present tense indicating the actual being in the present time as ***is***, for example, appeared to be less often used as one single element than we would expect it to be used in our today's language. This observation suggested a possible understanding of the present time as a natural state of permanent being with the time titan lingering in the still dominantly operating psychognostic mind. In the second place, the examples showed addition of suffixes to the names or adjectives to indicate their present timing in the meaning of being or doing, the functionality, rather than using the equivalent *is*, but by indicating the state of activity in those names and adjective in being differently or doing differently than being static. Finally, this differences in the states of being or doing in the present time showed a remarkable range of eight variable forms indicating eight dynamic states of the nouns or adjectives linguistically categorized as nominative, accusative, instrumental, dative, ablative, genitive, locative, and vocative,[27] expressing the time-related action and meaning modalities in the present time, without other worded evidence. Grammatically, this is referred to as inflection or conjugation of nouns or adjectives that are all expressed in the present tense. This observation indicates a tacitly accepted coherence of the action mode and its present time expressed together linguistically but without using the infinive form for being.

These forms of inflective changes practically play the role of single distinct additional conjunctive elements in our today languages making all the function words that still remain in the present tense of the actual present time in narrating sentences. Thus, the impression of still dominantly inherent time titan of the present moment of living, expressing precisely the present tense in action modalities in the worded language seemed natural in the archaic forms. The same is true in the present languages in some different modality with conjunctives or verbs again extending the continuation of the chain of thought primarily in the present time. And interestingly, this functional inclusion of the present time that initiated as either for securing continuity of the thought, or for saving the logicality of the stated facts, or for both, is particularly observable in the broader fundamental Indo-European root languages.

Finally the present tense of *being*, as ***am*** or ***is***, was relatively more frequent in contrast to all past tenses being least frequent in both Old Avestan and Old Persian[28] compared to present average language norms. This again tended to suggest the stage of dominant present time of psychognosis at early idiognostic expressions. The relative inconstancy of the use of the infinitive ***being*** (Avestan infinitive ***ah***) for just indicating the simple static state of being in the present tense (mainly third person), contrasting with the eight present-time dynamic forms indicating the active states of ***beings*** or ***doings***, also showed the firm tacit present time's more dominant expressivity in the relatively early idiognosis then than today. The archaic time thus appears showing an inherent constancy somewhat fixed in the form of present time and only expressed as such when radically needed for definite expression of the ***being*** state. Furthermore, making the nouns or adjectives to inflect, showing their dynamic working states with inclusion of the tacit present time, was in fact keeping with the simplicity of the basic two-frame models.

The fidelity of this expressivity to the structural foundation of the life commandment further shows that the basic biological expressivity trait also obeyed the commandment and followed its foundation in forming semions and moveons without needing to exteriorize the present time in psychognosis. In the worded idiognosis probably, initially the same protocol followed, naming the action as a whole (subject-action-verb-object), but keeping the functional state to process in the hidden present time, which changed in words from potential tacit and hidden status to be exteriorized gradually over living generations.

With this theoretical brief sketching of the course from silent to worded expressivity and from the hidden time titan to notable present time, one may think that there should have been tacit understanding of the being in the present tense in the communicating minds at that archaic time as a universal natural norm. Therefore, according to the meaning in the message, the present tense should have been explicitly used only in the two-frame basic presentation indicating uniquely the state of ***being*** in the actual real time for the static agent as ***am*** or ***is*** and only when needed for emphasis in the expression. Showing the appropriate inflected forms, and still in the present tense, served to indicate relations and dynamisms of nouns and adjectives keeping the limited two-frame models, agent frame plus time frame, to give the meaning. It follows that some earlier archaic time at the initiation of speaking, when the present time was indeed the only tacitly natural component of living, must have been sensed in the psychognostic mind but not openly expressed until idiognostic lingual itemization for the past and present was realized.

Accordingly, the pristine present tense was only exteriorized and used when confirmation of the ***being*** of the subject or action performed as ***doing***, in the obvious present time had to be firmly expressed as such. This reasoning exposed a wider scope to include a theoretical state of mind in hominids and early humans in which a monodimensional time and pure symbolic primitive concepts existed with the tacit belief that everything was only actual and real by its present time of being or doing. In that sense, every concept in reality should have been realized in the primitive mind as being composed of two parts—one being its symbolic presentation that was perceivable by the iconic (and later worded) code showing it, and the other, its inseparable innate present time harboring the ***doing*** that was tacitly there, without need for being expressed in static agents showing no dynamism and therefore no time change. But if dynamic, it would show the tacit functionality of the particular dynamism by conjugating the static agent's wording identity to show both the specific static meaning, the ***being***, for the time ground that is the present time and the time change for the specific dynamic change, the ***doing***, in that present time.

The course of events cited above justified considering more seriously the theoretical NET concept created for explaining expressivity expenditure of energy. In fact, this concept was responding to the natural question of what triggers expressivity by invoking the logical time-energy expenditure in assuming a mental trigger, the Navand that necessarily should be an intention-activated and emotionally charged starter. In this combination of theoretical essence, the only constituent that could eventually be demonstrated as tangible was the time, if it would fit properly the second frame of the two-frame models once words would come together with icons in symbol formation, filling literally the empty second frame. This prediction appeared to be realized by examples of linguistically present time usage domination in the old Avestan and Persian, as described, strongly suggesting the reality of psychoemotive and psychonoetic two-frame models and their constituents rendering plausible the action of such presumed trigger, the Navand.

The sceanario of a monodimensional present time appears interestingly reasonable for the initiation of time expression that is holistic and is in the tacit realization of the time titan, also forming the self-permanency concept as part of the basic symbolic concept of the bodily egocentric space time of the autos. In fact, neurodynamic energy used in concept formation for symbolic unnamed concept alone or symbolic and named concepts formed with the working memory has the working memory time as a component of the symbol formation. This situation can be assimilated to a spacetime of the symbol of which the psychognostic or the idiognostic realization only precisely shows the icon or the word respectively, but not the energy or its equivalent time that is transvested in the figurative being or doing expressive form. Although this hidden neurobiological permanent force (or its equivalent time) behind the mental realization of the entity that makes the concept is tacit and not showing in static concepts, it is essential in understanding of concept formation and is easily demonstrable in the deep structures of sentences. In other words, the names (the matter) are the codes singling out the direction of the neurobiological force (the energy) in algorithmic specification when describing an action (the outcome). In reality, the registration in memory and recall of the named entity cannot be achieved without the mental energy for the work, which legitimately covers the entire constitution of the semion or moveon, including the time. The holism, in fact, making out the symbolic meaning includes the agent (subject or object or both) and the neurodynamic processes with the energy and its carrier time. So naturally, the neurobiological energy used for expressivity predates the actual holistic concept realization and is invested in its significant identities, iconic or worded or both. This energy expenditure that remains unnoted in the static holistic concepts becomes noticeable in complex symbols in syntax, either in the form of suffix inflectional change in words or independent relay particles between words presenting effect of the energy-time complex in the Navand.

As every concept must use memory to be registered and recalled, both speechless iconic concepts with nonliterary status and worded literary concepts that also may have symbolic iconic meaning in general, use the same neurobiologic energy spent to produce and register them, which ought to include the energy spent as working memory, both implicit and explicit memories accordingly. What could possibly reduce the process only to working memory and the explicit memory in disputable memory research for worded concepts, as entertained in some writing,[29] incites doubt. As we know that nonworded holistic iconic concepts can have deep symbolic nonliterary meaning sometimes much broader in scope than implied with the words, the contribution of implicit memory to symbolic meaning can be major in the speechless psychognosis. This priority of precedence for the implicit memory has persisted through the time change from psychognostic speechless mind to our present eloquent worded mind, and meaning has dominated expressivity, finally reaching full-blown literary character. Furthermore, words did not initiate mental concepts of visual symbolic types that were already initiated by time-framed figures singly or in series and identified and interpreted figuratively without words. These nonliteral symbols encumbered the psychognostic minds of hominids and early humans with also basic psychonoetic values much longer than literary ones did with us. As both explicit literary concepts and implicit symbolic concepts do exist and as the explicit characterization by words followed the implicit functional nonworded symbols, therefore, the meaning must be accepted as the original dominant symbolic entity without needing words, irrespective of symbols remaining indefinitely innominate or becoming eventually nominated by words.

This aspect of implicitly originating expression is in line with evidence gained from Libet's conclusive data.[21] It forms an additional supportive evidence that predominantly psychoemotive and psychonoetic concept formation start in the speechless psychognostic mind in which a solid foundation of noetic structuralization already exists and is based on a universal model, which constitutes also the proof confirming and the axiom realizing the expressivity trait role in the Navand theory for the language formation.

REFERENCES FOR CHAPTER SEVEN

1. Goldsmith TH. 2006. What Birds See. *Scientific American*. 295:69-75.
2. Robinson A. 2003. *The Story of Writing*. Thames and Hudson, Inc. New York, NY.
3. Gibson EJ, Walk, RD. 1960. The "Visual Cliff". *Scientific American*. 202:64-71.
4. Yonas A, Granrud CE. 1984. The Development of Sensitivity to Kinetic, Binocular, and Pictorial Depth Information in Human Infants. In: Ingle D, Lee d, Jeannard,M, Eds. *Brain Mechanism and Spatial Vision*. Amsterdam, Nijhoff.
5. Yuille AL, Willman S. 1990. Computational Theories of Low Level Vision. In Oscherson DN, Kosslyn SM, Hollerback JM, Eds. *An Invitation to Cognitive Science. Visual Cognition and Action*. MIT Press. Cambridge, Massachusetts. London, England.
6. Land E. 1964. The Retinex. *American Scientist*. 52: 247-264.
7. Biedermann I. 1987. Recognition-By-Component: A theory of Human Image Understanding. *Psychological Review*. 94: 115-147.
8. Biedermann I. 1988. Aspects and Extension of a Theory of Human Image Understanding. In: Pylyshin Z. Ed. *Computational Process in Human Vision: An Interdisciplinary Perspective*. Norwood, N.J., Alex.
9. Biederman I. 1990. Higher Level Vision. In: *Visual Cognition and Action*. Vol. 2. Oscherson DN, Kosslyn SM, Hollerbach JM, Eds. The MIT Press. Cambridge, Massachusetts. London, England.
10. Mithen S. 1999. The Prehistory of the Mind. *The cognitive Origin of Art and Science*. P 110. Thames and Hudson.
11. Engel AK, Debener S, Kranczyoch C. 2006. Coming to Attention. *Scientific American Mind*. 17(4): 46-53.
12. Spelke ES. 1990. Origin of Visual Knowledge. In: *Visual Cognition and Action*. Vol. 2. Oscherson DN, Kosslyn SM, Hollerbah JM, Eds. P 117. The MIT Press. Cambridge, Massachusetts. London, England.
13. Brouillard P. 1998. Popular Psychological Tests. P 179. In: *Psychologist Desk Reference*.

Koocher GP, Norcross JC, Hill III SS, Eds. Oxford University Press. Oxford, New York.
14. Bonner JT. 1980. *The Evolution of Culture in Animals*. Princeton University Press. Princeton, New Jersey.
15. Marcus G. 2004. *The Birth of the Mind. How aTiny Number of Genes Create Complexities of Human Thoughts*. Perseus Books Group. New York.
16. Dawkins R. 1976. *The Selfish Gene*. Oxford University Press.
17. Greenspan SI, Shanker SG. 2004. *The First Idea. How Symbols, Language, and Intelligence Evolved from our Primate Ancestors to Modern Humans.* Da Capa Press. A member of the Perseus Books Group.
18. Savage-Rumbaugh S, Lewin R. 1994. *Kanzi : The Ape at the Brink of the Human Mind*. John Wiley and Sons, Inc. New York.
19. McNeill D. 2005. *Gesture and Thought*. Pp 247-249. University of Chicago Press, Chicago and London.
20. Dobbs D. 2006. A Revealing Reflection. Mirror Neurons are providing stunning insights into everything from how we learn to walk to how we empathize with others. *Scientific American Mind.* 17(2):22-27.
21. Libet B. 2004. *Mind Time. The Temporal Factor in Consciousness*. Harvard University Press. Cambridge, Massachucetts, London, England.
22. Amir-Jahed AK. 2002. *A Look at Thought in the Mirror of Speech*. (Text in Persian). Rahavard, A Persian Journal of Iranian Studies. 63: 50-61. (text in Persian).
23. Cook VJ, Newson M. 2001. *Chomsky's Universeal Grammer. An Introduction.* Blackwell Publisher.
24. Owens RE Jr, 1992. *Language Development. An Introduction*. Macmillan Publishing Company. New York.
25. Boyce M. 1992. *Zoroastrianism. Its Antiquity and Constant Vigour*. Mazda Publishing in association with Bibliotheca Persica. Costa Mesa, California and New York.
26. Hoffmann K. 1989. The Avestan Language. *Encyclopedia Iranica.* 3: 47-62.27. Razi H. 1989. *Avestan Grammar, Part II*. (Text in Persian). Fravahar Publication, Tehran.
28. Razi H. 1989. *Old Persian, Grammar*, Text, Lexicon. (Text in Persian and Old Persian). Fravahar Publication. Tehran.
29. Morris RGM. 2002. Episodic-Like Memory in Animals: Psychlogical Criteria, Neural Mechansms and the Value of Episodic-Like Tasks to Investigate Animal Models of Neurodegenrative Disease. In : *Episodic Memory*, Baddeley A, Conway M, Aggleton J. Eds. Pp 199-200. Oxford University Pre

CHAPTER EIGHT

NEOGNOSTIC STATES

FOUNDATIONS OF THE IDIOGNOSTIC CONSCIOUSNESS

Approaching the subject of the conscious state, or states to be more correct, requires defining consciousness. Such definition for a multifaceted subject that is consciousness in which not every facet is perfectly clear may hardly be integrally complete or correct. Attempting to solve the problem must first assure integrity by evaluating all factors concerning consciousness in order to reach a comprehensive conclusion. Next, the more difficult task is to realize the values of factors more decisive at the exclusion of others less significant to the variable states of consciousness to assure reasonable evaluation of a problem that is not mathematically solvable. To start with, we could say that consciousness can be regarded as a real time mental realization of each wakeful moment of life in focused concentration by the mind, realizing the self being in auto-hetero-gnostic control and capable of clear logical interpretation of moments in time. It may concern sensing, attention, intention, interpretation, and mental or physical action, and all this minus real time logic in dream vision consciousness. It is a part of the phenotype spacetime with its matter-energy-time components interacting in harmony with somatic biological conditions and realizing the whole processes constantly with the ongoing time. But consciousness is usually described briefly as a general state of wakeful awareness with perfect orientation in space and time. Opposing this general crude definition, the state of coma is regarded with different degrees of conscious to unconscious levels in a more elaborate way. In psychology texts, definitions of consciousness vary but ultimately come to separate two main spheres—the sphere of neurobiological mechanisms leading to consciousness and that of the philosophical causal factor for the mental experience referred to as the hard question in the puzzle of consciousness concerning the will.

Owen Flanagan has made an interesting allusion to the event of the chess match between the chess champion Garry Kasparov and the IBM Deep Blue computer, which ended in Kasparov's defeat.[1] This comparison of the high-level mental reasoning, using a logic of analysis and synthesis by serial arguments and algorithmic application in a conscious state, with a nonliving device able to do the same work but unconsciously, is the best example for the hard question, the causal factor, which remains undefined regardless of how thoroughly the mechanistic factors may be defined. The computer in fact does not have the will to win (or the joy of winning against the sadness of losing), which reflects the hard question, the unknown sine qua non, but yet it wins! This means that the hard problem, the causal puzzle, must be independent from the mechanistic logical guidelines reaching the solution, but it also is the origin for them. This realization establishes the veracity that the life commandment reflects the hard question with the pleasure-pain principle and not the logic per se that could be satisfied by a lifeless device. Therefore, these solution finder mechanisms in the computer should indicate unidirectional intermediary steps between the hypothetical unreal hard problem and the real solution, but in the mind, also from the solution once effected, to the real hard problem bidirectionally, understandably in a feedback supportive way.

Considering this elusive inescapable hard problem, if I have to formulate a definition for consciousness, I feel being forced to secure this real phenomenal entity of hard problem in the line of arguments with the supportive feedback effect of its outcome on itself, regardless of what it can be, a known scientific fact or a pure philosophical concept. In fact, a comprehensive definition cannot leave this relationship out. Respecting what has been so far discussed in biocracy reflecting biological manifestations in the phenotype by the bioforce, I would include a real spacetime entity for the phenotype including consciousness as inherent in its real time. This leads me to propose a definition for consciousness hopefully including both scientific and philosophical grounds.

Scientifically, many gaps exist in our knowledge, too numerous to allow logical connection, but we have one essential evidence, which is the evolutionary graded change that has made the structural anatomical ground for the consciousness to gain increasingly more neurodynamic capacity with proportional functional ability to give us the present conscious state. The states of biognosis showing attempted biologically definable differences of bioawareness, bioconsciousness, and psychognosis, even if not sharply distinct, do indeed show the merit of reality for the progressive change gained before our literary consciousness that I have named idiognosis. This undeniable trend of progressive change occurring in the phenotype spacetime on the one hand and the reciprocating effects of the phenotype spacetime with the environmental spacetime on the other confirm interacting of the two spacetimes in a causal way to assure the trend of change. Such changing trend conditioned by autoreactive processes, if reaching equilibrium, is expected to reflect evidence of those processes in the final equilibrium. The hard question therefore must have been elaborated during the change along the different progressive stages leading to consciousness and must have been also reinforced continually. This is indeed a fact that needs be reconsidered to be incorporated in defining consciousness on both scientific and philosophical grounds.

The overall considerations consequently lead me to propose a definition for consciousness based on its natural history and summarized in the following principles:

1. Consciousness originates by autos' bioforce in the form of biognosis to bioconsciousness transcending to psychognosis, through auto-hetero-gnosis, and is reinforced by modular selfs transiting from psychognosis to idiognosis. Initiation of consciousness starts changing the time indefinite matter-energy uncertainty to recognizable time definite type. It expresses definiteness physically and mentally through the expressivity trait to oneself and nonselfs.
2. Consciousness works in the phenotype's spacetime of energy, matter, and time using the NET system of expressivity trait engaging sequencing in real definite extendable time by memory in the wakeful state and in the holistic time indefinite (time titan) visions in dreaming.
3. Conscious recognition must be regarded analogous to *EITHER* as against *OR* mechanism in the open *EITHER* and *OR* of the subatomic quantum physics' spacetime, the phenomenon being changed to *EITHER* alone as against *OR* alone by the conscious will in focused attention. Conscious realization of the visual iconic figures by self(s) can be analogically assimilated to the physical photonic matter and that of the meaning similarly to the energy of light in physics.
4. Consciousness is limited in the *EITHER* versus *OR* choice, but gains unlimited extension when the material iconic visual detection, combined with the abstractive meaning by words, makes real the totality of matter-wave analogy elaborating the interminable abstractive conscious mind of the immaterial biology.

This state of mind commonly called consciousness is included with the other Gnostic states in the classification given in chapter 5 (Table I) in a generic outline for Neognosis. From that outline, autognosis was discussed in previous chapters and especially in relation to the pillars of emotions in chapter 5, mainly evidencing bioconsciousness. Psychognosis, as we have seen, was proposed for what becomes conscious in the mind in terms of its psychic value, like all affective sensations of different sorts including the four archaic sets of emotions described as pillars of emotions (Table II). It was further elaborated as forming the fundamental building blocks of the nonworded two-frame formulas for meaning by the semions and moveons discussed in the last chapter.

For idiognosis, the term is meant to include all codifiable items by literary wording and retrievable through explicit memory, which can be learned by choice, or created for communication purposes by necessity including such figurable items like in arts and science in general, in short any word-coded and recallable item as conscious content with meaning, concrete or abstract.

Based on considerations for the principle of biocracy in the face of evolution, we can understand that the conscious state has developed as a natural selection following psychognosis and bioconsciousness, which themselves followed the simpler state of sensing in bioawareness and even further back in archaic sensing that I have named paleognostic state. These staging sequences, theorized to explain progress from an initial tropism to autonomous reflexive acting and to the controlled expressive sensing and creative abstractive thinking, appear deterministic. In reality, the fundamental truth is the fact that so called unconscious state in the most primitive life forms has changed into the state of perfection in the most refined relationship of the living organism with itself and its environment. This change shows that sensing mechanism has evolved into primarily implicit consciousness, the psychognosis, and the explicit consciousness, the idiognosis, matching the needed functions with incomparable adequacy through the evolution. This course of events certainly appears deterministic, a determinism that has been dirigible by the hard problem with its self-maintaining supportive feedback on the bioforce principle responding to outer environmental impositions. It is, in fact, determinism in that sense, the sense of persisting principle of the bioforce in DNA providing expressive adaptability with the chance of self-supporting determining assurance for unending progress.

Reversibility being impossible in evolution, determinism seems clearly acting on the human mind with the time definite logic of facts versus no facts and *being* versus *nonbeing*. This phenomenon, interpreted following time indefinite quantum logic of simultaneous *fact* and *no fact* and *being* and *nonbeing* of uncertainty quantum principle, in essence, seems permitting periodic reversibility in a sequential phasic course of awake conscious states alternating with sleeping consciousness showing time definiteness and indefinitness and also with phasic continuity permitting biological life. So our state of conscious realization seems to reflect our deterministic phasic continuity that we see with the product of our biologically condensed time in material ontogeny and in immaterial memory. This reality relates to our autobiography a necessary deterministic image in the condensed time that does not exist out of biology and is visualized in iconic immateriality. It is this condensed time that gives the biological ontologic material soma and the abstractive nonmaterial psyche. This subject justifies scrutiny of Gnostic thoughts initiated by *will*, which are the ultimate immaterial products of the materially built conscious power that needs be explained through neurodynamic energy exerted in time, just as the electromagnetic waves are explained by their matter-energy interactions expanded in time.

The decisive adaptive trait in evolution perfecting sensing interpretation can be tested in studying comparative neuroanatomy as examples of progressively more refined brain structuring in the vertebrate species. This is particularly clear with skull endocast impressions of language centers in *Homo erectus*.[2] More importantly, however, it can be also tested functionally in the human neurophysiology. In this connection, the question that can be crucial is how important and decisive the archaic sensing with the original autonomy, here named paleognosis but called unconsciousness in today's common reference, can be consciously active in the present human neurophysiology? This question has been partially dealt with in the past two chapters in relation with dreams and with discussion of culture in animals. It will be further considered in this chapter in establishing *le droit de cite´* for the simultaneous presence of psychognostic and idiognostic states under neognostic title, or as so commonly called consciousness in short. The subject in essence also exposes the hard problem, the state of conscious will that must have been originated in its nucleus in the biological autonomy. Any answer attempted will have to include the hard problem in the substance of autos reflected through self to define it in consciousness. In line with the logic of biocracy, the will must be searched in autos as it must be reflecting bioforce binding the energy and mass. The will can be indeed regarded as the conscious realization of the innate autocratic drive. Although it cannot be identified in substance just as any insubstantial emotional feeling when consciously recognized, it is still revealed in the conscious sphere as a reality in self. So declaration of the will is by self, but its origin is in autos, and we can search for the will in autos not only philosophically but scientifically as well. Therefore, my discussions to follow will focus on significant data from respectable neuro-somatic investigations available on the subject. To facilitate comprehension and avoid confusion, I will use the words *unconscious* and *unconsciousness* in their familiar meanings.

THE GHOST IN DISGUISE

The sense of self is deeply rooted in our minds, and when consciously acting, unless being psychotic, schizophrenic, or else, we well know that we are the agent doing what we decide to do but no one else, and we are responsible for the good or bad of our acts. The point of discussion to follow is precisely the fact that this appears not to be entirely true! Some well-conducted neurophysiologic experiments show the opposite and indicate that initiation of our conscious acts is, in fact, unconscious and precedes our personal conscious awareness before it confirms that we are performing the act so that, paradoxically, we may not be the agent doing what we think we do. This paradox is akin to a ghost in some disguise replacing us and needs just a little more explanation.

Pioneering experiments by Penfield and Roberts on the surgically exposed human brain in mid-twentieth century,[3] followed also mainly by the work of Jasper and Bertrand,[4] opened the way for subsequent similar studies that provided a wealth of information on the subject of sensory and motor activations by specific localized cortical stimulation.[5] These and other similar studies using transcutaneous cortical recording by electroencephalograms (EEG) and direct cerebral cortical recordings in neurosurgical interventions on patients cleared many puzzles and answered some mind-boggling questions on the functional specificity and coordinated brain activities, but our unknown inner decision maker, the hypothetical ghost in disguise, remained masked.

In relation to intending to perform an action, studies done by Libet and associates[6] form the fundamental basis for understanding the conscious act and its components. The primordial interesting revelation of results of these and similar studies by others,[7] partly confirmatory and partly controversial to Libet's results and interpretations, is the fact that all intended mental procedures seem invariably composed of an initial unconscious component and a subsequent conscious end point, taking a time generally called neuronal adequacy time or neuronal correlate of consciousness and imparting doubt to the subject of free will.[8] In other words, what we call conscious invariably starts unconsciously by the actual neuronal activity, which is undetectable to us as we only see and comprehend the entire process as conscious.

The actual experiment by Libet was primarily designed to check the timing of conscious initiation of a motor act in relation to what was known as readiness potential (RP) in works following in the same field of investigations, initiated by Penfield and Roberts. RP had been shown to precede motor action and could be recorded by EEG before the action itself could be recorded on a myogram. The question was whether the conscious will to act was also synchronous with RP or was it asynchronous and in what order in timing. Libet had to record the initiation of RP on EEG and the action potential of the muscle group doing the movement (M), consisting in a wrist flexion, by electromyogram (EMG), but had also to integrate somehow the initiation of the subject's will (W) in starting the action of wrist flexion.

Recording the time of the initiation of the will, reported by the subject being examined, necessitated registering it in a way that could not cause another RP confusing the RP of the motor action to be recorded on EEG. This was done in Libet's experiments using a different way, basically perceptual and not motor, by instructing the subject to note the position of a circling light spot running at known speed of one circle every 2.56 seconds, noting the light position secondarily at time of deciding to act, but reporting it after action was completed.

Experiments were conducted forty times for each participant and recordings averaged. As some subjects reported that some of their intentions to act were not immediate prior to their acts, the RP times recorded varied from longest 1000 milliseconds (msec) to shortest delay of 500 milliseconds on EEG before the movement recording showed M, which appeared on EMG. The decision time to act (the will), W, appeared about 200 milliseconds before M, so that recordings invariably established an initiating RP prior to intending voluntarily the motor act by W and ending in M. The sequence followed the order of RP (1000 to 500 milliseconds before M) to W (200 milliseconds before M) to M.[6]

A general interpretation of the briefly sketched experimental results above, excluding divergent details in the opinions of critics,[7] but emphasizing the overall basic agreement on the initiating unconscious modus unjustly considered conscious is inevitable. This interpretation can plausibly confirm that any so called conscious act is in reality composed of an initiating unconscious modus prior to consciously realized recognition in the mind of the performer and prior to actual evidence of the completion of the action.

Interpretation of the illusory acceptance of the voluntary act as a bona fide willful conscious decision opens two main directions of debate. In one with dualistic inclination, interpretation may consider the timing of RP preceding W and M as an indication of an independent action to volition, a ghost act, a clear Cartesian view. On the other line of thought namely proposed and supported by Libet, the natural course may indeed be the way expected and shown by the result, which evidences the ground preparation RP and the effect of volition W to naturally occur prior to M. The initiation is of course unconscious but the precedence of W to M is conscious and indicates the intention to go ahead. However, W does not indicate uniquely a YES, it does also leave time for a NO. In other words, in the 200 milliseconds from W to M, a vetoing can still take place, which in fact had been reported by some subjects in the experiments. In addition, the timing of M could be consciously retrospectively recorded in mind in a retroactive way as the initiating event.[8] So in brief, the retroacted process could be interpreted as a commodity arrangement performed automatically by the conscious state as, for instance, in the head up conscious retinal presentation of the real head down retinal figures in vision or more elaborately analogous to color interpretation of the mind from wavelength physical retinal perceptions.

Another way to interpret the timing sequences of the conscious neuromotor process for an explanation to justify precedence by unconscious initiation is to refer to the concept of the time titan that I have sketched in chapter 4 and as implicated with the metaphoric invisible present tense in the last chapter. As the process in its entirety is one action that is intended and therefore endogenously activated, it cannot be indicating the environmental timing and can only reflect the autos's time titan, which is always unconscious as it is the inner eternal time unknown to consciousness. It only becomes consciously known when it becomes exogenous to autos, once shown in a mental image, concept, or action realized by the conscious self. Therefore, it is only natural that any conscious action to be known as ***conscious will*** *have to be exteriorized from autos to become known by self as an exogenous entity leaving the realm of unconsciousness and entering the sphere of consciousness*. The timing for conscious sensing is necessarily after the transfer is made, but the entire transfer is one continuous process that could ultimately be recorded as either unconscious or conscious, if unconscious could ever be recorded consciously! In this explanation, every process is naturally and correctly evidenced and no commodity action is needed in principle if the self could similarly record unconsciousness.

In another series of experiments, checking the evoked potential (EP) on the cortex that develops in response to a peripheral liminal electrical stimulus to the skin, Libet's findings indicated that the EP occurred about 500 millisecond before the subject sensed the stimulation consciously and reported it subjectively. If a comparatively liminal electrical stimulation for producing the same subjective feeling was applied directly to the cortex, EP did not occur but the time for consciously sensing and reporting the feeling had to be extended with a chain of repeated stimulations for about 500 milliseconds before the conscious experience would be produced. To check the effect of the route difference for the stimulations to induce consciousness, further recordings in the neural tracts from the spinal conduction to the medulla to thalamus were also made in patients having electrodes placed in subcortical regions, and Libet's findings again showed similarly a delay of about 500 milliseconds whether the normal route of conduction through the subcortical channel was used or the abnormal direct cortical stimulation was effected.[9] In these experiments, which show the same time delay from initiation of the stimulation to the conscious experience, the time delay is not noted by the subject as it is uniformly unconscious, and again, a presumed retrospective mental process reports the stimulation and effect as one whole action, which is a sensory type and is from an exogenous stimulatory source. In actual fact, again what is unconscious obviously is not, and what is conscious is reported. Is the ghost's action, the hard question, included in this unnoticed time?

The outstanding clarification that comes out of these experiments is the fact that an important unique specificity of mental processes exists in conscious timing for both purely sensory and mixed sensory-motor actions after an unconscious starting. In the more complex situation, whether the stimulus is from the environment and exogenous in reality, like electrical stimulation to the skin or to the cortex directly, or endogenous by mental conscious decision making, which is in reality endogenous in the mind's primordial autos, but exogenous to self, the consciousness in the self follows the unconscious start. The conscious effects are, in fact, all exogenous to autos, and once represented by self, they are in an irreversible modus. Autos, being the archaic totipotent ruler in autocracy, presents all unconscious and emotional senses in the indefinite time titan, and uses the unlimited progressively increasing scale of pleasure-pain axis of the life commandment. In the simpler type, we are facing perceptive conscious sensation from an outside stimulus arousing conscious attention over the unconscious subthreshold stimuli, and in the more complex type, we are dealing with an inner attention from autos, which is still not conscious but becomes conscious on the way to an intention leading to a motor action. So in the final analysis and regardless of the route for the stimulus transfer to the brain, the essential revelation is that both perceptive (attention only) and decisive (attention-intention) mental processes start unconsciously in the subcortical to cortical structures and become conscious from the unconscious type by processes in the cortical structures with a delay of about 500 milliseconds : *a time between autos calling before and self-responding after irreversibility.* This phenomenon in reality represents a constant time difference in the transfer from the psychognostic time titan to the real time. Is the call from autos to self the oracle of the hidden ghost? Or is there a more reasonable explanation?

Reflecting on the reciprocal relationship of the limbic system with the neocortex and considering the transfer of the nervous impulse through the subcortical nuclei, mainly limbic and unconscious, to the neocortex, potentially conscious, we detect that the change into conscious reporting of skin sensation or intended will is independent from the type of nerve actions being a peripheral action potential or a cerebral evoked potential for attention-only mental work or readiness potential for attention-intention mechanisms. It seems either to be dependent on a direct feedbacking between the limbic and the cortical functions or an indirect one through cortical association areas referring the sensation back to the skin or realizing the intention to act. The feedback is between the autos and the self so to speak, initiating with limbic activation through the normal route of brain stimulation, or even abnormally by direct cortical stimulation, which causes the sensation to be felt on the skin—in any way, it is showing a direct relation between limbic's autos and neocortex's self or an indirect one through neocortical association areas. So we can presume that this feedback time must essentially remain confined within some time limit of neurodynamically transfer process that must be equal to the delay between unconscious and conscious states. This state of affairs confirms the solid evidence for the unaccounted time and its hidden content that has been named the hard question, as indeed looking at consciousness without seeing its background leaves many questions centered on the will. The hardest question being the essence of consciousness is ultimately the expression of autos's bioforce with the change from uncertainty of the unlimited quantum time scale to certainty of biological sensing for, and attention-intention on making the choice for that scale change. The will, therefore, inherent in autos, is expressed by self as yes or no, consciously recognized and declared as a conjoint decision output: the ghost in this interpretation is therefore autos but wearing the mask of the self.

The comparison of normal and abnormal routing of the stimulus to cause the conscious sensation also reveals the possibility of nerve-conduction time difference that is not to be neglected. In Libet's experiments, the site of the peripheral stimulation either close or far from the head, although indicating 14-50 milliseconds difference in conduction time, did not influence the 500 millisecond delay in recorded cortical events before conscious feeling, neither did lower or higher frequency of subliminal stimulation applied on the brain. In other words, the time effect on consciousness production was constant with all subliminal stimuli on the brain that had to be prolonged in a chain of repeated stimuli from any site (proportional in intensity for cutaneous or cortical stimulation) for the duration of about 0.5 seconds.[9] Therefore this time, constancy seems to be the single prerequisite time dimension in the human mind for consciousness to appear as a realized will. This time dimension, so-called neuronal adequacy time, or the neural correlate of consciousness already cited previously, seems to be a single evident preparatory time, but possibly including a minimum of two component times, a neocortical association time and a cortical-limbic feedback time. The range of this time period averages 500 milliseconds before conscious awareness according to Libet's experiments,[9] and therefore, this time range can be regarded as constant time laps from psychognostic autos to idiognostic self that I will call Libet's constant "L."

The evidence for this feedback between limbic system and neocortex, in addition to dream scene mechanisms discussed previously, can be seen in the measurable time difference between the unconscious and the conscious states, without neocortex and with it, or in reality between a state of biological processing that uses its own indefinite permanent time with no need for consciousness, and a second one which needs more than the all inclusive biological time and must have its own measurable time. That difference is impossible to define except indeed in the form of the Libet constant. As the two states of mind can only exist interdependably in a symbiotic analogy, they must balance their time scales. Again, we can invoke the explanation of time titan and presume that what it does in the autos unconsciously is fundamental and conscious sensing occurs after the Libet constant delay, which is borrowed from the psychognostic time titan. We can realize now that the reality of this time delay, with its functionality assuring limbic-cortical relaying in the unconscious time indefinitness, bridges the scientific evolutionary evidence with the deterministic philosophical finality assumed in the ultimate time indefinitness of abstraction with extended consciousness: a bridge between indefinite eternity states through material biology.

The change of indefinite to definite time granted the power of perfect changeability from synthetical to analytical mental handling of items by the phenotype for safeguarding the egocentric position of the self and the autos. In fact, it is evident that the biognosis, the uniformal conscious ground of the *infinit monodimensional present time* of the time titan, from the very paleognostic initiation dealing only with the inner spacetime metabolic changes to mindful psychognosis and idiognosis, still did not have to cope with the precise actual present time, as all sensory and neuromotor actions could be carried out at one time scale (the inner permanent present time continuing without break), except in acts related to external environmental spacetime changes imposing variations like fight and flight. In essence, bioconsciousness and psychognosis with a primitive working memory and implicit memory could be the ultimate in monodimensional time handling only within the actual present time compatible with the unconscious permanency of eternal time titan, a state in which processes are essentially handled in parallel by the right hemisphere with a free fluidity connecting them. This state of biognosis has existed at one long evolutionary time with the predominant right hemispheric handling of activities by the bicameral brain[10] fully serving the inner eternal time titan. But with neocortical development to the stage when precise timing in sequencing and a more elaborate memory, especially explicit declarative memory, became overwhelming, then left hemispheric function with an analytical time scale related to conceptual semantic items prevailed for that task to be accomplished. Exiting the monodimensional consciousness by the agent of autos, the self, and matching inner time titan with environmental time change then became inevitable, allowing the process to take place with neognosis, not to replace bioconsciousness and psychognosis but to supplant them and serve the biopsychon as the egocentric judge realizing selfhood opposed by the external world. All this is being based on fully interchangeable synthetical holism with analytical partialism in mental interpretation.

AUTOS - SELF - WILL

Biocracy has but one time scale to use, the biological time for biological processes that are arranged in an invariable sequencing for basic SR by chemical, metabolic, neurohumoral, and hormonal mechanisms and in usual normal locomotion. This sequencing is intrinsic to the nature of biological processes, autonomous, immeasurable, and not needing consciousness necessitating change from time indefinite to time definite type; it is a course of action within the time titan to perpetuate the infinity of time titan through reproduction. On the other hand, with evolutionary impositions necessitating adaptive changes to secure external time realization, this holistic time system was no longer sufficient, and time analysis came to play its role in biocracy with auto-hetero-gnosis, the individualization of the self and the establishment of the conscious time definite scale. Thus, this Gnostic state relayed the inner biological time, the time titan, and the independent environmental time to alert and allow biocracy to save its biological balance in the face of external changes. Evidently, with developed central nervous system, complex metabolic, neurohumoral, and hormonal processes allowing adaptive seasonal and periodical changes on the one hand and locomotion and fight-and-flight responses needing accurate rapid timely reactions on the other, a new mechanism was needed to connect environmental time effects with corresponding short-term and long-term somatic variations so that everything would be kept in continuing balance. This necessity incited the evolutionary adaptation to establish a more adequate system of harmonious work between internal and external worlds. Thus, reaching full consciousness, the idiognosis became essential to serve living needs in collaboration with psychognosis and to exert control on it when forced by real time priorities mainly in social living.

The most important processes, in fact, to be kept in balance were related to emotions, the reactive responses by the biocrat to peripheral environmental effects, all perceptive senses evoking feeling and psychologically charged reactions of various degrees and intensities. Some of these mental processes, if becoming quite hostile, obviously resisting external forces of injurious nature to self or justifying a frank fight, threatened the mental balance of the self positioned between internal norms of autos aiming for supremacy and satisfaction, and opposing exigencies of ego and superego imposing on self by the early societal rules. This formed the states of mind in psychognosis that, now after establishment of idiognosis, had to be given second priority especially in social settings with increasing society norms. So in essence, idiognosis, aided by explicit memory extension was formed idiomatically by the extensive intertwining of experimental additions from emotional meanings attached to primary natural iconic frames in abstractive forms, with the norms of logic in the elaborate mind, but with main control by the pleasure-pain balance.

The essence of time-related idioms is the real present time, which is outside time titan and multidimensional in terms of being fragmental unlike time titan, which is a constant monodimensional continuous present time. Any factual event in the real present time is an idiom realized or created by our *self*. We live with our real present time idioms in our wakeful activities if consciously attending to all events. Our biological permanent present time, our time titan, reigns in our sleep giving us intermixing visions of all types of events in our inner present time. Our conscious self is the central permanent idiom of the real present time, which cannot lose its idiomatic time frame and firmly keeps it separate from all external spacetime effects and other past and future time idioms, even from the most autos-oriented self in the past or in the future as modular selfs are involved. Idiognostic consciousness, in fact, considers each moment's present spacetime characteristics of the dominant self. The conscious self is in reality the idiomatic center of the real time as autos is that of the infinite time titan. But what is the limit of the real present time for the conscious self if different from the present time of autos operating with semions and moveons of holistic values?

To answer this question, a reminding remark should be first made by reiterating that the model for the self to be conceptualized and acknowledged accordingly is identical with that used by autos based on semions and moveons of two-frame forms presenting the carrier-frame for the agent and the time-frame for the effect, respectively. However, it is different from autos' situation by the fact that the agent is presented by an identifying worded code, a graphic symbol with meaning, which provides additional clue to the agent's identity precoded by known icons, facilitating its recall. So in short, the graphic symbols facilitate passing from the indefinite time to the definite one. Additionally, the word to appear in the time frame can now precisely give the meaning related to the agent's specific action that can be shown to remain in the present time or to extend to the past or to the future. So in essence, the present time length for both autos and self seems to be equal as a segment of consciously realized time: the present moment of recall that includes the sequence of unconscious autos acting and conscious self recognizing.

Viewing the question for the self alone, the plausible solution is by finding out beyond what length of time after one attention-intention task is completed, a new one is engaged, and by measuring that length of time. Considering the findings of Libet's experiments, the 200 milliseconds between the conscious will (W) and the action shown by myogram (M) is the mean time for such a defined segment of time. Knowing that self, realizing the conscious will, is preceded by the unconscious time of about 500 milliseconds, we should accept the total of a mean 700 milliseconds for the self's working memory completing a task of attention-intention type, which includes the Libet's constant time. Thus, the basic time segment of the working memory for psychognosis must be about 500 milliseconds and for idiognosis 700 milliseconds. The role played by the implicit memory as working memory in psychognosis is basic in the neurodynamic process of attention-intention to be considered psychognostic until realized in worded form as idiognostic. This finding opens up a horizon of further revelation to appear in what theoretically is a chasm but practically a bond in times between autos and self: the time for the unified will to be sensed. Thus, it becomes evident that the psychognostic constitutional foundations including the life commandment with pleasure-pain principle and the basic two-frame models used in all semions and moveons operate unchanged in both psychognostic nonworded and idiognostic worded consciousness using the same basic implicit memory. Even when the worded codes for all elements used, concrete and abstract, are available, which only appear when expressivity must be shown with words, the basic semion-moveon principle with the fundamental two-frame models are the invariable operating norms. Furthermore, the will (W), which is in essence a moment of time permanency of self-recognition of ***to do or not to do***, naturally initiates psychognostically in the limbic system to be recorded by self 500 milliseconds later; this is the time for the ghost to appear personified in the will, so to speak, under the mask of self.

It has been shown that harmonious brain activity, more often in the form of rhythmic synchronous neuronal discharging, characterizes consciousness.[11] This baseline rhythmic synchrony, called gamma waves, seems to be the ground for sensing conscious selfness. The will to be sensed probably forms when the limbic system's neurodynamic charges reaching the neocortical zones for the self's abstractive frontal and spatial left parietal areas,[12,13] respectively, elaborating abstractive and physical positional selfhood, are activated and react to the received stimuli adding to the harmonious united synchronous response. Thus, it is likely that a reinforcement of these areas reacting to the limbic discharge during the 200 milliseconds, while the self-response is being finalized, enhances the sensing of the will power to be experienced. Positive feedbacks for regulating the limbic impulse in harmony with stimulatory or inhibitory contributing associative brain areas also likely work in coordinating the final voluntary output between limbic autos and cortical self. This time seems to be the specific time for the will's conscious realization in a binary process setting definiteness at full functionality in real time and seems to be the essence in confirming the selfness for yes or no decision making. This reciprocal autos-self feedbacking seems to operate maximally in the wakeful state essentially on logics in real time, and minimally in sleeping, mainly on affects and in time titan.

The will, from the actual neurobiological evidences and the evolutionary documents in hands, seems to have been psychologically elaborated from the autonomous archaic autocrat. Psychognostic mental interpretation of holistic semion-moveon significances are essentially synthetic functions generally handled by the right hemisphere, but the executive manifestation of motor type response is generally controlled by the left hemisphere's frontal lobe. Endocast findings of the Broca's motor language center in Homo erectus in the global motor area of the left frontal lobe seems suggesting some specific motor refined act satisfying the holistic semion-moveon connecion to have been represented initially in this region. This anatomic and functional evidence in reality reflects the autocrat's expressivity manifestation in refined motor responses naturally confined in the motor area of the frontal lobe to become motor language center of the future. Furthermore, it also evidences the refinement of motor responses evolved from the archaic sensing-reacting to percepting-responding and to autonomous willful action. The foundation of the will is evident in this course of action by the bioforce effect in the autocrat and its biological evolution, transcending the auto-hetero-gnosis to secure its recognition by the conscious human self. This is self's own *ipso facto cognito* property oblivious of its real origin.

More precision related to self and will, along this line of thought, can be brought from other sources of information with combination of synthetic holistic right hemispheric functions and related left analytical hemispheric work performed simultaneously. One such example is the functioning of the surgically divided hemispheres for treatment of epilepsy in otherwise normal humans as was outlined in chapter 3 with the patient who could play his own left and right hemispheres on a mutual guessing game, showing perfect integrity of his holistic right hemisphere opposing the analytical left hemispheric action. This patient's self and will were authentically unchanged and remained the same as before the operation, explaining intact independent reciprocity of both hemispheres but connected through relays with the limbic system. Further examples could be cited in analytically synthetic outcomes with scientific constructions, poetry, graphic art, and music for example that except for possible memory deficits or neuromotor inconsistency occasionally indicating differences in performance, the constancy of self related to will is invariably secured. In these instances, concentrating willfully on an artistic production seems to bring out again resources from autos and the unfathomable depth of psychognosis that would not be possible without solid self-will identity. In this consideration, a not well-known but potentially independent power of the mind originating by ego action and the manifest autonomy of autos seem to coact through selfs and express a kinetic leap from the potential capability that is offered to autos to achieve excellence in artistic realization.

The matching of icons and times in dreams included in theoretical possibilities for logical or illogical combinations of content frames versus time frames now can be explained by the presence or absence of the autos-self bridging of time that is seen in actual observation. The bridging seems to be more often absent in REM sleep of deeper type as opposed to more often present in non-REM of less deeper sleep. The amount of neural energy spent in dreaming should be theoretically limited to the Libet time equivalent for a dreaming scene as no directive autos-self axis is used like in real time wakeful states. The fundamental life commandment application in mental work with pleasure-pain principle and the basic two-frame formulas used by semions and moveons can all be holistically handled by a known minimal constant quantum of neurobiological energy spent for each logical end of attention-intention work in psychognostic and idiognostic consciousness hypothetically equal to the Libet's time constant energy equivalent. In both psychognosis and idiognosis, the basic unchangeable quantum of energy to be spent to trigger the neurodynamic process of an attention-intention type should be the same energy equivalent of the 500 milliseconds Libet's constant, with the added 200 milliseconds for the worded expressivity in idiognosis.

In neurodynamic conscious domain, the basic energy spent should be the psychognostic time equivalent, realizing the achieved somatic steps with a sense of satisfaction or elation (pleasure-pain principle), about equal in duration to the shortest time for the two-frame model of life commandment to be completed. In consciously estimable scale, two practical situations can show examples mimicking either the short-lived intense suffering from discontinuous pains of neuralgic type or the enhanced psychedelic pleasure of the orgasmic time, the second example being seemingly in solid relation with the logic of SR.

GENESIS OF IDIOGNOSIS

Did idiognosis really start by a special mental capacity independent from psychognosis is a natural question for many who consider the talking by humans as an absolute sign of not being animals? Regardless of this question's absurdity, it deserves the clear NO answer for the record.

Before language to start and verbal idioms of any kind to appear, the basic pattern of perception—attention and attention-intention were solidly formed with the early concepts of meanings attached to actions taking form in the mind as early iconically based mental idioms. The meaning of the archaic mental images, based on cause-effect mechanism, was certainly noted and recorded, though in bioconscious or psychognostic rather than in any more elaborate abstract form. The primitive concepts of actions thus formed were most likely from the more frequently observed events, and more importantly, from their stronger explicit image memories with stronger implicit emotional contents. Therefore, whatever communicative idiom could have been formed and used, it would have basically represented a causality event from an iconic frame together with a time frame always as related to the egocentrist autos. It can be presumed, naturally, that the content frame expressed the most significant part, the emotional meaning of the action observed, but the time frame the least significant one, the common always present time of the inner time titan, the psychognostic present tense. Thus, theoretically, we can presume that the actual product, which was the primitive concept of an action taking place, was formed by the working memory and both implicit memory and the early forming explicit memory, matching the old archaic life commandment of logic and happiness. This would be in reality an actual idiom, but in terms of mental representation, it would be exemplified by the semion-moveon models primarily without vocal worded equivalence.

So idiognosis was in reality psychognosis exteriorized from the inner permanent time titan frame constantly present, to an outer time frame of inconstant present time, a time that was no longer permanent and interminable. This exteriorization naturally served to firmly establish the self, initiated in auto-hetero-gnosis and the meaning of changing of outside phenomena to it. With idiognosis, the nucleus of idiomatic thinking, the self, was exteriorized early in the archaic iconic symbolic language and with it most every closely related item to it to be phonetically expressed in the monodimensional present time (check the word *mama* in the last section of chapter 11 for further evidence). The question of self or self(s) was discussed in the third chapter in the section of *Situs Solivagus* primarily trying to use *Solivagus* singular identity matching the autos. However, the essence of self to be identified must be based on its exteriorization from autos, and this can only be understood with the idiognosis along with and because of psychognosis.

The exteriorization process included another equally important change in the frames of percepts. This change occurred in both the content frames and the time frames. It gradually altered the emotional meaning of the content frame bringing the meaning out of its bona fide natural emotional form into a structured form by mental creative processes, being artificially colored, augmented, reduced, or even zeroed, like in social dealing norms or in art and science artifacts. The changing indeed not only altered the content frame with its emotional ingredient being changed but also the time frame from the titanic permanently present time to the real ephemeral present time and potentially to both past and future times. The time frames could now have a significant enhancing priority role in items of artificially constructed elements, in languages, scientific symbolism, artistic creations, and in society rules and civilized behavior. These changes took thousands of years to give us now the impression of the consciousness being evaluated differently in emotions and feelings as compared to its application in scientific fields. The symbiosis of the emotional and nonemotional human animal submitting to the rules of biocracy, struggling a way out for also ultimate happiness in material life, or with artistic or scientific achievements, is the product of the long history of paleognostic to neognostic change. The final exteriorization of the natural psychognosis in its artificial byproduct which is idiognosis is but a useful complementary addition.

FROM ART TO SCIENCE

The very initial transposition from psychognosis to idiognosis in mental ideation and external representation was a real transition over a long period of time. It took place initially in figurative expressions using bodily means of gesticulation, grimacing, vocalization with primitive rhythmic singing and drumming expressions and possibly even drawing and painting art. In this connection, the discovery by Barham of more than three hundred fragments of pigment in a cave in Zambia, believed to be between 350,000 to 400,000 years old[14] and highly suggestive of body coloring or cave coloring for ritual purposes or both attaches particular significance to the expressivity trait dominating the eventual use of verbal language. The rationale behind this assumption is twofold. In the first place, the simplest form of a concept in the mind using the two-frame formula for symbolic expression required a content frame, which could include bodily signaling, vocal signaling, drawing signs, or all combined. Naturally, concept of an icon in the visual reality or visional imagination could be strong enough because of its meaning, to be exteriorized for example as a drawing more impressive with colors like in ancient cave drawings. Such exteriorization by the actual psychoneuromotor coordination resulting in bodily message enhancement with colors, or as an art product, is actually the simplest showing of an idiomatic idea with holistic structure and meaning in the present time of the artist's autos initially, transposable into the present time of self by recall.

In this type of configuration reproducing exteriorization from psychognosis to early idiognosis in figurative form, the outstanding finding is naturally the absence of any time frame in any form. Later, when symbolic figures establish and simple figurative idioms most certainly representative of mental images start to show constant realities like sun, sky, moon, animal types, etc., naturally there is yet no time concretization except the tacit present time, as the mental life is likely running in the monodimensional time of the all inclusive present. In fact, we only witness time representation in symbolic form when it is linguistically clearly identified by the proper idioms for it. The earliest such evidence in record may be a scored piece of bone interpreted as indicating the lunar cycle, dated to forty thousand years BC.[16] Thus, idiognosis in reality started in the psychognosis and was exteriorized first in the wakeful present time permanency to be later defined with idiomatic frames of its content that remained longtime in that *monodimensional present time*. Differentiated past time from the present, only became possible by special recognition for the time frames, not so much because of the figures in the associated content frames than because of the figures' affective importance and by memory differentiation from pure implicit to explicit type. Recalling coded contents by language helped abstraction, which allowed separate segments of the time frames to be localized consecutively. This whole system evidently worked effectively based on semion and moveon in their simplified mechanisms.

In relation with the time governing the actual production of the visual art as opposed to auditory art, a point to be empahasized is the fact that the visual art production enjoys continuity of time despite interruptions in the production completion, but not the auditory art. It does not suffer from other events coming in contact with the self, causing interruption of the self from its motivation behind the production, and therefore is a ground for permitting additional idiognostic presentations such as different figurative items and signs to be included with the changing time. In other words, visual art by virtue of its extendable production time separated itself from pure psychognostic ground and became a lasting means of reporting early concepts eventually leading to scripture. But auditory art, in singing and drumming for instance, being strictly time dependent could only express immediate psychognostic affective vocalization and remained most holistically fidel to the psychognostic affective expressions.

Painting, in fact, is actuating an extrapolation from an external visual perception or from an internal visional one by fine psychomotor coordination producing an external graphic expression that usually satisfies the self and possibly others as well. The stimulus is the most affectively significant perceptual or imagined figure model that serves psychologically the commandment of logic and happiness triggering the semion-moveon neurodynamic mechanistic system and extending it. Masterly skill in copying or original production will further add to the positive feedback in skill perfection and in reinforcing the logic and happiness commandment. If the stimulus is internal, in an attention-intention act of deciding to exteriorize a concept in visual art, the entire process repeats itself in sequence similarly with a content that mainly exposes the artist's emotional charge. This is in essence the initiation of the idiognositic mechanism, which can go far from its basic psychognostic ground and produce a host of imaginable visual idioms, as well as opening a vast horizon of communication with figures.

Creation of art in reality is the clearest example of harmonious teaming of the psychognosis and idiognosis based solidly on the life commandment and the stereotypic semion-moveon mechanism. The nucleus of graphic expression in fact is probably as old as the *Homo habilis* and his craft. The answer to the old question of what beauty should be and how should it be defined, for instance, can be found in how closely the life commandment is followed in the production of the artistic work or the view on the beauty as nuanced in psychognostic to idiognostic cultures and through the ages.[17] Here we find the initiation of conventional factors in influencing the production of an idiom that is an example of one fundamental and basic psychognostic origin in expressive productions.

We can witness the evolutionary progress of artistic performances in *Homo sapiens* in examining the course of the visual and auditory-vocal arts. In such an examination, we may surprisingly find some constancy in the fundamental foundation of the musical art in terms of rhythmicity and vocalization, which, in spite of all changes occurring through the millennia of time, has remained basically the same in its emotionally endowed and generally more pleasant and enjoyable character to the average listener. Recent studies on diffusion tensor imaging (DTI) also support stronger preferential myelinization of conductive auditory pathways in the brain.[18] This emotionally oriented aspect of the auditory art, possibly with some cultural ritual contents reflecting totemic archetypal meaning, like special ceremonial rituals with singing, is the most that has remained or can be inferred from analogies observable in some archaic tribal rituals today. On the other hand, in visual arts only a small number of our ancestral heritage is saved intact to reveal the passage from psychognosis to idiognosis in clear form in archaic prehistoric painting.[19] The bulk of evidence seems to be with the hieroglyphic figures that typify the initiation of natural transitional emotive passage of psychognostic beliefs from full morphism to gradual amorphism in letters in the history of pictographic languages.

The transfer from psychognosis to idiognosis in its most significant presentation in today's form could have caused a loss of emotional meaning of the life, artificially splitting the logic from the happiness principle in the life commandment and developing almost an independent logical realm, were it not for the inner anticipatory sense for expecting happiness to come as an ultimate gain in completing even the pure logical work. The change of course has been more significant in the visual than in the auditory life. Indeed, both visual and auditory functions, as already mentioned, have essentially two sets of values for both relaying information and inducing sensation that can be, but are not, necessarily equal. Either art form can theoretically have greater or lesser emotion inducing effect, pleasing to displeasing. However, the auditory vocal or musical perceptions (or worded expressions), are predominantly emotion prone and emotion producing. It is impossible in fact to separate logic and happiness in the auditory art but only partially with vocal lyric expression and the meaning it can give with the communicated words to musical themes. However, the content of a visual art can induce a forced experience by an enormous emotional visual effect, or none if such effect is totally absent, and without any language being used. Furthermore, auditory arts impress by a series of consecutive time frames bringing a full emotional meaning in chain, but the visual art is one single time frame with one single unchangeable emotional effect, like some artist's bizarre dream paintings,[20] or a totally nonemotional symbolic content. The first one impresses by successive chain content presentation with emotional meaning blank to vision, the second is one single visual frame that can become blank to emotion. The first one necessitates multiple frames in continuity to produce the emotional exteriorization and cannot be segmental without losing its emotional charge, but the second one can be as simple as a single frame that can change from full to empty emotional meaning. The musical art, if accompanied by lyrics, can add tremendously its emotional harmony. Visual art, in contrast, is devoid of lyrics and human voice and can easily contain emotionally empty symbols, and this is how it originated the basis for idiognostic noetic idioms in the fundamentally emotive psychognosis, ending in abstract scientific symbols and signs.

This difference in the expression exteriorization of the idiognosis from psychognosis in art, in reality the origin of noetic idiognosis, has been gradual through the ages, and the transition has witnessed a progressive gain in idiognostic nonemotional character with a loss in psychognostic emotional aspect of idiomatic items. As this is in fact the reality and as the idiognostic symbolization can reach near purity status of emotionally neutral scientific idioms being routinely used, the life commandment of logic and happiness has lost its happy significant half with advancing scientific cultures forcing greater times to be spent with emotionally empty mental work. Unless a rewarding goal is envisaged in the scientific sense of satisfaction, naturally no mind would voluntarily undertake an emotionally empty task.

Craving for joyful events and all pleasure-seeking activities, no matter at what price and what consequences in the emotionally buoyant youth, for example, often without consideration to ps and pi discussed in chapter 1, is not infrequent in human biocracy against which imposed codes in idiognosis of the society, the social consciousness, is often powerless.

WHAT MADE THE BIOCODES

When we reflect on the whole subject of creation, life, evolution, mind, and consciousness, we end in realizing that in so many unknown facts related to all of these realities, only one is evident and that is a continuity assuring a bioactive force going on with the assumed infinite time. But even earlier than this bioactive force continuity as a pristine code is the energy-matter relation in which is the prime fundamental coded reality. The biocode modality of that fundamental immortality principle in the gene, DNA and its relentless evolving course seems to be as impervious in its relayed continuation in time as ever, and in this, it is the coding process that is pivotal. What indeed made this coding? My answer to this question here is exactly the same as in concluding the thought experiment described in chapter 2: bioforce with its reciprocal relation of the material and nonmaterial components as a model. This coding, in fact, is a real biocode; it is in essence similar in every nonbiological and biological element, from electromagnetic radiations, photons and ions, to DNA and the whole animal life series. The word *biocode* is self-explanatory and serves to define what is needed to know about the steps of biological submolecular, molecular, enzymatic, cellular processes, and the greater scale physiologic reactions altogether. These functions of natural biological processes are regulated by the natural biocodes replicated from the bioforce model. In fact, the reciprocity of the material to nonmaterial components of the bioforce is evidenced in the bipartite sequence of cause and effect frames in all biology. These natural biocodes reflect not only natural expressivity inherent in emotively based psychognostic semions-moveons holistic concepts but do the same in their subsequent literary language codings. They would expand to include memes and will ultimately lead to infinite abstraction creation of the immaterial biology.

Conscious biocodes, expected in exteriorization of idiognosis from psychognosis in examples of artistic creations, would naturally express the holistic meaning of the content of frames of images in psychognosis related to the time frame, which is the uniform infinite time titan. As such, the biocode is in its natural form, expressed by the action of self, in a tangible visual art reproduction or a musical tone, as examples that would become idiognostically labeled and reproduced by explicit memory, making conscious holistic biocodes in extended forms with motor or phonetic (gesticulation or voicing) or both expressivities.

Explicit conscious biocodes were needed essentially for two reasons in the conscious biopsychons' lives. The first being the need for the self to be able to recall mental images by a tag on them more easily as single analytical parts rather than as holistic iconic figures and to communicate internally with one self for adjusting other mental icons in concepts and inferences comparatively. The second is to communicate with other conscious minds using the same biocodes. These biocodes, formed already internally by psychognosis, were exteriorized by visible signs and vocalizations in various idiomorphic presentations exteriorized from the psychognostic speechless mind. I will expand on this aspect of idiognosis in the next two chapters dealing more extensively with language. In this chapter, I still have to discuss the basic principles involved in the production and use of mental idioms and concepts to be processed in consciousness.

To itemize idioms, concepts, and thoughts and to bring them from their psychognostic foundation out into idiognosis in conscious order, mental mechanisms activated by self operate in two parallel lines of conduct. In the first one, the general principles of the life commandment, innate in psychognosis, are used as the main lines of the protocol borrowed from autos by the self, and in the second, attention-intention direction line is used for the transfer initiated in psychognosis to be evidenced in idiognosis consciously with a constant transfer delay of the Libet constant time between autos and self. This delay from psychognosis to idiognosis that is a natural part of the operation and uses psychognostic time titan is yet counted as conscious because of the conscious will being counted as part of the self, although self is operating in the actual real time and not in the infinite time titan. Both lines of conducts have been reviewed earlier, but the second, the attention-intension one, needs more elucidation.

FROM CONCRETE TO ABSTRACT

Ideation can be thought of as simple conceptualization that can expand further and, therefore, can be regarded as a chain of processes that must have a starting point, a continuing length, and an ending point. In theory, the simplest idea that must have a meaning by definition would be made of either a solo single itemized concept forming a static idea like any single name or a dynamic idea like expressed in a verb or a function word or a minimum of two such single concepts demonstrating changing from one to the other or both unified with indicated change implying time effect. Again, in theory, the starting unit and the ending one could be similar or dissimilar, but by way of logical rules, an idea in the mind is based on the meaning it generates, or the meaning that generates it. In either equally agreeable definition, what the initiation is based on cannot lead to an extension without change and cannot complete the ideation by itself alone except in a static single idiom or symbol producing a simple solo concept of concrete or abstract static nature. Thus, in essence, realization of the single static idiom represents constancy in time, a static event, an indication of present tense. Therefore, the principle in conceptualization leading to ideas dictates that concepts are static or dynamic in relation to time.

1. In the single static concept, single words (names, pronouns, adjectives) remain unchanged in the monodimensional present time.
2. In the single dynamic concept (verbs, function words), dynamism is inherent in the word expressing the concept, not without an inclusive tacit time representative for the change, which implies tacit subject, object, or both.
3. In more than single concept idea, the starting conceptual element (usually a name) needs a minimum of another element (name, adjective, verb, or function word) related but distinct in content meaning to induce dynamism in idea formation either remaining in monodimensional present time or extending the time dimension.
4. In all such concepts forming a dynamic idea, ideation process may require a minimum of one relay element between the items to represent dynamism or time change (verbs or function words).

The mental principles for coding to form ideas presented on these foundations can be immediately noted to be the same familiar principles that govern in life in the two-frame structural basis of code-forming life commandment and semion and moveon formation. Indeed, coding the content frames of meaning or the time frames of change and reaching the results by the mental processes constantly follows the same principle. The coding rules in idiognosis are in essence the same rules used from paleognosis through psychognosis that have provided the biological and physiologicl scale processes for ages in accordance with the biocratic life commandment principles. There, nothing more than the concrete biological elements formed the coded idioms themselves and in the permanency of the constant present monodimensional unchangeable time titan. But now in the idiognosis, mental tags must be applied to any element to be exteriorized from the psychognostic realm or imagined anew, including the time changing factor, in order to be mentally manipulated at will and by simple recall. The processes here take the form of attention, evoked or willed, for single constant adynamic monodimensional conceptual ideas, or attention-intention, again evoked or willed, for ideas of two or more conceptual elements with evident dynamic changing from monodimensional to extended pluridimensional time.

As an example, when a name is heard, attention is evoked by the auditory perception and may or may not be followed by an intention to an action to be performed, and the process is of course conscious, realized in the mind with the actual first frame of the two-frame model, to be either followed by a second frame indicating an action or no further frame if no action is implied. The first frame contains the name in the present time of perception, with an expected second frame for time to follow that just shows statism for the name as static. The first frame is in reality followed with a second frame presenting no change (empty frame or no frame). If attention is willfully applied to simply remember an appointment for instance, the conscious recall with an inapparent intention leading to a mental work that is just memory reinforcement is a dynamic change in an extendable present time with or without executing the intention. The mental reconstruction if succeeds is similar to the first example except that intention causes dynamism implying representation by a second frame for the time used to exteriorize the name from memory into actual conscious present time. If attention is followed by an apparent intention, for instance to change an appointment, the entire process is evidently conscious as a completed mental work and is again in the extendable present time. Realization in the mind includes both first and second frames for the content, the appointment, and the changing to be made respectively. This basic conceptualization using the two-frame model naturally includes either the tacit statism of the content of the first frame revealed unintentionally with no second frame to follow or intentionally, calling for a rechek, necessitating the second frame actualizing conscious idiognosis regardless of motor execution for any change, and the process is completed with the completion of the second frame. If attention is focused to call a name, to calculate a small number, or to look at a target for more detail, in short for a mental or physical single aim, dynamic changing in the present extendable time is confined in the second frame in which the time extensibility is to the limit of the working memory. In all these examples, dynamism effectuated through attention-intention process is framed in the time frame theoretically from potential to real completion of a task, in static perception-comprehension only or in dynamic attention-intention-completion engaging the expressivity trait using NET explained earlier. Triggering concepts fill the first frames, dynamic time change the second frames, and the tacit present time, the psychognostic time titan, is exteriorized with the Libet constant time plus the idiognostic addition as explained. These examples reflect the situation with the inapparent subjective self as the agent whose identity is not expressed in any frame as being ipso facto part of autognosis. But we understand that an objectivated heterognostic subject will occupy the first frame when self is exposing heterognostic dynamism through its autognostic expression.

Similar examples could show that any single conscious attention can operate alone or can be followed with an intention to form single or chain succession of ideas, presenting time dynamically extendable to the permissible limits of single or sequenced working memory according to the mental task planed.

In these examples, the process is a simple basic feedback from a stimulus leading to a change with a final equilibration by mental energy spent to eliminate the stimulus and reach stability. This is precisely the model of semion-moveon as symbolic unities used in psychognosis without words, and basically similarly in idiognosis with words. In all examples of willful attention followed by an intention to act and the actual completion of the action, the same principle is governing and the chain of the ideation may use more than a single action, but following the same protocol to reestablish mental equilibrium. All instances of the attention only processes are of the type described in the experiments of Libet provoking evoked potential (EP), and all examples of attention-intention in achieving an evident physical or mental work are those characterized with readiness potential (RP) as remarked earlier. Finally all these processes, commonly interpreted as conscious in idiognosis, in reality start unconsciously on the semion-moveon principle in psychognosis with Libet constant time to become exteriorized in words in idiognosis. Recent ideas gained from advanced imaging of cerebral cortical activities in conscious states indicate all these possible mechanisms, engaging groups of pyramidal cortical cells in large networks of millions of cells, able to change patterns of assembling coalitions according to mental tasks performed, and in millisecond scales.[21]

In conclusion, all idiognostic ideations with coding providing the language in final analysis are not possible without the psychognostic ground rooted in the psychognostic speechless mind and without the psychognostic initiation in the time titan with the known time delay to idiognosis. The mental work using these biocodes is firmly dependent on this time portion of Libet's constant that must belong to either the conscious real time, which cannot be possible according to the experimentally tested evidence, or rather to the infinite time titan of autos which seems to justify the appropriation by inference deduction shown with the additional 200 milliseconds time for self-realization that follows.

In the next chapter, I will extend the presentation of idiognostic coding in relation to the content frames and their meanings in ideation in schematic figurative formulation and in conformity with the commandment of logic and happiness, using the pragmatic forms of the attention-intention concept.

REFERENCES FOR CHAPTER EIGHT

1. Flanagan O. 1999. Consciousness. In: *A companion to Cognitive Science*. Bechtel w, Graham G. Eds. Pp 176-185.
2. Mithen S. *The Prehistory of the Mind. The Cognitive origins of Art and Science*. P110. Thames and Hudson.
3. Penfield W, Roberts L. 1959. *Speech and Brain Mechanisms*. Princeton University Press.
4. Jasper H, Bertrand G. 1966. Recording with microelectrode in stereotaxic surgery for Parkinson's disease. *Journal of Neurosurgery*. 24:219-224.
5. Libet B, Alberts WW, Wright Jr EW, Feinstein B. 1967. Responses of Human Somatosensory Cortex to Stimuli below Thresholds of Conscious Sensation. *Science* 158:1597-1600.
6. Libet B, Gleason CA, Wright EW, Pearl DK. 1983. Time of Conscious Intention to Act in Relation to Onset of Cerebral Activity (Readiness Potential): The Unconscious Initiation of A Freely Voluntary Act. *Brain*. 106:623-42.
7. Haggard P, Newman C, Magno E. 1999. On the Perceived Time of Voluntary Actions. *British J Physiology*. 90:291-303.
8. Libet B. 1999. Do we Have Free Will? *J Consciousness Studies*. 6(8-9), 47-55.
9. Libet B. 2004. *Mind Time. The Temporal Factor in Consciousness*. Harvard University Press. Cambridge, Massachusetts, London, England.
10. Jayne J. 1990. *The Origin of Consciousness in the Break-Down of the Bicameral Mind*. Houghton Mufflin Company. Boston.
11. Zeman A. 2002. *Consciousness, A User's Guide*. P 294. Yale University Press.
12. Feinberg TE. 2004. Not What, But Where Is Your "Self"? *Cerebrum*. 6(3):49-62.
13. Newberg A, D'Aquili E, Rause W. 2002. *Why God Won't Go Away*. Ballantine Books. New York.
14. Greenspan SI, Shanker SG. 2004. *The First Idea. How Symbols, Language, and Intelligence Evolved from our Primate Ancestors to Modern Humans*. P 159. Da Kapa Press. Perseus Books Group.
15. Lemonic MD. 1995. Stone Age Bombshell. *Time*. 145(25) 49.
16. Kunzig R. 1999. A Tale of two Obsessed Archeologists, one Ancient City, and Nagging Doubt About Whether Science Can Ever Hope to Reveal The Past. *Discover*. P 84-91.
17. Heath R. 2001. *Sun, Moon, and Earth*. Wooden Books. Walker and company, New York.
18. Fields D. 2008. White Matter Matters. *Scientific American*. 298(3):54-61.
19. Romm S. 1992. *The Changing Face of Beauty*. Mosby Year Book.

20. Geertz H. 1995. *Images of Power*. Balinese Paintings Made for George Bateson and Margaret Meads. University of Hawaii Press, Honolulu.
21. Greenfield S. 2007. How Does Consciousness Happen? *Scientific American*. 297(4): 76-83.

CHAPTER NINE

TRANSFER TO INFINITY

THE LIMITLESS SYMBIONT

In this chapter, I will expose and discuss the foundations of idiognosis and its expansion in relation to the symbiont it forms, the basic ideations in philosophical meaning it expresses, the scope of conceptual meaning it can encompass, and finally, I will sketch the ground possibilities for word formation and noetic development to follow. These generalities are unavoidable now in order to grasp a firm understanding of reasons for the language to appear when we reach the actual mechanisms of word formation in the next chapter.

The first chapter title in this book, "Bioforce in Symbiogenesis," and especially the section title "Symbiosis in Destiny" meant to emphasize, crystal clear, the universality of the bioforce reflected in two different worlds, physics and biology, uniting them in symbiosis in the human mind by creating the mental life. The same symbiotic meaning can be applied in much closer analogy to idiognostic creation of concepts and ideas of linguistically coded category along with "The Psychognostic Speechless Mind," titling chapter 7. These two modalities of expressivity, being basically of one and the same essence, are bona fide biological entities and share one foundation. Thus, "Bioforce in Symbiogenesis" seems to have progressed toward facilitating unification of the body and soul in the course of evolution.

In the first chapter, I was considering the physical world versus the biological world: the environmental effects to autos from all nonautos entities. In psychognosis, these effects quantitatively and qualitatively could be synthetically identified and interpreted, and in idiognosis, they could be analytically identified and phonetically coded for recalling. They would make the stimuli for originating the mental ideas. It is true that both psychognosis and idiognosis contribute to make up the mind content, the entity that can be referred to as the symbiont that is indeed the created unlimited mental world. This world is yet never completely expressible by idiognosis alone. Already in psychognosis and in spite of the near complete discerning of the exogenous nonautos world by semion and moveon intervention, there is unquestionable evidence for the limitedness of the holistic procedures with symbolic presentation to grant full expression. We have to accept that all holistic semantic conceptualization by semions and moveons gives crude meanings in psychognosis at the prelingual coded stage. By transfer extensions from synthetic to analytic means of interpretation, metaphorization of various types, and possible new formatting mechanisms with words and syntax, comparatively significant gain will ensue in idiognosis. However, neither the initial conceptually formed holistic meaning in psychognosis nor their subsequent naturally coded linguistic forms with all ultimate changes in the worded language of idiognosis can alone assure perfect expressivity due to the unlimitedness of the mind's nature. The reason resides in the affective meaningful content of the mind and its dynamic reality with the life commandment acting on every perception. Considering this togetherness, the symbiont that is shaped by both psychognosis and idiognosis is the idiognostically worded world that can present items of imagination on the basic models but in free abstraction. This powered idiognosis can theoretically expand the horizons of the living mental ideas and their metamorphosis beyond imaginable frontiers because it disposes of both iconic psychognostic and worded idiognostic identities for its constituents and with added power of abstraction extending into the world of metaphysics. In this capacity, an interchange between psychognosis and idiognosis can operate indefinitely, disguised in immaterial biological substrate of words and both natural and conatural items of emotional and noetic types can contribute to the richness of mental ideation more and more approaching biological immateriality.

The linguistically constituted idiognosis is, in fact, a world of potential infinity expanding into fields of unlimited scope and variations. It is not just a separate mental living as a symbiont, but a source made of wondrous self-replicable abundance of varieties of cultural products. To give a brief focusing overview, the symbiont is a metaphoric expression itself making other living metaphors by extension assembling to form mentation, depending on psychognostic frame rule but acting with a free will. Mentation, in this respect, is an all-inclusive ability of meaningful visualization of concept forming ideas, either concrete and quite clear and precise or abstract and somewhat imprecise, working as one functional organ making the body and the soul of the symbiont. This functional organ evolves from the simple sensing to perception, comprehension, and interpretation finally forming the semion and moveon models for future worded ideas to be constructed. Words could be regarded to constitute the consumable elements for this organ to function holistically and maintain the system. The energy itself is the initial neurodynamic energy spent to actualize the meaning in each word as the result of expressivity by Navand, and the word itself serves as the carrier of the energy (audible or visible form) called Namaa in the Navand theory of languages defined in chapter 7, which is to be discussed further. Namaa is the only perceptible (visible, audible, or otherwise sensed) concrete element of the entity named, either naming the icon or making the lexical form. In this process, the chain of Navand-Energy-Time (NET) as one functional quantum of energy used in one attention-intention neurodynamic process is crucial in triggering the expressivity process. The neuroperceptive-psychointerpretive mental systems follow the proper chains of procedures to potential limitless expansions beyond concrete limits into indefinitely formulated abstractions beyond the realms of matter and energy. This leap from the material to the ultimate immaterial biology, we must remember, started in reality in the pristine autonomy with the bioforce in its initial material form.

In philosophical sense, such possible expansion in reality relates the mind of the psychognostic autos and the mentation of the idiognostic self escaping eventually all material boundaries reaching back to the immaterial origin after a material journey. In fact, psychognosis operating through holistic semion-moveon concepts reflects the axiomatic life commandment as an immutable model. This model, together with idiognosis introducing subunits in the forms of phonemes and morphemes enhancing flexibility with expansion of applicability for the expressivity trait in abstraction, opens the doors to a world of abstract infinity. Such change needed for extendable use of ideation, no doubt, could never have been achieved with the ideation remaining in the realm of psychognosis and limited to semion-moveon restricted extent. In fact, chain constitution of these models in psychognostic expressivity using bodily signs and vocalization together could sufficiently extend the comprehensibility of expressive evidences alone before the evolutive stage of toolmaking, but not with composite toolmaking and thereafter,[1,2] when more subtle worded indications were needed to perform fine work. In any event, regardless of when and how language initiated the extension in ideation by the words coming to code the archaic models of semions and moveons exceeded far beyond the psychognostic possibilities to further allow the nuancing variations in ideas to be established. This showed what did take place, rehearsed, and expanded ultimately beyond earlier limits.

Perhaps, we could continue calling this state of idiognostic affairs simply consciousness as we have been doing in cognitive sciences, but evidently, consciousness encompasses the entire scope from the pristine autonomy in the bioforce to the holognosis in the metaphysics and is not only idiognosis. When faced with different aspects of consciousness that have been lacking precision in our expressions, we are not able to answer all questions. These aspects of conscious states, suspected and witnessed in animals or evident in ourselves, like nostalgic qualia of some sort, realization of some symbolic understanding psychologically but vague, or apparently clear concepts in metaphoric terms, though evidencing conscious character, are not equally conscious in degree of expressivity. Even using elaborate language and extensive wording varieties, all we achieve is sometimes an approximation to what our conscious states really tend to tell us. We create words for descriptions, but we cannot see our conscious states crystal clear in those words, and we naturally satisfy ourselves with the shadowy image we can makeup of our own inner conscious feeling and knowledge. In fact, idiognostic mentation can expand unlimitedly in unrealities more on apparent logical relation of abstract concept but cannot replace the psychognostic mind with the realities related to autos and emotions. Keeping the words *conscious* and *consciousness* for what they can mean, but recognizing that what we are doing is an impossible exact labeling to show the truth is the very indication that we are constantly missing our deep truth. This hidden truth of an affect that is never exactly what the word says, or like the concept of infinity that even the most clear mathematical idiognostic coding cannot bring to integrity,[3] relates to our nucleus of bioforce in our autos that hides the uncertainty principle. Reflecting further on our inner undefinable truth, thanks to our idiognostic ability, we should realize that there can be also truth in realities beyond ours that may appear irrealities.

Neurobiological development with the changes leading to idiognosis has served somatic biological processes in parallel along their evolving structural anatomical and physiological specifications. This parallelism in function is constant and has been so throughout the evolution in biognosis from paleognosis to neognosis. In idiognosis, this constancy remains, applied in phonetically created codes for concrete and abstract mental entities to produce a new life, an exteriorized artificial living, in appearance seemingly out of pure biology and beyond somatic needs and possibilities. This occasioned the inappropriate inference that has induced dualists to propose the erroneous dualistic theory. In reality, idiognosis is an extension of psychognosis and does not justify any dualism except in extended function. In fact, evolutive adaptation has permitted neurobiology to allow the self to see self items in exteriorized entities, making new artificial worlds of abstracts in philosophy and metaphysics: a situation akin to the metaphoric example in chapter 4 of the two mirrors facing one another, showing unlimited reflections as one reality, but in extended scope of pure inconcrete beings beyond reality.

Psychognostic ideas are in essence stereotyped, making holistic complexes more constant in form, focused and limited in expansion. They reveal much affective charge as a whole action effect in general. However, idiognostic ones, based on words for symbols that present a nucleus of more or less precise subtle meaning but a connective halo of other acceptable inductive meanings according to their types eventually make an exponential expansion of ideation chains with necessarily less dominant affectivity and of relative exactness only. Such idea types are mainly exemplified in description of feeling, abstractive interpretation, literary work, artistic expressions, and finally in generalization and metaphysical domain. This aspect of the ideation justifies more examination and scrutiny.

In order to organize the study of the symbiont making the idiognostic world, the two components of such study namely the material (constituent building elements) and the building method (mechanisms of construction) need examination and explanation that ideally should be handled clearly separately. But this subject of conventionally regarded preconscious as against conscious and more exactly named psychognosis in contrast but leading closely to idiognosis concerns iconic matter versus worded matter with embedded functions rendering distinct separation of material from methods impractical. Discussions concerning the change from psychognosis to idiognosis as a whole blended process will follow in the following two chapters. Here the material and the method will be considered together but following reasonable distinct lines of explanation as much as possible.

IMAGES IN THE MIRROR OF MIND

The origin of ideas in the psychognostic mind started mainly with visual perception and followed with mental interpretation for realizing every external environmental figure and its spacetime identification, meaning, and attributable value in potential relation or actual action concerning the egocentrist autos. This basic idea formation was discussed in chapter 7. The chronologic gap from that psychognostic type of ideation to the very first worded declaration expressing a definite conceptual idea historically recorded is not known. What I have found to possibly represent the oldest referenced formulated major conceptual idea we can isolate from antiquity records in logical realization of belief with expectation of outcome in the human mind is the famous Zoroastrian axiom ***"righteousness leads to happiness,"*** discussed with the ego principles in chapter 4. From that time (second millennium BC) to the present, the subject of idea in the dawn of philosophy that should be legitimately called Zoroastrian, to Greek flourishing, and to the present time, has indeed changed face and has been expanded essentially from an initial subjectivist approach to an objectivist view and again to a subjectivist one, but with many metamorphosed intermediaries. Philosophical thinking goes as far as to include even the extreme nihilistic subjectivist possibility that nothing would exist outside of us, an absurd *ipse dixit* not in line with human good sense to say the least.

Regardless of these objectivist and subjectivist variations in views and contentions, it is to be pointed out that the image forming the idea has been the real subject of debates in all these variations and not so much the idea by itself. In other words, imagination regardless of its mechanistic mental realisation has been the primary object of debate. A detailed discussion of imagination, reviewing the background philosophical history of the subject from Platonic and Aristotelian time to the present, respecting the bond of imagination with idea by Johnson,[4] concludes to a theory of imagination that should be at least based on categorization, schemata, metaphorical projection, metonymy, and narrative structure. These prerequisites are considered fundamental as minimum necessary basic components to solid realization of imagination. But in my discussions exposing the visual perceptions and interpretations detailed in chapter 7, which ended in meaning presentation and symbolic concepts' structures in mental imagination, I passed over those steps more in biological than philosophical terms.

The philosophy on image interpretation and on the idea itself as image contingency with its variability according to conditional external and internal associated factors looks upon the problem from somewhat a different angle. From that angle, attributes of autos and then self in interpretation of images to become ideas may be different in terms of the external realities and internal meanings because of the autos and psychognostic interpretations being basically pure biological needs responded by pure affective reactions. But those of self with idiognosis, including aspects of literality in logic and judgmental understanding, go far beyond biological frontiers. This consideration is certainly valid, but only in secondary conditioned states consequent to literality expansions. In terms of the fundamental expressivity that must be manifested either psychognostically, idiognostically, or both in the interhuman relations, the image-idea axis for interpretation and action is precisely the same for both psychognosis and idiognosis and seems to follow a basic algorithmic pattern. Continuation of the analytical differentiation of the ideas from the holistic form down to atomic parcels, the phonemes and morphemes of spoken languages will build the world of words with their metamorphosis. This has become most evident in extended ideations and expressions in philosophy, and yet a philosophical salient debate of the image-idea relationship is not completely settled.

Whether the reciprocity between imagination and ideation in the mind is a full-scale infallible one or is fallible and leaves points of differences in degrees from one to the other in both ways is not of our concern at this time. It is clear that the basic expressivity from psychognostic to idiognostic models is genetic and, therefore, respects an allowance for adaptability on the genetic blueprint to produce the final outline,[5] certainly not excluding adaptation to new horizons opening by the language. The changing conditions in literary idiognostic expressions naturally add refinement to the basic psychognostic models keeping the foundational algorithm but may intercalate elements of idea shaping and extending the expressive possibilities and variations to extreme expressive abstractions extending into metaphysical architectural structuring.

An interesting new philosophical viewing that can fundamentally change ideation and abstract imagination reinforcing classical determinism in the production of idiognosis is the concept of anthropic principle.[6] This concept, as explained by Gardner, is centered on the physical constants and laws that appear to suggest a life-friendly cosmos from a mere compatibility to a natural contingency of life as part of cosmos, to a state of determinant functionality akin to a cosmic DNA engendering life cofunctionally through the friendly physical constants and laws. This strong level of assertion in the anthropic principle goes as far as to assure ontogeny for the life, accepting a blue print for life replication. Accepting such assertion naturally seems to solve many problems particularly that of the idiognosis providing for the symbiont and its immaterialism with unlimited abstraction as an ultimate fulfillment reinforcing the life extension in immaterial form. We will have the opportunity to reexamine such radicalisms in the light of further evidence in our last chapter of the book.

ELEMENTS OF EXPRESSIVITY

Expressivity, as I have said earlier, is a trait developed and refined through the evolution. Its nucleus is in the biocratic tropism and dynamism. It changes from the archaic reflective-reactive manifestation with changes that occur in the central nervous system development in primates to reach a well-defined mental expressivity, evidenced with global bodily manifestations in action and vocalization. Expressions by gesticulation or in voice changes, intonation, shouting, crying, etc., as well as in words and meaning, all work in harmony with physical bodily expressions. Mental expressivity is assured by constructive expressive formulas making the expressive ideas. Processes to exteriorize the expressions are affection-related and affection-directed throughout life. In humans, the archaic expressivity reaches the sophisticated level of presenting two types whose origins can be already detected in the primates. One is spontaneous, reactive, not under voluntarily aimed control, like all pillars of emotions that engender expressive behavior. The other is intentionally aimed expression to be distinguished as initiated intentionally and being voluntary, possibly controllable, but still affectively influenced, also detectable in primates but used in interhuman communications so clearly.

In the human mind, and as reflected in present languages, there are generally five types or categories of expressions that are exposable in interlocution, each being initiated by one primarily affective mental force (Navand, which originates the expression intended to reach an end). These are the five classes of the intentional ideations that make declarative, imperative, interrogative, exclamative, and nominative sentences. It is clear that the affective force behind these types of expressions is generic, as a general directional mental force voluntarily applied for an intention to be accomplished, which is under the Libet constancy time value that evidences in my interpretation the neurodynamic unitary factor NET explained in chapter 7. NET, in my belief, originates from autos and in the psychognostic speechless mind engenders full expressivity including the bodily language expressions and the voice together, showing the meaning of the attention-intention process by all demonstrable affective cues. In idiognosis, NET acts similarly but uses the affective meanings in words on the same identical models of semions and moveons as detailed in chapter 7. The main communicative interlocutionary five classes of expressive forms will be essential in indicating the intention of the expressions. These generic meaning formations giving the fundamental expressive sentences may also include more subtle specific affective senses, which will be included with qualifying indicator words without altering the main structural models. These secondary affective meanings can be included either frankly and directly by the true meaning of the qualifying word or in transposable meaning of words. For example, in the following sentences:

1. This is a beautiful spring day.
2. What a beautiful spring day!

The first sentence is declarative and confirms the being of *a beautiful spring day*, which is the meaning intended to be expressed in the declarative way, itself forced by the affective effect of the beauty imposing auto- and allo-confirmation by expression to complete the intentional process. The second is exclamative, but confirmatory. Both expressions confirm *a beautiful spring day*, but the second sentence has additional specific affective impression asserted in its different form. Surprisingly, the second sentence also has no linguistically defined verb *is* as is seen in the first sentence but includes the verb in meaning and expressing the idea even more impressively. This effect has been realized by a meaning transfer to the word *what* and with the addition of the exclamation sign (!), thus making a change from declarative to exclamative affection expression.

The mechanisms making meaning transfer possible use direct processes by the actual word meaning similarity or indirect ones transferring meaning to the extended present time in *is*, the time titan model in psychognosis and from there to other target words. In fact, the second sentence in the conventional complete form should be:

3. What a beautiful spring day is this!

In which the meaning still includes the *being* by *is* for *a beautiful spring day*, which *being*, the tacit *is* carries to *what* in sentence 3 where the self is addressing the exclamatory sentence to itself being aware of the tacit *is* that disappears in sentence 2. This example indicates that meaning transfer in the worded language can use the basic time titan as a normally inherent but inapparent transfer medium in *is* and also can be expected to expand with the increasing number of words using direct meaning-action unitary significance between words as in sentence 1 in direct way or when a change of intentionality in the sentence type is realized, making indirect transfer as in our example 2. Of these two mechanisms of transfer, the indirect time-related one seems to be the primordial type, which is operating in the psychognosis in the realms of autos but also usable in idiognosis. In fact, again, the intentionality change making the declarative sentence vary to exclamatory type is by the present time relay between *is* and *what*. The direct way by similarity of the worded meaning, on the other hand, seems to have become evident in idiognosis using words for time expression, although possibly rooted in fuzzy meaning and ideas in psychognosis.

It is important to realize that the origin of this possibility of meaning change in ideas that takes place in idiognosis and is demonstrable by the structure of sentences has been established in psychognosis. The basic two-frame semion and moveon models exemplified by the life commandment are in fact the same used in idiognosis. I have discussed this subject in chapter 7 where the unworded semions and moveons only show the holistic action in the present time titan as a unified abridged symbolic meaning about an act that is being done. Even there, each such symbolic meaning of the action being done is, in fact, representing a bona fide verb meaning that can be seen to achieve a result, which per se may be a single meaning, but as a performed action, it is applicable to multiple possible performers or agents, subjects or objects. This implies unity and constancy of the meaning for the action showing the constant result but encompassing variability of meaning for different subjects and objects. A scene of various predators (subjects) hunting and the outcome of the result in various hunted animals (objects) would be a typical example. Thus, connection of the meaning of one action to different subjects and objects is a natural semantic conditioning fact. If the action and the result can also include some variability in subtle ways in attacking and evading efforts for the chased animal, the observer's semantic interpretation is certainly not restrained to one constant frame and can interpret the relation of the constant action type to expand relative to subjects and objects allowing interchanging in the perceived meanings. So the normal ground for meaning transfer is experienced in a natural way and such transfer is plausibly operational by both direct and indirect ways of transference (active and passive verb ideas) earlier outlined. Each such experience is also recorded in the present tense, the present time titan from biognosis through psychognosis and idiognosis, using the natural connectivity provided by the permanent present time. Thus, metaphoric mechanisms can be said to have originated, and tacit metaphor by implicit meaning have existed before worded expressions to mean passive and active verb forms.

Such tacit metaphors in psychognosis are real and are detectable in treacherous acting of primates, in vervet monkeys for example with specific cries for specific alarming for predator sighted that sometimes are used to send away others from located food sources. The treacherous acts are a psychognostic manipulation, implying production and assertion of untrue meaning and forcing the false comprehension of that meaning in others by using the expressive sign or the phonation for that meaning. It is basically a process breaking the bond between the spacetime reality and its normally related component meanings. This particularity of extreme meaning change, as well as the singularity noted in spoken jokes, is metaphorically originated already in psychognosis. In essence, the nucleus of a symbolic idea in its holistic type or in its modified, subdivided, detailed idiognostic application is subject to variations and permissive expansion presenting a real scope of meanings that should be recognized and studied.

The origination of forms for the expressive ideas in the holistic symbols or in the broken down worded ones, which seems to be founding the language gradually over a long period of time, starts with the transfer of ideas. A transfer from the speechless tongue of psychognosis to speaking idiognosis extends the elements and the scope of expressivity precisely by the origination of meaning manipulation, expansion over time, transfer in different active and passive verb forms and metaphors, and production of a field of applicability of parcels into whole based on new directions and rules.

THE SCOPE OF CONCEPTUAL MEANING

Meaning related to any concrete object is more or less clear on what image of that object is actually formed in the mind though it is certainly not without possible visual errors and not the same in all minds alike. For abstract facts, imaginary objects alone, or objects and functions, meanings become variably defined in the minds depending on the interpretation and understanding both of the object and its related functional potentiality and also its affective mood incitation in states of mind. Most descriptions and definitions for conceptual meanings agree that a central core meaning is usually understood by the mental interpretation for the symbolic concept entity as general reference among communicating minds but a wide range of variable contingent side meanings accompanies the central core.

Further, the scope of meanings in imagination may be much wider than what the visual expression indicates. Imagination, just as may be experienced in dreams, has unlimited presentations with practically no conditioning restriction in meaning except in line with life commandment and psychodynamic balance as shown in chapter 6. It is also restricted to be naturally within the infinite dimensions of the all inclusive time titan, i.e., in the continuous present time even in extended dreaming with scenes of past events. Such dreams reinforce the extendable present time naturally to the past according to deep memory span, but never to any future, which is not recognized in the biological time except in the immaterial biology of abstraction. Imagination in the wakeful state can be conditioned according to the life commandment of logic and happiness, being factual and logic if in full awareness or less logical and more fanciful in daydreaming on the pleasure-pain axis. But restriction within the scope of the present time (time titan) is generally realized. This restriction is because of natural limitation of meanings attached to imaginations that are under the biological directional limitation of mental interpretation to guide the meaning on the two-way axes of autos to environment or self to nonself. This is dimensional limitation concerning the egocentrist autos relation to environmental stimuli, an important point that will be further discussed later. This limitation can be viewed as a constant replica of the early biocratic tropism in the *situs solivagus* reflecting the neurodynamic function limitation to adapt to the new Gnostic expansions from psychognosis to idiognosis.

As the psychognostic interpretation of the environmental stimuli leading to semions and moveons are based on the axis of autos to environment, so is the nucleus of the conceptual meanings being directed to and from autos. In idiognosis, conceptual meaning symbolized in words follows the same pattern and, with more variability, widening the scope of meanings and their changes due to the fact that holistic semions and moveons move into possible extended metaphoric scope and new subdivided forms in words and more so in phonemes and morphemes combinations. This change brings about expansions of the field of expressivity and the governing rules for the fusion of the atomic elements (phonemes and morphemes) and molecular synthetic combinations (words and sentences) to realize the desired substance effect: a unified meaning intention in the expressive idea.

A clarifying touch on linguistic means of meaning transfers is needed to expose how the implicit psychognostic meanings could have been interpreted to be acceptable to match an environmental item. I will start with metaphors in linguistic terms that define words with meaning that can replace other words' meanings usually by analogy. The metaphor intends to make a desired change in the expressive meaning forcing more attention on the part of the interlocutor to grasp the analogy that often exceeds the meaning of the replaced word with promoting decorative or demoting pejorative emphasis intended to produce the effect. Classic examples given for metaphors by Goatly[7] demonstrate these facts, such as in the following examples:

- Life is a box of chocolates.
- Virginity is a frozen asset.
- The past is a foreign country.
- Thatcher was Reagan's lapdog.
- Women who remain housewives turn into vegetables.

These examples clearly show what metaphors can induce intentionally in terms of meaning transfer in expressivity. Interpretation of conventionality of metaphor usage classified by Goatly indicates five degrees for metaphors from dead to alive. An estimate of the normal routine daily use of metaphors gives the figure of 5.7 metaphors per minute of talking,[8] indicating their important role in idea expressions by meaning transfer. It becomes clear that any metaphor has an inner semantic charge and an outer extendable halo in the chain of concepts. The continuity of communication uses metaphors routinely either in identified concrete examples or in tacitly understood abstract meaning. Metonymy is another means for meaning transfer and usually involves acquisition of new additional meanings by one word over an original defined limited meaning. For some investigators there is no fundamental difference between these two major linguistic means of meaning transfer.[9] In other words, conceptualization with accepting meaning change for an item according to needs for expressivity, in essence, involves the same functional mental mechanism and provides the same analogical result irrespective of the means used for the transfer. Meaning transfers also use other subtle linguistic means functionally effecting change from whole to part and from exaggeration-expansion to reduction-minimization and vice versa.[10] All these transfer mechanisms could have an older root in psychognosis with tacit implicit memory codes before becoming functional in the explicit memory with worded codes.

This brief exposure of the basic meaning transfers that ultimately indicate essential similarity and uniformity in neuroperceptive-psychointerpretive mechanisms suggests a significant operating factor in transfer mechanisms. That factor is usually seen without being observed. It is the relativity in meaning for concepts that faces the monodimensional present time, the time titan, naturally subjecting the meaning transfer to rules of the life commandment. On this foundation, subsequent psychognostic models and ultimate linguistic formations will be built, and extension of mental life will be allowed to go theoretically and practically unrestrained. This extends the relativity scale of meanings in the mind that are nonstatic, inherently dynamic, and show analogy with the uncertainty principle of the near-immaterial subatomic quantum world. This fact seems suggesting another confirmation that bioforce principles would be also controlling the immaterial biology and its energy expansions in literalism to the far-reached world of metaphysics.

FOUNDATION OF IMPLICIT PSYCHOGNOSTIC CODING

How metaphorizing initiated in the human mind in extending a meaning from one static or dynamic idiom to another seems to need more precise explanation for a more understandable conclusion, and the answer is to be searched and found in the more fundamental psychognosis before actual idiognostic worded language developed.

Let us first see the possibility of the original ways of symbolic meanings to be coded in the biopsychon's mind that must have been formed naturally on the axis of autos to nonautos environmental objects. These meanings, no doubt, must have been formed through auto-hetero-gnostic processes between external environmental entities evaluated in the light of known facts related to the egocentrist entity of the biopsychon. Using comparative gross functional characteristics of environmental items by the egocentrist autos as similar or dissimilar to oneself must be playing the pivotal role. The best naturally available standard to the biopsychon's speechless mind for such comparative evaluation is, in fact, the biopsychon's physical body as whole or part that could be used as symbolic imaginative links between the biopsychon and the environmental various single entities. Any such link could be normally presumed to be formed in the speechless psychognostic mind based on auto-hetero-gnosis allowing comparison of the detected nonautos environmental animate and inanimate figures with the familiar own physical whole or parts. Such comparability is sensed to initiate identifying whatever functional similarity may be detected relative to one's own physical whole or part in the category of function-attached meanings. The process effectuates, in fact, a primitive tacit identification and naming in a holistic way and without sharp precision in a nonworded metaphoric mechanism of meaning assimilation on functional basis. This identification must have started in the broadest sense of functional categorization, animate as against inanimate for example, referred to autos' own functional possibilities. The construction of reality in humans as based on the pioneering work of Piaget may be regarded fundamental in this connection.[11]

Comparability in size and number implying also contingent functionality appears as a perfectly natural neuro-perceptive possibility leading to neuro-interpretive facilitation to primitive holistic concept formation in psychognosis as models for future use. Two such fundamental concept formations indeed have been demonstrated to naturally occur in the remote evolutionary history firstly in function-related concepts and secondly in size or number-related ones.

The fundamental function-related psychognostic concepts as examples are SR based and directed by life commandment, and compare the environmental entity-action events as good or bad in relation to the egocentrist autos' physical body as a whole. This archaic concept is the most holistic by implying only the meaning of the action regardless of subjects or objects involved. This model originates the more defined urgent semions and moveons by the presumed mechanism involving neuroperceptive, psychointerpretive, and neurodynamic processes. In psychognosis, these models seem effective with the less precise and more general functional meaning of good or evil effects of the environmental events. These additional elaborations in concept formation direct the related aspects between environmental spacetime changes and the autos in its egocentric status to follow the relativity axis of life commandment of logic and pleasure-pain principle.

The second simplest form and also plausibly the earliest concept formation contemporaneous with function-oriented types in the speechless psychognostic mind is that of number one for unity, a conditioning functional unitary value apparently irrespective of size. Let us take autos-self in its egocentric position as a unitary element facing another single nonself solo entity of any origin and type. The psychognostic sensing from the nonautos unitary element to autos' comprehension could have any imaginable effect except that it could not possibly disturb autos' sense of integral unity as the feedbacking is between two unities not changing the numerical value of one, a value of innate understanding for autos-self. However, facing two or more same nonautos entities, regardless of any other feedback effects, reflects on the autos' unitary value, inciting the sense of numeral difference with nonautos items of more than unity. This is the basis of the phenomenon of subiting elaborated by Lakoff and Nunez[3] that can be used for judging metaphorically between the unity and more than unity. This subiting is an act of innate archaic cognition that we can categorize as biognostic or certainly psychognostic. It is a conceptualization-metaphorization at the same time in which numerality (one) is constant but size may be variable. This is the simplest concept that can be at the stage of symbolic-semantic conceptualization in psychognosis that can originate metaphorization by extending from one to more than one, prior to any code for numbering, and is usually limited to interpret only three similar items or perhaps four.[12] The process seems to be a perception-interpretation in parallel. Experiments in infants of preverbal age have indeed confirmed this simple subiting calculation to be present prior to language acquisition.[13] Also, experiments in conditioning rats to press upon a lever a selected number of times to get food when hungry have shown their numerical capacity for number duplication in pressing a number of times, equal or close to the number required.[14] This is a bona fide example of symbolic-semantic concept with simultaneous comparative judgment, which is basic, together with metaphorization of psychognostic type, and needs no linguistic coding. The concept is innately present in autos-self, the unlabeled initiating self, not being communicable except by expressive gesticulation prior to words.

This basic conceptualization, incorporating meaning in numbers before any explicit code for numbers is formed, indicates the fundamental possibility of concept formation without phonetically coded semantics. It confirms that sensing applied to items for identification must be preformed by autos in psychognosis and then used by self in idiognosis on the basis of familiarity or unfamiliarity to autos or self in implicit or explicit memory. In fact, feedback effects to self from all nonself elements, quantitatively or qualitatively, can be interpreted and identified as such, and then coded for recall. The evidence for this much discerning in the psychognostic mind prior to explicit worded code formation seems unquestionable. We have to accept that all fundamental semantic concept formations started in psychognosis at the precoded stage before literality and by metaphorization based on biologically meaningful facts related to autos, then changed to the coded status by self in idiognosis. Thus, neither the initial uncoded metaphorization in psychognosis nor the subsequent coded metaphors in idiognosis are really arbitrarily formed, and in fact, they reflect affective meaningful aspects of items to autos and self and in line with the life commandment and with innate metaphoric variability of pleasure-pain principle analogous to quantum uncertainty. Leaving this very clear explanatory example of innate numerality in psychognostic conceptualization, I can now approach the other spacetime effects on autos-self concept formation. According to this natural rule, the origination of semantics in its nucleus that initiates a stepwise semantic growth, starting in psychognosis and leading to code formation in idiognosis appears to only need four basic detecting-defining systems (DDS) in the mind to identify spacetime relations for

1. ***Orientation,***
2. ***Quantity,***
3. ***Quality,***
4. ***Time.***

Finally, internal system of synthetic functional capacity for interpretation of the detected-identified data is the key in the whole conceptualization. This last system that uses neurodynamic energy and mental problem-solving mechanisms is the agency that gives the final forms of semions and moveons in psychognosis on whose models the ultimate worded holistic concepts and subdivided derivations will be formed. This last innate problem-solving system, elaborating the final interpretation, uses functional details of neuropsychodynamic factors, practically unknown, which set the basis of relativity meaning to autos from each of the four categories of nonautos fields. This complex of unknown relativistic system in the mind has only one known guideline, that is, the life commandment. Further discussion on this functional aspect of the mentation in the transfer from psychognostic to idiognostic expressions should be through reviewing the Navand theory of languages as it could relate to these four DDS systems, but mostly to dynamism related with time, and secondarily with idiognostic coding by words.

Initiation of Navand theory development in chapter 7 introduced the idea by forming the working ground, but what will establish the compatibility of the theory in harmony with psychognosis gradually changing into idiognosis and its actual applications in linguistics needs further detailed examination that will be done with further elaborate discussion on idiognosis.

These four DDS spacetime related systems and the axiomatic archaic relativity scale of the life commandment form the fundamental constituent ground for idiognosis. The mechanisms used necessarily respect the personal preferences of autos to be decisive in priority coding in idiognosis. The foundation of mental coding, which is initially iconic and implicitly recognized as categories and later defined as subsystems with divisions in sharper definitions ultimately reaches explicit coded status with words in speaking expressivity. Our conceptual categories identifying basic structural linguistic elements, i.e., subject, object, verb, effect in their psychognostic primitive forms with symbolic meaning that make and keep a bond between autos and its environment serve the trait of expressivity, which must go on expanding. Implication of DDS, the four system categories, serves the natural estimation of autos' environmental meanings only as related to the egocentric position of autos in the restricted four systems' functional capacities that provide the significant meaning values to autos-self implicitly. The foundation of autos' functional psychognostic mental activities without explicitly identifying codes, and the self's with explicit idiognostic codes, can be assumed to be logically based on the principle of tacit concept metaphorization from the egocentric position of autos and therefore self.

Probably, as cited earlier, the first idealistic remark to this effect should be referred to the work of Piaget in relation to assimilation and accommodation of mental experience in children.[11] As we will see later, the tacitly stereotyped holistic semions and moveons induce metaphorized forms of labeling by meanings most acceptable to autos.

Perhaps the best logical way to approach the problem is to start with the primordial unitary entity in the psychognostic mind, that is, the autos, and to consider this autos' central standing unity to nonautos items sensed and comprehended as unknown unities, more or less, in terms of numbers (divisible quantity, implicit group quality), size (indivisible quantity, implicit unitary quality), spatial location (orientation, direction, closeness, remoteness), and timing (immediacy, nonimmediacy). This stage of basic sensing would be what we can call the symbolic-semantic conceptualization, the same that has given semions and moveons, the holistic meaning categories. This category range of a core meaning that could expand must be regarded as the initial source for the metaphors to be formed. This is the first step of neuroperceptive-neurointerpretive work prior to the stage of expressing the sense by words that we should preferably call the stage of semantic-holistic conceptualization prior to semantic-phonetic conceptualization. These psychognostic stages are all pictorial according to evidence from neurocognitive studies and dream research, as we have previously discussed, and are all emotionally conditioned.

THE MATRIX FOR THE IDIOGNOSTIC COMPENDIUM

In conclusion, implicit psychognostic coding for recognition of nonautos items by autos is based on the factor of relativity interplaying with comparability of nonautos entities to autos through neuropsychodynamic recognition checking the four categories of nonautos spacetime elements. These four mental systems for *orientation*, *quantity*, *quality*, and *time*, achieving detection-interpretation for the egocentric autos, make the matrix for concepts to be initiated and extended with the evolving expressivity trait. The evolution has shown the importance of this foundation in the ultimate explosive linguistic expansion witnessed today.

Of these four DDS fields of topographical evaluation by autos-self, I take the most elusive one first which is the time. Time is both elusive and essential by its nature that is immeasurable, but grants measurability. The immemorial time, which is innate in all life forms in autos and which I have named time titan, is theoretically the incalculable infinite time in physical and philosophical interpretation. The memory time, on the other hand, is the foundation for comprehending the self positional standing of autos and the nonself elements of the environment to it, making implicit memory in psychognosis. Autos' neuroperceptive-psychointerpretive system complex is thus able to evaluate nonautos' items dimensional and functional meanings along and, in the infinite scale of time, covered by autos' implicit memory or the autos-self forming agency. This scale of time is primarily limited to the present time covered by the working memory in its defined form and is changeable to an extendable present time by each recall. The mechanisms involved provide an implicit coding for recall when the many elemental entities in topographical positions relative to autos are detected by their iconic figures. In this way, the semions and moveons are iconically recallable without needing names. In fact, the recall is served by explicit visual memory as can be exemplified according to geons mental pictorial meanings explained in chapter 7. This realization is naturally confined to the present permanent monodimensional time.

The main inseparably associated element with time, to be coded in psychognosis, is action effect. Action identification uses complex metaphorization as related to orientation, quantity, and quality the other three metaphorically semantic fundamentals. The action could be a simple assertion of the state of being with implicit sense of present time, being tacitly included for either an autos inner sense or nonautos entity presence. It may indicate reciprocal actions between the nonautos entities with distinction of actor and object in iconic forms and even with an inner sense of good or evil result in line with the pleasure-pain principle. Action identification includes perception of moving displacement in spacetime with an implicit sense representing a form of symbolic-semantic metaphorization that has been called topokinetic memory.[15] All these are being psychognostically sensed as iconic-coded concepts whose biological significances set semantics of their affective values directed to autos. Extension of the meaning to autos by actions of two nonautos entities to one another can be interpreted as metaphoric introjection affecting autos or extrojection in generalization to other nonautos. Whatever effect observed can be taken as an independent nonautos symbolic-semantic concept with metaphoric interchanging between entities involved. This level of sophistication is still in the realm of psychognostic possibilities affecting autos with iconic memory to be further established in idiognotic self and with codes in explicit memory.

Time identification integrally, and not as a segmental part of the infinite time scale harboring an event, is based on the archaic biological sensing of the permanently continuous time titan constantly present through biognosis, bioawareness, bioconsciousness, psychognosis, and idiognosis. This time is the present time at any moment, within the working memory span implicitly, or by recall as the present holding the past. It is an inner sense as expressed by Kant, like all cognitions occurring as an inner sense. This fundamental basic time, which can accommodate extensions to past and future, is essentially the static sense of stable time presenting a feeling of *moment permanency*, an ***ephemeral instant memorial time, a memory of immemorality, an instant of immortality***. This is the metaphoric representation of the fundamental uncoded sense of time in psychognosis, which, if coded with an iconic item, is tacitly known and extendable. Even in idiognosis with concepts of actions, motions, and the metaphoric plethora of various time-related meanings, the time as an independent reality is not definable but is definable in the form of the state of being of the event in question. This time may be included tacitly in idiognostic expressions without separate codes along idiognostic codes expressing various occasions with affections, or it can be expressed in action description or separately in coded form referencing its abstract meaning. Metaphoric meaning of time to self is what perception of sameness for the self incites in contrasting sensation with variability of nonselfs and with visual frames changing constantly in the mind. The ultimate meaning is uniformity of a single ground for all various changes observed in consecutive episodes progressing unidirectionally. Metaphoric projection of this unidirectionality with irreversibility sensed by the self is basic and makes the foundation of both implicit and explicit memories. These two types of memories that could be also called psychognostic and idiognostic, respectively, are fundamentally the same in regards to time. In fact, unidirectionality of events in time, allowing detection and recording sequences consecutively and irreversibility observed that sets the limit of no return making divisions in time are fundamental basic metaphors that generate all subsequent time values to be differentiated.

Orientational concepts in psychognosis are based on the recognition of autos' spacetime to autos itself as central, comparing the environmental spacetimes of every nonautos entity in vectorial relation to autos. Thus, conceptual models are sensed to bear the meanings of the three-dimensional sites, forward-backward, up-down, and right-left orientations to the central autos. To this simple static spacetime orientation for iconic coding, the dynamic effects in terms of near-far and motional speed will be naturally added and integrated by the neuroperceptive-psychointerpretive problem-solving system of psychognosis, assuring the normal functioning of the living organism. We can remember such example of effect in the shortening safe space of a rapidly approaching object causing withdrawal of the head in infants described in chapter 7.

The quantity distinction as mentioned earlier entails the holistic conceptual togetherness versus separation of size and number and seems to be initiated with the number and not with the size. The size normally forms contingency of a whole integral unitary entity presenting no contrast to the similar tacit autos-self unity to be detected prior to the more distinctly presenting number contrast. The size quantitation involving volume difference is basically function-related in psychognostic perceptual interpretation in relation with body parts and their functions in the egocentrist autos. In this way, natural psychognostic holistic concepts for size related function form in relation with the body parts. This subject will be discussed further in chapter 11 in relation with the respective early naturally formed names.

Quantity perception and expression in terms of number, in fact, seems to be the earliest psychognostic-idiognostic natural coding with numeral subiting, as discussed earlier, restricted to one to four similar items. This is a clear example of metaphoric conceptualization showing the differential sensing that actuates the meaning of more than unity contrasted with the innate unitary sense of self. This metaphoric sense also includes the extendable meaning of variable different idiognostic mathematical sets.[13] Extension of it in continued increasing or decreasing differences and their meanings is metaphorically basic and is semantically realized in psychognosis in unworded forms. Further extended steps, forming the foundation of mathematics with all abstract notions in mathematical thinking, will follow in idiognosis with the appropriate coding made and used to that effect.

Mathematics, in fact, forms the only precise universal language that, in the passage from psychognostic uncoded subiting to the coded format in idiognosis, can use the psychognostic crude foundation but without metaphoric conditioning to express the most subtle abstract ideas with an unfailing precision. The reason is of course total logic and objectivity developed with it by the restricted unitary quanta in sequencing order to plus or minus. It is essentially the unitary basis for its foundation in all minds that enforces the differential stepping in increasing or decreasing order to be consecutive and without direct relation to psychognostic affectivity. Thus, the only idiognostic coding of uniformal and universal applicability is the mathematical codes with the metaphoric mechanism operating in them minimally as strict relay that leaves mathematics immutable as a language of pure logic.

Parenthetically, it is of interest to note that mathematical consciousness with its unfailing pure logic, when coming to the concept of infinity, may lose precision and accept approximation short of any real solution.[12] Such, however, should not be regarded as the correct conclusion in the living world for creation is in reality all integral. Numerality that cannot constantly show integrity of the creation is artifactual as it imposes time definiteness in the fundamentally time indefinite world of creation. The hypothetical reciprocal feedbacking of energetic nature of the two fundamental components of quantum string theory, which is an integral complex as with positron and electron of photon for example, interpreted as infinite, and the mathematical metaphor of infinity seem unsolvable realities. Mathematics as an early idiognostic product preceded physics about two millennia indicating this false dead end puzzle of creation that is now no longer valid. This reality that creation has its own logic of beyond exactitude of the mathematical logic, as we so evidently find in the uncertainty principle of quantum physics, reveals that indeed God may play dice against Einstein's belief, but weighted dice, of course, that have matter, energy, and uncertainty.

The last of the four fundamental systems of nonliterary conceptualization (psychognostic type) is with metaphors that create the sense and the concept of quality. This inherently relative scale conceptualization with meaning of good versus bad in a grossly generic categorization and, with all possible extensions of the contrast-based sensing, form in psychognosis on the biological significance of the metabolically balanced state that is targeted to autos. In this sense, the biological value is archaically elaborated by the energy with metabolic changes as described with the autogenetic egos, later extended to frank pleasure-pain principle. In this sense, the biological value of every process enhancing the balance as a positive feedback is the root for the metaphoric *good* to autos and any contrary effect with a negative feedback is that of the metaphoric *bad*. This contrast-based metaphoric foundation is as old as biocracy and can be said to be aged about four thousand million years on earth. All the archaic forms of paleognosis to neognosis have actually served biocracy on that unconscious metaphoric basis. In psychognosis, this metaphor reaches its complexities forming meaningful preconcepts in the face of sensed variations of the environments in addition to pure metabolic factors, with proto-egos and their impressions on autos in line with biodynamic life commandment. In idiognosis, further additional metaphoric meaning of egos as concepts of knowledge, power, purity, precision, righteousness, honesty, etc., with contrasting opposite concepts and more abstract pleasure-pain meanings associated with various affects will appear and form the dominating mental atmosphere of social living where norms are all metaphors of rights and wrongs.

FORMATIVE NOETIC MECHANISMS IN PSYCHOGNOSIS

The aim in this chapter was to describe the preparatory ground and conditions facilitating model formation for understanding idiognostic coding by words. So far, the basis of what components could be used in model formation and in what relationship has been discussed. In this last section, I will include what I believe is fundamental to indicate how the elaborated components described and theories exposed could relate together and fit into functional models that have practically served nonworded concepts in psychognosis and will apply to idiognosis.

My discussion will focus on the affectively dominated holistic concepts and the role of memory in their elaborations that form knowledge iconically coded in psychognosis. This knowledge could be about a general fact, a reality known to the hominid (daybreak, nightfall, etc.) called noetic, or it could be a knowledge by personal experience (escaping a predator, breaking a leg, etc.) called autonoetic. In conventional terms, the first one is knowledge of semantic memory type, and the second is episodic.[16] The episodic memory includes some time distinction related to the past. The basis for both types is referred to as perceptive memory,[17] which is essentially what is confined within the working memory time to be encoded. This terminology sounds somewhat artificial in the sense that the reality of the fact known is the same, the knowledge, and the memory is also functioning to actualize the same essence of knowledge. But understandably the difference is in the different ways leading to actualize this knowledge—one being of autos origin and the other of independent sources.

All autonoetic facts are affectively oriented. The affective charge, though variable, is what gives the meaning to the event, which, in addition, is time specific. These autonoetic, psychognostic, recallable pieces of knowledge on iconic basis have both the semantic and the episodic specification in memory terms in an inseparable way. However, their encoding and formation have been interpreted differently. Tulving has considered the process to include an axis of perceptual-semantic-episodic steps in series and has named the system SPI for serial parallel independent functions involved in coordination furnishing respectively perception identification, knowing in the present, and remembering from the past.[17] Simons has advocated a multiple input, MI, model in which the serial restriction from semantic to episodic memory is excluded.[18] Tests in all related experiments have been invariably carried out linguistically, which can openly miss the affective meaning of iconic knowledge not handed with words at all or imperfectly. In my belief, the psychognostic knowledge of iconic conceptual type is the most sincere and straightforward in meaning; it is meaning episodic or episodic meaning as it is essentially egocentric rather than allocentric in abstraction with secondary reference to self. Semions and moveons in autonoetic sense include the most precise semantic content possible in a holistic concept whether at the perception-response level or at episodic level remembering a past event, but in both, their affective charge is the sine qua non factor. In other words, separating semantic from episodic memory in psychognostic interpretation does not appear to make sense. Formation of an inside world of symbolic identification of causality effect as *knowledge* in early humans, before language development, must have had a respectable richness of autonoetic data and inner vocabulary of protosemantic-episodic memory gained knowledge.

Another question that must be clarified concerns noetic information in the same psychognostic situation. So far, I have considered the point of initiation for concepts of environmental entities and events in the psychognostic mind to be on the egocentric autos. Could an allocentric understanding by mental projection also function at this early stage? In simpler words, could abstractive power have had reached the level to allow imaginative autonomous sources of external origin to form in psychognosis? Or said differently, could metaphysics as pure abstraction have had a root in psychognosis?

In the human mind with language and double coding of concepts both iconically and linguistically, allocentric representation of facts and assimilated knowledge on that basis is unquestionable, and existence of a metaphysical world is a reality in the conscious mind. The reason is precisely the impersonality the worded codes can grant to abstract entities or to concrete ones with imaginable potential or real functionality and movement. The mechanism is psychological and uses introjection of the external world elements as embodied selfs and recognizing that they could also be external autonomous powers looking upon the internal one. This level of abstraction seems usual in the well-developed normal human mind. The role of double coding, iconic and linguistic in facilitating this introjection-extrojection axis, affection-motivated or oriented, is sensed to be potentially significant in forming a linguistic code for an emotionally imaginative unreal icon. Furthermore, the long-term memory for events and their spatial context in this mechanism seems to be playing a crucial role. Unfortunately, comparative anatomical studies to show significant variations in the long-term memory substrate of temporal and parietal lobes and the hippocampus of the brain in primates and humans or detailed comparative studies on human damaged and undamaged brains are not available. Were they in hands, they could be valuable, although still not infallible with psychological differences in confirming solid inference for abstractive imagination toward metaphysical concepts in psychognostic speechless mind. A reliable study for such meaningful information still would have to be organized and performed. However, studies of functional magnetic resonance imaging (f MRI), in fact, have shown, to some extent, the simultaneous functional activation of the long-term memory substrates in the brain, interpretable for egocentric and allocentric spatial viewing with difference in deficient function compared to normal controls.[19] The results indicate the importance of neocortical long-term memory sites and hippocampus's cofunctional earlier contribution, together supporting today's status of long-term memory in neurophysiology. Good sense would also accept some long-term memory to be functional in the iconic psychognostic mind with sufficient extent to include allocentric abstraction accepting external unreal power sources basic in metaphysical imagination. If this assumption is in fact credited, then it enforces the view that double coding, iconic and linguistic, continued on and influenced the preformed foundation of metaphysics in psychognosis to develop its present status in the human mind. No doubt emotion based psychognosis implies the likelihood of possible imaginative items for the nucleus of such presumed metaphysics to have been formed. A priori, idiognosis completed the creation of an exteriorized self out of autos already having started in psychognosis that assumed allocentric projected functionality in addition to the shared egocentric one with autos, allowing inimaginable abstractive power gain but with diluted affective psychognostic charge in an expanding idiognostic field of increasing impersonal immaterial abstraction.

Considering the issue of the affective charge, here is the evidence once more for a long continued inappropriate practice of using the word unconsciousness globally with all affectivities it includes in contrast to consciousness with substantially controlled or diminished affective content. The added linguistic coding assuredly enhanced the psychognostic symbolic knowledge in long-term memory, giving it double identity by words but not really double affectivity charge and indeed restricted the affective exposure in the limited capacity of single words with limited metaphoric scope. The extent of abstraction certainly expanded incalculably by words for the self-ego effects promoting impersonality in abstract ideation using linguistic formulas. The change was in reality from psychognostically defined affective holism to materially divisible idiognostic metaphors with less defined affective dynamism. Fundamentally, emotion-based motto idiom, not really doubling the emotional charge but changing it from what it was, often reached less defined level. This change seemed to reach adynamic *stato* idiom level (static nonaffective scientific items for example) without motto dynamism and *motto* content for the self-ego entities. In fact, overwhelming association of affect with holistic psychognostic concepts must have limited the scope of neuroperceptive elaboration to more rigid algorhithm and more stereotypic topokinetic memory whereas with words, these limits would become looser including representations of nonaffective types and confusing metaphoric effects: a change from confined but intensely directed emotion to a limited level with about uniformally expanding vagueness. We can say that psychognosis centered a great emotional charge to limited channels forming dynamic motto concepts. Addition of words bringing labels to these emotionally charged concepts facilitated spread but attenuated the strong expressive emotions. Double identity by word codes given to psychognostic concepts certainly helped explicit expressions but reduced the implicit affective charge to some degree as the worded codes could never reflect the entire emotive contents of the coded materials. This fact explains the necessity for and the inseparability of intonation from the spoken language. Furthermore, new concepts of scientific type, affect-empty, emotionless, diluted *motto idioms* (*stato idioms*) gradually reducing dynamic strong expressions in civilization are expanding in our main expressivity field, that is, the language.

Perhaps to understand more clearly what psychognosis is, and what are all Gnostic states that were labeled unconscious (often even including the entire psychognosis) before it, is to realize the inseparably bound affective charge in the holistic concepts in semions and moveons by their mere iconic meaning summarized in psychognostic holism. This affective charge, autos-centered, is invariably present in and inseparably bound with every item in psychognosis. We could say that psychognosis has only motto items (emotion-dominated meaning) that would take stato forms with words in idiognosis still hiding some affective charge as stato-motto or newly formed pure stato items, stato-stato, without any affective charge like in scientific terms or acronyms. In this sense, the foundation of abstraction acted as the nucleus of impersonality expanding into immaterialism with only the substrate of memory being its biological basis. However, the abstract world of biological immaterialism in idea formulation into philosophy and metaphysics again assembles the essence of human affectivity inherent in the ever lasting life commandment with logic and pleasure-pain principles. Immateriality of metaphysics, if interpreted as beyond relativity and consciousness must be in fact devoid of material stato idioms. The fundamental expressivity trait based on emotion and emotive energy is indeed imperishable in the immaterial biology and expands to ultimately find new fields on the basic algorithm of the life commandment as we will see in this book's last chapter.

REFERENCES FOR CHAPTER NINE

1. Mithen S. 1999. The Prehistory of the Mind. *The cognitive Origins of Art and Science*. Pp 126-127. Thames and Hudson.
2. Greenspan SI, Shanker SG. 2004. *The First Idea. How Symbols, Language, and Intelligence evolved from our Primate Ancestors to Modern Humans.* pp158-159. Da Capo Press. A member of Perseus Books Group.
3. Lakoff G, Nunez RE. 2000. *Where Mathematics Comes From.* Pp158-175. Basic Books. A Member of the Perseus Books Group.
4. Johnson M. 1990. The Body in the Mind. *The Bodily Basis of Meaning, Imagination, and Reason.* The University of Chicago Press. Chicago and London.
5. Marcus G. 2004. The Birth of the Mind. *How A Tiny Number of Genes Creates the Complexities of Human Thought.* Basic Books. A Member of the Perseus Books Group, New York.

6. Phipps C. 2006. A New Dawn for Cosmology. *What is enlightenment*. 33: 42-48.
7. Goatly A. 2000. *The language of Metaphors*. Routledge, New York.
8. Gibbs RW Jr. 1992. Categorization and Metaphor Understanding. *Psychological Review*. 99(3):572-577.
9. Traugott EC, Dasher RB. 2002. *Regularity in Semantic Change*. Cambridge University Press.
10. Campbell L. 2004. *Historical Linguistics. An Introduction*. MIT Press, Cambridge Massachusetts.
11. Piaget J. 1937. The Construction of Reality in the Child. *In the Essential Piaget*. Gruber HE, Vorech JJ(Eds). Basic Books. New York 1977.
12. Lakoff J, Nunez RE. 2000. *Where Mathematics Comes From. How the Embodied Mind Brings Mathematics into Being*. Pp 19-21. Basic Books. Perseus Books Company.
13. Butterworth B. 1999. *What Counts: How Every Brain Is Hardwired for Math*. Free Press. New York.
14. Mechner F, Guevrekian L. 1962. Effects of Deprivation upon Counting and Timing in Rats. *J Experimental Analysis of Behavior.* 5:463-466.
15. Berthoz A. 1998. Parietal and Hippocampbal Contribution to topokinetic and topographic Memory. In: *Hippocampbal and Parietal Foundation of Spatial Cognition*. Burges et al. Eds. Oxford University Press.
16. Gardiner JM. 2002. Episodic Memory and Autonoetic Consciousness: A First Person Approach. Pp 11-30. In: *Episodic Memory. New Directions in Research*. Baddely A, Conway M, Aggleton J. Eds. The royal Society. Oxford University Press.
17. Tulving E. 2002. Episodic Memory and Common Sense: How Far Apart? Pp 269-287. In: *Episodic Memory. New Directions in Research*. Baddely A, Conway M, Aggleton J (Eds). The royal Society. Oxford University Press.
18. Simon JS, Graham KS, Galton CJ, Patterson K, Hodges JR. 2001. Semantic Knowledge and Episodic Memory for faces in Semantic Dementia. *Neuropsychology* 15: 101-114.
19. Burgess N, Becker S, King JA, O'Keefe J. 2002. Memory for Events and Their Spatial Context: Models and Experiments. Pp 249-268. In: *Episodic Memory. New Directions in Research.* Baddely A, Conway M, Aggleton J. Eds. The Royal Society. Oxford University Press.

CHAPTER TEN

THE WORLD OF IDIOGNOSIS

CONSTRUCTION PLANS

The last three chapters have covered material with discussions clarifying the point that we had reached our natural consciousness before words were created, and that our mental living is all conscious in terms of time indefinite and time definite consciousness and is not totally dependent on our language, which initially developed to better serve our expressivity. That expressivity, as a trait inherent in our biological affectivity, secured better reciprocal interhuman relations in our initial small hominid herds and continues doing it in our present civilized structures.[1] But our mental activity was also used for other nonaffective purposes as well as evidenced by psychonoetic activity described in chapter 7 and by the nuance between noetic and autonoetic realizations noted in the last chapter. This is clearly why words were developed in the first place and then how, by extension, they proliferated naturally (or unnaturally!) to so many senses and nonsenses in the orthodox biological meaning in our advanced enormous civilized societies.

This contradictory implication of extensive literary expression can make an interesting philosophical debate for or against extension of idiognostic expansions as such expansions increase stato idioms and decrease motto idioms, leaving less lively incentives to face more insignificant one in the mind. To make the point, an inciting remark from a story in one intuitive fiction from the era of fiction writings on outer space visitors is up to the point. Briefly, it describes a *psychognostic* biopsychon traveler from another planet visiting ours. The traveler sees a human on a bench in a park reading a news paper. *She-he-it* approaches the earthly man, smiles at him, and offers him a beautiful heavenly flower with inebriating scent. The earthly man, perplexed by the visitor's encounter, but delighted by the flower's heavenly scent, puts down the news paper and grabs the flower. The visitor throws away the news paper and nods his head at him in silence to forget the paper and be happy with the flower scent.

This inciting story shows that over and above the natural biological needs to be served by our expressivity trait, in the eyes of a psychognostic mind, the world of words has expanded explosively and extended to unimaginable fields, and what was controlled by us in the service of our feelings has gradually exerted control on us, imposing a self or selfs on our autos with so many unreal and unnecessarily created items, so obliviously normal in our conscious norms neglecting the essence, and puzzling to naturally minded onlookers. Let us hope that the natural mind of humanity will never dissolve in the idiognostic artifacts.

To return to our main task, let us first recognize that the constructive plans for idiognosis must be distinguished in two functionally distinct categories. The first is the plan in which the crude iconic matter of the psychognostic origin is mentally manipulated to create a semantic world. This plan activates memory steps; it is intrinsically neurointerpretive and includes several unknown factors with the most mysterious one being the initial mental trigger named Navand in chapter 7, which is responsible for the whole action resuming the expressivity trait. The second functional category plan is the one in which an auditory identifying sign, to be communicable, is added phonetically to the inner iconic meaning of the private and virtually noncommunicable nature in our standards, which is outlined in chapter 9 in relation to autos in its egocentric position. These two plans differ in importance—the first one really builds the private mental structure in the mind with potentially unlimited power of extension and potential exteriorization, and the second one makes it practically available to be shared by other intelligences.

The private plan uses an internal language, which is uniquely binary type, based on pleasure-pain principle of *either-or* indication. It constructs the very early means of behavioral expressivity. It may go to include altruistic consideration in the governing autocratic logic, which again obeys the pleasure-pain principle, but with the altruism securing satisfaction ultimately for autos. This subject can be debated according to different points of view. However, biocracy is primarily concerned with supremacy of autos, and biological autocracy is a reality whose manifestations in idiognostic experiences also follow the life commandment principles. As such, the very private psychognostic expression versus less private idiognostic one may vary in terms of more pleasurably axed egotropistic private decision making, opposing the more open logically axed allotropistic considerations. Now, psychognostically viewed, the master is still autos, and self's ultimate idiognostic expressions safeguard the same pleasurably dominant meaning to autos, but an autos projected out to allomorphic equivalents in the arena of idiognosis and social norms as target of altruism. Without idiognosis, such projection would have been quasi-impossible.

THE PRIVATE MEMORY DOMAIN

The first plan, already initiated and established in psychognosis, is naturally memory dependent. It is clear that memory as a whole is involved in the process of perceptive interpretation including iconic code relation to meaning, and whether any specific type of memory is more instrumental in the entire process of formation, encoding or recalling, and using of the iconic idiom to be known and used is a secondary matter and is not precisely clear. In this connection, it is interesting to note that indeed some degree of private, inapparent, autonomic memory effect in the brain regulates time intervals at organic cell levels.[2] According to this memory activity that can be basically related to the circadian effect in the brain, gene expressions can show variation in the somatic cells, like in cardiac and hepatic cells, that would show periodical short-time variations during the twenty-four-hour cycles. These variations keep interval timing in the organs. The regularity of these changes seems to function on feedback basis of neuronal synchronized discharges between cortex and basal ganglia autonomously and tacitly.[2] This can be regarded as a basis for the implicit memory type periodicity that could initiate explicit realization if consciously recognized. Thus, the foundation of the autonomous innate private timekeeping in terms of regulating cellular manifestations like in the heart or the liver, genetically controlled[2], seems to be a definitely valid fact. This degree of memory specificity that can be regarded as some privately controlled capability for specific cells, though not clearly detectable in the abstractive domain, can yet indicate similar mechanisms to be possibly operational in abstractive psychognostic to idiognostic concept identification. It works with the respective cellular networks particularly in the brain like in Mirror Cells of the left frontal lobe.

Memory as a functional capacity that can be extended may naturally show variable specificities of functions. These varying aspects have engendered apparently different memory types that accordingly appear to impose legitimately precise names and functional specifications when it comes to study the general role of memory in biology.[3] However, the only plausible variation of functional differences of memory types is in short-term versus long-term biological functions. These two basic types of memories are definitely distinct based on solid anatomic and physiologic grounds of limbic and cortical neuroanatomic and neurodynamic foundations respectively. The passage from the short-term to long-term memory types is essentially from an unembodied emotional significance of the initial unsecured synaptic network of the limbic impression to an embodied secured cortical semantic value for which the networking is reconstructible if triggered. This passaging-through mechanism is inherent to expressivity trait and binded to NET trilogy and secures the embodiment of the emotional effect with the event of the iconic scene, or word meaning, thus securing a recallable semantic value. The initial step may be purely psychognostic with emotional meaning of iconic scenes and may be said to essentially engage the implicit memory. The actual steps for securing the initial ephemeral networking into a fixed one may be said to engage episodic memory function. The final storage of the purely iconic scene significance of psychognostic type that is privately recallable as with semion-moveon mechanism, if added with words, is practically recallable to the communicating self and can be said to have used semantic memory.

One way in which a schema can be proposed for showing the exteriorization of psychognostic meanings into coded meanings, to acquire distinct semantic identification, could be formulated in terms of memory mechanisms from ideas, to specified meanings, to distinct codes in time sequences as:

Implicit Memory Source	Episodic Memory Source	Explicit Memory Source
Meanings of Sequences =>	Changing Sequential Meaning to =>	Changing Timed Meanings
In total Holistic Time	Timed Meaning	to coded meaning

This hypothetical representation of changing face of the meanings in biocracy from the early meanings of sequences of inner metabolic steps of energy building and expending, to more stereotyped sense of timed itemized sequences and ultimately to coded itemization, can be said to certainly have originated with the implicit memory and ended with the explicit semantic memory. The timed meanings in codes in the last stage is in essence demonstrating the actualization of the final interpretation of the idiom, which includes the evidence of the matter (iconic or words) and the actual present time of it (permanent present time titan, or the real present time). This is the stage of semantic realization of the idioms—icons or words or both. This plan seems to be basically functional in psychognosis in the iconic visual world based on its holistic concept.

It should be emphasized that the private plan, structured on memory basis, is holistic and contains realities iconically initiated, iconically meaningful, and iconically recallable. These fundamentally figurative concepts are single grounded, i.e., they are only identified by figurative emotional meaning for each associated icon-action or icon-result that are realities kept in memory. They are not to be voluntarily recalled as they have no reality beyond their present time of being except at the first registration or by recurring based on external hints for recurrence, which revives the emotion associated with the meaning of the figurative scenes and recalls, bringing out the totality of the concepts, semion-moveon type.

It can be said that these materialistic iconic concepts are all real experiences for which psychognostic neurodynamic mechanisms forming semions and moveons act automatically when a recall may actualize. The private memory domain is thus strictly visual-visional and essentially concrete, keeping and realizing concrete holistic concepts of real-life events in psychognosis. The same basic model then will secure the idiognostic explicitly realizable facts with words.

THE PUBLIC ARENA

The second plan uses the primordial models formed in psychognosis and exteriorizes the semantic content of concepts in communicable phonetic codes. In this process, the second plan seems establishing continuity between the actual spacetimes of the communicating minds possibly by virtue of the double idioms, icons and words, securing a more immediate relay and as long as the communication is running. This spacetime relay is tacitly sensed but is not so realized or realizable with the speechless psychognosis as it is with the worded language in idiognosis though both serve the expressivity trait and both are equally using the fundamental two-frame model described earlier.

To further support the basic concept, psychognosis must be recognized as the solid ground that not only provided the initial materials and models (autos controlled) for serving the biological expressivity trait but actually acceded to the satisfactory level of expression to serve the psychognostic speechless mind. This provision allowed perfection in the language construction, gradually through millennia, to become more fluent and to be more practical in responding to the increasing needs for the expressivity to be served by idiognosis and for the developing environmentally conditioned and socially imposed self. However, basic materials and frames for constructions were the same old ones as in psychognosis and were on a binary biological system of *being versus nonbeing*, using the usual two frames—one for the matter and one for the energy (or time or motional meaning change of the matter) as described in past discussions.

These models had just the static iconic elements filling the first frame for the matter, visual figure entities that could change or cause change in scenic presentation, and no figurative representation in the second frame for the time or its equivalent energy housed in the invisible time titan. The second frame with no figure content, showing the time permanency in its implicit invisible form in psychognosis, was in reality the ground for the potential dynamism of the first frame's figure to occur, a dynamism proving the reality of the energy being spent, but depending on time that assures the expenditure. The static or dynamic state of the first frame's figure could not be seen without the second frame proving statism or dynamism. In idiognosis, the iconic representation changed gradually to icon-word in the first frame for the matter. This was the only change necessary to occur, the frame for the time titan, however, being left untouched, either remaining empty in the case of a static unchangeable item like a name or becoming filled with the identifying word for the present permanent time (for instance with *is* in English) confirming the static stance or with a word indicating dynamism (any verb form, or conjunctive word as function words). This aspect of basic formative structuring of the psychodynamic and psychonoetic concepts in psychognosis was treated in chapter 7 (Table III) and will be again discussed further with the Navand theory of languages in this chapter.

The actual construction of idiognosis initiating phonetic labeling to identify the iconically known meaning reflections in the mind will be properly detailed in due time. Its mechanism of neurodynamic nature however needs appropriate elaboration before approaching the phonetic labeling steps. Let us just clearly say that the fundamental neurodynamic initiation is emotional and does not start out of the blue but follows deep valued emotions between close members of the human family, namely, parents and children. Remnants of these archaic emotions and bonds between parents and children can be vividly witnessed in today's societies and are reflected in the similarity of name codes identifying mother, father, me, etc., in Indo-European languages as an example. The type specific phonations in words that are used for identifying static states with names or adjectives and dynamic states with verbs or conjunctive function words are not my concern at this time. This subject will be outlined later in discussing linguistic roots from known cognates.

To objectivate the explanation for the second plan of construction for idiognosis, which is on the same foundation as in psychognosis, but with word labels, I will use the spacetime concept. The foundation of the single idiom formation is by a mental mechanism using the working memory for perception to encoding for the new idioms or working memory for recalling old idioms. It is essential to symbolically formulate the construction of an idiom in its basic psychognostic form in which the matter for identifying the idiom is iconic. This model for the single idiom should be constant and usable throughout the steps to come for new concept formation, visually iconic nonlingual, iconic lingual, or noniconic absolute lingual of abstract types. This process can be shown in an equation in which the idiom to be formed is presented in a spacetime presentation, with the word-code as the content matter and the dynamically invested time material as energy in the following form:

$$\text{Idiom to be formed} = \text{Matter} + \text{Energy} + \text{Time}.$$

In this classical spacetime formula, the matter is the only known element mentally defined by a name code and, therefore, conventionally known. The energy is known as the Navand energy used mentally to encode the idiom. The time value is unknown, but may be regarded equal to the time the working memory uses to encode and register the idiom, or the time commensurate with the energy used. We can summarize the steps as

Idiom to be Coded (IC) + Working Memory Time (WMT) = Coded Idiom (CI)

Or in simpler form:

$$IC + WMT = CI.$$

And we can see that CI, as an element of the spacetime, can be written

$$CI = \text{Code Label} + \text{Energy} + \text{Time}.$$

And as energy and time can be regarded to be reciprocal conditional figures of one another, we can write

I. CI = Code Name + Time.

The equation restates in essence the same two-frame modal form for concept formation initially formulated for the life commandment in chapter six. The essence is the stimulus by the Navand energy working through time, giving memory, like *recirculating biology as moments*. In fact, the time continuum for the dynamism to occur is accommodating the energy of the stimulus by the attention provoking name code either from external environmental or internal mind sources.

The essential difference of this plan from the private plan for concept formation is the fact that triggering recall is voluntarily possible by the explicit memory at ease through the coding of the iconic concepts. Every worded concrete concept is thus instantly recallable using the memory ground, which now is functioning as an accessible semantic memory. In this way, two fields of free ideation practically open in the mind—one idiognostic with an internal expanded field to be checked and rechecked at will and unlimitedly, without needing a triggering external repetition to reexperience a passed event, and another basically private psychognostic through which to possibly also induce and control a similar effects in other minds that use the same idiognostic codes. Idiognosis thus opens the doors to a new world of unlimited concrete and abstract experiences and unlimited travels in time in talking to self and to others both directly in precise literary terms and indirectly in metaphoric abstract meaning.

The crux of the matter is now the inner responsible autonomous will. This autonomy, in whatever substance or form or principle, is certainly only made realizable but not formed by idiognosis. We could say that its realization through pondering by self, per se, indicates that its mere recognition may be proof for it to be nonself. This of course is one side of the puzzle, which can be easily solved accepting bioforce, biocracy, and autos to explain the puzzle. The other side is that idiognostic world exposes all concrete iconic concepts with identifying words as if they are reflected in a mirror that can only give coded reflections by words instead of real concrete figures or the real contents, therefore every abstract concept also will have an unreal image of word. This point cannot be denied as indeed words have created immense fields of metaphors by meaning extension and, above all, by abstraction, and all these abstract concepts have no concrete representatives or have ill-defined shadowy figures at best. The only practicality remains the reality of conscious will, which invariably follows a known lapse of time to become realized as conscious and thus can only reflect autos. The segment of time, which I have called the Libet constant already discussed,[4] must be regarded as representing the time value liberating the energy by Navand initiated through autos with will that is recognized by self only after it is realized.

IDIOMAS IN THE MIND

In the formula **I** for the coded idiom, time represents the energy in equivalence. This energy in the trilogy concept of NET is in reality a more comprehensive form as was shown in chapter seven for expressivity trait. It serves coding by its originator called Navand, and is actualizing the coding or recalling a coded idiom. It is borrowed energy from the basic time titan ground energy of the balanced metabolic state. The energy as the metabolic neurodynamic fuel guaranteed to be permanently available in the balanced metabolism and the time as the inner unlimited time titan are both inherent basic parts of bioforce in biology. Both are used in psychognosis initially in starting symbolic semantic implicit coding with semions and moveons prior to idiognostic explicit semantic phonetic coding. As idiognosis follows psychognostic steps, it actuates explicit individualization of coded idioms, which change from the psychognostic symbolic form made in the permanent time titan into singled out specific idioms by word labels obviously without separating from their time matrix, which is the present time, and will be reflecting the same in the span of the working memory. By choosing the time rather than the energy as representative to simplify the formula, not only the semantic serial implication in the coding process fits the time continuum, its essence as "*being*" indicating the time as the expected content of the second frame will also be naturally matched by worded examples as we will see.

This two-frame model equation seems to be the fundamental basis for code application in mental work, in perceiving, forming, or expressing all single idioms. We can see that we must include an inherent energy part with each idiom thus formed, and indeed, there is no mental idiom formed with no energy spent. This energy particle whose equivalent of time is represented in the formula I may appear unknowable at the first glance, but remembering Libet's constant universally operating in any and all conscious attention-intention processes, we are forced to admit that the working memory must include that constant in acting to form or to retrieve the idiom. We can assume that the initial WMT for the encoding process includes Libet's constant to accomplish the task, but for subsequent chain recalls under normal conditions of current semantic memory availability, perhaps all the energy needed may be equal to WMT minus Libet's constant with significant energy saving.

The process of attention-intention to provide for conscious recognition and manipulation of idioms is a one-time starting process to follow the chain to the intended end. This is consciously secured through explicit memory without using Libet's constant time anew between steps in continuation of the working memory to reach the intended end in one attention-intention process. In the normal process, the course of events goes on as if memory ground functions as a holistic exchange facility between the time titan and the real time and allows repeated usage of idioms, expanding the added mental life with the minimum of energy and time spent by the working memory. The initial neurodynamic energy expenditure and the time usage by the working memory that includes Libet constant are thus minimized to the strict necessary for each attention-intention protocol. In this way, the serial processes of mental ideas turning into words can keep going on in idiognosis most effectively and economically continuing in chain without reusing the entire working memory at each repeated stage for connection of idioms but with its face value minus the Libet constant. This mechanism may theoretically include the elemental connective idioms developed in language (all function words) as auto-triggers for extending the action of Navand, the very initiating trigger that can be assumed as Libet constant initiating and securing relays without repetition between starting attention to completing intention.

THE HIDDEN DYNAMISM

The richness of the new material in form and meaning in idiognosis is astronomical. The main reason for this immeasurable richness is an automatic exponential potential of self-proliferating items resulting from the mutual enhancing effects of the speechless psychognosis and speaking idiognosis in combination, which can feedback one another in a theoretical but possibly reciprocal way, reflecting on the mind that harbors both. In fact, the psychoaffective and psychonoetic nature of concepts described in chapter 7 can form expandable fields of idiognostic presentations mainly by metaphoric meaning transfer to cite just one endeavor as witnessed, which is the literature. This richness in form also includes rich dynamism in formulated meaning, which, in essence, reflects the power of the expressivity trait that works in two distinct forms—unintended tacit expressivity and intended directing expressivity.

In the first form, unintended and uncontrolled, like in daydreaming, the individual meaning of iconic concepts now also identified with words is per se a potential origin of spontaneous interchanging and expanding of meanings in the mind with baseline energy expenditure assuring a potential ground of expressivity. If we remember that the visual impressions are not controllable and both visual and visional imaginative automatic serial scenes are rather the rule, we understand the reality of the scenic cinematic extensions that may continue in our minds uncontrollably as it is also true with music and rhythmicity in general. The field of language, by the metaphoric meanings alone and with relays made in the mind between iconic only, iconic-lingual, and lingual only items, may extend profusely, practically without keeping limit. This richness and dynamism is inherent to the new situation and is autonomic in the state of potentially enhanced permissiveness in meaning relays through linguistic formulation and with or without eventual directing intentionality leading to open expression.

In the second form, the minded expressivity, which uses semantic exteriorization to transfer intent is actively responsible for the extent of forms and dynamisms of idioms. Based on this reality, the intent (the will through autos) actualizes the potential expressivity and originates Navand as the innermost mental integral power wested in the agent that is self. The intent therefore principally can be in the form of directed affective type, initiated with motto content as discussed in the last chapter, or a nonaffective stato initiated without motto content, mainly handled by self permissively. The targets will thus include affective idioms and concepts or mixed affective-neutral or neutral types, going back and forth in mentation, freely or elected to be directed and expressed. So in this line of consideration, the new material could be significantly dynamic, emotion containing like in all affectively charged expressions, or it could be static and neutral as in most symbolic scientific wordings or could be mixed. The change from unintended to intended expressivity is fundamental in exteriorizing the NET trilogy, which in analogy changes the subatomic quantum uncertainty principle of time indefinite nature into the directional selection and fixation of time definite type.

Dynamism detected in the visual speechless mind in psychognosis by iconic changes must be reflected in words in idiognosis. A priori, it appears that this may not be achieved with just the two-frame models making a module and may necessitate another frame for some symbol for the dynamic expression by words, but this possibility is only noted to be rejected. In fact, the frame for the matter, the identified worded name and the frame for the time through which the dynamic effect takes place are the only frames for these two precise functions, and if the final result is to be shown including the dynamically changed meaning, whatever symbolic word must be added has to be included in the second frame for the allocated time. The additional element is in reality assuring time relaying. The evolving mind from psychognosis to idiognosis, in fact, solved the problem without introducing a new frame by adjusting the names and the adjectives with inflections in the same allocated time for the second frame as was seen in chapter 7. In this way, dynamism that needed worded representation was expressed by inflections (usually a vowel suffix change) as the simplest novelty introduced without a radical change in the model. This solution left intact the basic two-frame model that was established in psychognosis as one constant primordial principle. In the idiognosis continuing in the early monodimensional present time era, with multitudes of applicable meaning expressions to be exteriorized by words, the mind created function words and verbs. These words with intent or action charge, naturally using the frame of time and showing specific dynamisms and relational connection with logical sequencing, replacing inflections of nouns and adjectives, prevented additional frame formation. Thus the basic two-frame model was safeguarded. Accordingly, motto idioms, stato idioms, or mixed types could still be used on the same old principle of binary origin (matter in first frame-its being or nonbeing in second frame) serving expressivity. In this setup, names and/or adjectives could be served by verbs and function words with dynamic precision and logical sequencing, expressing the holistic concise meaning of psychognostic understanding in a more evidenced way. This expressivity power gain could certainly affect both psychoaffective and psychonoetic capabilities, but it increased incomparably more the noetic one quantitatively. The evidence is clear, comparing the realization of the state of mind for qualia versus science that hardly allows qualitative comparison, but shows unlimited quantitative expansion for the scientific items.

These formulary two-frame model representations schematizing idioms formation and storage include Libet constant in real timing. In reality, they carry the process of attention-intention to initiate new idioms or for recognition and manipulation of idioms that are already consciously formed but are to be recalled. The process of formation and storage starts with working memory (including Libet constant) and that of recall for forgotten idioms would use the same process and time whereas the recall for current idioms theoretically follows in chain to accomplish the intention using just one initiating Libet constant itself equivalent to Navand energy value. Here, to gain a more comprehensive view, a recall of the notion of the quantum energy, as discussed in chapter 3, in the auto-hetero-gnosis is appropriate, reinforcing basic quantum rules.

Working memory seems to actually make the connection between the implicit and the explicit memory for completing the attention-intention process so that time accompaniment for the process to be initiated and completed would just use the Libet constant once with the working memory extending the Navand effect through the conscious work to accomplish the task. Or we can say that working memory trigerred by the Navand energy equivalent to Libet constant starts in psychognosis establishing the connection with the time titan, and the rest of the process in idiognsis goes on as in actual psychognosis in a time permanency but in implicit automatic time-defined sequencing. The process may be likened to an electrical key switching on all lights at once in a circuitry in parallel connections. Thus, for each attention-intention process in idiognosis, the working memory starts with the Libet constant with Navand allowing one quantum of energy to be spent for the action completion. In this view, the quantum is attention-intention ruled, and in its biological nature, it may vary with each global unit of work in accord with intermediary steps involved but using the span of the working memory extendable within its normal range without interruption or undue repetition.

THE FORCE OF MEANING

Further insight may be gained considering the network circuitry of neuronal connections in the brain and the quantal energy expenditure at synapses keeping with the concept elaborated by Walker.[5] In accord with this author who also refers to the work of Katz and Milhedi[6] on synaptic energy transfer, there seems to be an actual quantum mechanical tunneling of electrons through the synapses firing with release of neurotransmitters. The result that is stimulatory to the post synaptic neuron and may also turn to be inhibitory through intermediary channels thereafter—thus evidencing inherent duality on the one hand and assuming a defined number of electrons (up to nine) tunneled for a defined synaptic transmission on the other—is frankly quantum-natured. The great number of 10 billion neurons and 23.5 trillion synapses of the brain then can assure the probabilistic uncertainties of quantum physics[5] in simultaneous synaptic firing. This consideration lends credit to the notion that the mind seems to have a 200 milliseconds time for conscious yes or no to an action started unconsciously according to Libet's results discussed earlier. As such, psychognosis does not need but the basic Libet constant time or Navand energy for semions-moveons construction, but idiognosis-actualizing self with yes versus no decision would need Libet's constant plus additional 200 mesc to clear its position with words.

The crux of the matter here is in the meaning elaborated originally in the mind from semions and the conclusive reaction to it synthesized in the mind with moveons, the first inciting the second for significant alerting meaning to autos, but if not alerting, it indicates status quo and no moveon formation. The basic fundamental two-frame model of static or dynamic matter and time change indication for each will be unchangeably operational in the idiognosis as well, but with additional work. In both frames, the element of the meaning value is important in revealing the identity of the material as the subject initiating the change, or object evidencing the change, indicating the cause and the effect of the change observed in a crucially clear way. But the weight of material identification versus meaning of the action as primary or secondary importance may be more ponderous for the subject of the change than for the object of the change or vice versa, or perhaps entirely for the meaning of the action itself at the expense of no significance for either subject, object, or both. Thus, the ending effect of the change may only symbolize the meaning of the action once completed, and the change itself giving the meaning would be the primary theme that seems to be fundamental in the holistic psychognosis, not the agents (subject/object) at either ends. This, in fact, should be our conclusion believing the mirror neurons functions reflecting the actual action in time, the motional physical change of both self and nonself body movements in repeated reflections and registrations.[7] Thus, the action of external physical nature of the cause (stimulus) and its time change sequences to complete, and the self reaction to the action again in terms of mirror neurons reactivity in the brain elaborate the holistic action significance, which is of prime importance in constructing semions and moveons. All these steps must be elaborated in idiognosis by using words with all subtleties of the literary meaning included in expressivity.

Dynamic relation between subject and object, established in psychognosis, is therefore clearly subordinate to the main fact, which is the action, but becomes evident in the expression in idiognosis in transitive or intransitive forms of dynamism related to agents. The role of the meaning for the demonstrated dynamism result is, in fact, the crucial one and is primarily time dependent. It is what the sequencial visual impression of events is in the mind and in which iconic figurative dependence of the meaning to either a subject or an object is not a rule. This dependence is intrinsic to the meaning of action. In fact, a visible subject or object may not be even sensed in the biologically interpreted natural environmental changes like light or dark or cold or warm or even in wider seasonal changes that all show the state of being for a dynamic change with its mentally interpreted meaning but not the invisible agents involved (subjects in retrospect, or objects in prospect). Accordingly, expression of dynamism in the environmental changes may be fundamentally change-ruled and secondarily agent-ruled. Dynamism, in fact, is the change per se with or without evidence of any agent involved.

In terms of applicable logic, directing the essence of action significance as the crucial meaning determining either an anterograde causality inductive logic interpretation, subject to object, or a retrograde deductive one, object to subject, it can be operating if identifiable agents are included in the reasoning. But the principle of the two-frame models imposes that the action starting and ending in the first frame is being allowed and carried by the second frame as a whole; therefore, both subject and object, if any, cannot be separately carried in the second frame, which carries all effect only as a whole dynamism in time. Thus, the only possibility would be for the action to include both subject and object either at separate ends of the effect or at one end as object that could only be the subject itself or better as an action-imbedded object. Thus, transitivity, passivity, and impersonality can be explained. This basic psychognostic feature of the speechless mind shows itself in the variability of verb meanings and models of their worded usage in idiognosis although the principle is uniformly change-dominated meaning for dynamism in the expressed concepts.

The meaning related dynamism in expression, maximum in motto idioms, is intentional and, thus, under some mental norms. These norms represent primarily the orientational relationship of the autos in its egocentric position to nonautos environmental items as seen in the past chapter, which determines the affective meaning of the item and secondarily, the affective permissiveness of the society's restrictions. For motto idioms, the plan of construction in essence brings the biological meaning of living moments from the archaic biocratic type of holistic nature, significant to autos, into the idiognostic sphere preserving the core but with some surface change. This process, initially operating in the time titan, reaches the status of independent time significance with memory, serving the intermediary phase, and finally that of being label-significant and time-significant in the completed idiognosis. In this change, the idiognostic mental presentation obeys the two-frame modular process to assert functional stance of either static or dynamic meaning as related to time, making them solid inseparable parts of the present time. The present time of the actual conscious realization that gives recognition of the idiom as nonself and accompanies each meaning and the whole chain of concepts and ideation in the entire mentation is still in the inner time titan transiting into the created real present time (biochronogenesis) with the working memory. So in actual fact, changing static to dynamic status of the idioms is the factor that introduces steps in the time titan allowing transfer to real time that is ultimately expressed in idiodynamic verbs as will be seen in the next chapter along discussions on the Navand theory of languages.

Motto idioms in idiognosis reflect an affective meaning, and the time material as the present moment reminisces the time titan. Explicit recognition of self in idiognosis, instead of implicit recognition of autos in psychognosis, does not remain a holistic unity; it is also affected, and self(s) can be formed just as time can be divided. This dynamic change establishes the modular self(s) with variable functional capacities in the application of idiognostic activities with motto, stato, and mixed idioms. Thus, full attention-intention focusing on various aspects of affective, artistic, or scientific activities are made possible in modular forms. The steps could be shortcut mentally for most modular self(s) in the numerous particular fields of art and science linguistically coded that, for brevity, could all be grouped in the three categories of motto, mixed, and stato items in decreasing affectivity order. However, in final analysis, the force directing expressivity, regardless of means assuring expressions, is unidirectional, allowing the evolutionary principle of irreversible progress to continue to no visible end with time definiteness.

This trend of relentless forward moving is remarkable for keeping unflinchingly secured in an infallible constancy. This fact is further evidenced in the transitional change from psychognosis to idiognosis in which the primordial holistic expressivity on semion-moveon construcion is clearly saved with the unchangeable two-frame models for concept formation in spite of word fragmentations affecting holistic psychognostic concepts. This fragmentation, it should be pointed out, seems theoretically illusory, as fragments can only be significant if making a whole. The guiding rule is based on meaning regardless of contributing parts formulating the meaning. Thus, it appears that the primordial essence of the bioforce in its progressive irreversible character reflected in expressivity is obeyed in expressive modalities and guides every step in the forward course infallibly, reflecting its integral principle of immortality. This principle and its axiomatic nature are governing all biological transitions in evolution. In expressivity by idiognosis, we must accept, the addition of words has not affected the basic integrality and the energy balance in any way jeopardizing this axiomatic forward continuation of the biological life to its unknown destination. Indeed, the plethora of words in idiognosis seems to be unreal, similar to spread pieces of a jigsaw puzzle that hide a reality that must be suspected and searched to be found. The only reality is in the meaning formed by the holistic two-frame models allowing concept formation in psychognosis and idiognosis, using neurodynamic energy and revealing vast scopes forming the ultimate immaterial biology that serves metaphysics: Abstraction through time.

THE NAVAND THEORY OF LANGUAGES

The original investigation leading to the Navand theory of languages was done in Persian, Old Persian, and Avestan languages,[8] hence the word Navand, meaning impetus in Persian. The investigation was described in part in chapter 7 with the speechless psychognostic mind in what concerned the foundation of the expressivity trait shown to form psychoaffective and psychonoetic fundamental models for holistic wordless concepts. The prototype patterns were formed in a two-frame model—the first frame to contain the cause and the second the time for the process to give the result. It was argued that the cause frame containing the nonworded iconic figure of the agent should be always coupled with the time frame, and the causal figure could be static or dynamic, forming a static or dynamic concept. In both types, either the cause remaining the same in the static state without demonstrating any change and coming out in the extended present time as a constant uniform meaning or making the transfer of the meaning change in dynamic concepts to be shown as the result, the time frame did not show a figure for time but for the result as a state of being. In both cases the time itself was the continuum of the time titan and did not show iconic figure representing time, it only carried the meaning of the effect, static or dynamic. This presumed process would result in the holistic psychognostic speechless sense either in a summarized meaning of a static concept in the present time, with whatever the intrinsic meaning of the unchanged figure in the first frame would be (morphon), or would give a dynamic concept that would show the dynamic meaning change (semion) and carried through the second frame to give the final meaning format (moveon) again naturally in the present time.

This scenario in the speechless mind appeared to be the way expressivity trait would operate, and the concept of NET was introduced to explain the energy expenditure, started by Navand, necessarily accompanying the operation. The concept defined the intent of the expressivity trait by Navand as the trigger agent from autos. The neural energy spent was therefore understood to be unknown. The time, however, appeared one of the elements in the NET trilogy (Navand, Energy, Time) that could be potentially demonstrable and testable to establish the soundness of the hypothesis. It could be shown by inference in the psychognostic speechless mind referring to the time titan as the plausible biological tacit permanency of the present time, which could radically indicate "*being,*" asserted in words and concretely confirmed if the time frame would actually carry this worded concept for time as *being*, *am*, *is*, etc. With the worded concept indicating the time, not only the entire expressivity trait operation could be solidly proved including its contributing matter (icon and/or word) and energy (time) to the process, but the fundamental chain of the evolution leading to this stage of the speechless mind and the following speaking mind would also find confirming support. All the basically established models of expressions could then be comprehensively understood and accepted on a solid foundation.

According to the basic rule of the Navand theory,[8] every lingual item as a word of any category will include the two fundamental contributing elements to its construction: (1) the lettered symbol as a carrier, which identifies the meaning it exteriorizes; and (2) the carrier's functional energy (abridged in Navand from the NET principle) as the force of expressivity. Thus, Navand theory of languages postulates that every word in the speaking mind is formed by two inseparable representatives:

1. Identity representative with the word itself called Namaa (aspect in Persian)
2. Energy representative with the dynamism in the word's meaning called Navand (impetus in Persian)

And we can summarize this idea in:

II. WORD CONCEPT = NAMAA + NAVAND.

In this formula that uses the Persian words from the initial work on the subject,[8] Namaa is from the Indo-European common root, meaning name and symbolizing the iconic representative, if any, mentally and audiovisually. Therefore, Namaa can contain from negligible to full significance capacity identifying the meaning of the item. Navand, from the NET complex, summarizes the power of energy-time in the meaning transfer in either single static concepts like single names, or dynamic concepts with change accompanying the subjects at the beginning, or the objects at the end of the dynamism. Navand expresses the inner autonomous mental dynamic component of each and every word in language as the word's charge of energy sine qua non for the meaning to be produced. It can carry presumed component of the unconscious formative foundation of any concept initiating to be consciously realized in the mind (with Libet constant), with its identifying word that may have no precise material content like interjections *ah*, *aah*, *oh*, *o*, *ouch*, etc., or a word with clear conceptual meaning like every concrete object to be actualized and exteriorized.[8] Thus every word in Navand theory has both Namaa and Navand that each can have the respective charge of meaning and dynamism from a minimum to a maximum in an understandably inverse reciprocal balance—minimum Namaa significance with maximum Navand effect (highest emotional charge like in interjections) and minumum Navand effect with maximum Namaa content (highest noetic charge). The first group includes mainly initial psychognostic meanings with their transition to idiognosis and englobes all function words including verbs, and the second represents words of the transitional phase and late idiognostic formation covering all content words. Namaa, appearing initially as just the vehicle of the Navand expressing emotionality, will change through the transitional period from psychognosis to idiognosis to lose emotionality of autonoetic nature gradually in favor of the content that becomes of nonemotional noetic type to reach today's scientific symbolic figures.

The idea of a biological agent representing expressivity trait, to engender realized mental expression in semions and moveons as iconically structured holistic concepts in psychognosis, eventually reproducible in idiognosis, seemed plausible and worthy of further expansion in the Navand theory. The NET trilogy then was envisaged, which provided reasonable satisfactory explanation, but the agent Navand, which was the key figure in the whole process, remained purely hypothetical. With Libet's work clearly confirming the premonitory unconscious timing for all attention-intention mental processes in an invariable constancy,[4] the theoretical Navand found experimental support. So in summary, the theory that was initiated on deduction based on mental realization exteriorizing the concept by incriminating the unknown mental energy through time called Navand, ultimately found its convincing experimental proof in the Libet's constant preconscious time. This initially psychognostically based setup clearly uses the energy equivalent of Libet's time that I consider a quantum represented by Navand triggering the working memory. In the Navand theory, the holistic semions and moveons of psychognosis are initiated by Navand through visual stimuli of iconic nature and in idiognosis by words with audition, vision, or both.Dynamism by Navand operates in two different ways; the basic way is by sustaining the matter of the concept in the time titan continuum and transposing it into the real present time using implicit-explicit memory. In doing this, dynamic modulation is transferred from inapparent time titan to the present real time. In this basic way, Navand dynamism serves to ascertain the neuroperceptive conscious recognition of a concept induced as first time impression or recall from memory exposing the Namaa of the concept to be exteriorized from the inapparent present time into the actual present time. This dynamism is tacitly in the present time without needing confirmation by an indicator in worded form of the *being* category. The mere conscious actualization of the concept is necessarily contemporary to its expression and is naturally in the present time needing no additional confirmation. This basic dynamism works in two modalities and accompanies every conscious voluntary idea, which may include mental actualization of the Namaa (name of a concrete or abstract item) by the

Navand. In the first modality, the basic dynamism distinctly forms the ground stage for all ideations to start voluntarily by Navand with Libet's constant and to stop voluntarily without going further just finishing the attentional change. In the second modality, the process continues in the present time to accomplish an intentional work in chain.

In its second form, the Navand dynamism introduces *being* in words indicating potential or real progress in the present time. In doing this, changes of Navand dynamism, from the basic ground of static immobile type into a variation inflicting moving dynamism, secures changing forms of the initial material that is the first originating Namaa by adding qualificative words and intercalating *being* in its present tense with *is,am, are* in English or equivalent words in other languages. This brings out verbally the dynamic change in mental actuality. In this second type of dynamic action, an embedded dynamism may become included by Navand using *doing* in worded expressions, thus exteriorizing special functional character of the initial Namaa in addition to its actuality of *being*. This dynamism forms the foundation of the monodimensional transitional time era of psychognosis-idiognosis for an unknown extended period. In fact, this monodimensional time period of an extended present time seems to originate in time titan and exhibit signs of conscious recognition when realized between the past and the future transiting from psychognosis to idiognosis. A tangible example became known to the world with survivors of tsunami disaster of 2004 known as Moken people or Sea Gypsies well studied by Ivanoff.[9] These people have no concept of time, no word for hello or goodbye indicating a beginning and an ending, and no abstract notion of need (want) as they only visualize the action of giving or taking in their expressions. Changes from such dominance of a uniformly reigning present time, only showing time change through action, take long time to allow meaning specifications to be eventually introduced in the syntax with function word and various verb types as we will explain in the next chapter.

In complex dynamic concepts of more than one name as in adding conceptual meanings and in syntax, Navand serves to continue the chain of the ideation with the verbs and function words. The formula II is in essence the same as the formula I shown for the idiognosis construction. The foundation is invariable and constant, but changes occur in the mental interpretation and transition into linguistic formation, presenting some subtle practicality or peculiarity in actual linguistic application. Thus, the Navand theory of languages establishes the general foundations of linguistic rules for expressivity trait from the archaic times that seem to have been formed on a solid unchangeable pattern in all prelingual preparations in the psychognostic speechless mind. Navand theory establishes and explains beyond possible doubt the universality proposed by Chomsky[10] as an inherent fact applicable to principles and also can explain particularities and variations in spoken languages as I will have the occasion to briefly check in relation with Chomsky's universal grammar in further discussions.

REFERENCES FOR CHAPTER TEN

1. Mithen S. 1999. *The Prehistory of the Mind. The Cognitive Origins of Art and Science*. Thames and Hudson, New York, NY.
2. Wright K. 2006. Times of our Lives. A Matter of Time. *Scientific American special edition*. Pp. 26-33.
3. Baddeley A, Conway M, Aggleton J. 2002. *Episodic Memory. New Directions in Research.* Oxford University Press.
4. Libet B. 2004. *Mind Time. The Temporal Factor in Consciousness*. Harvard University Press. Cambridge Massachusetts, England.
5. Walker EH. 2000. *Physics of Consciousness*. Basic Books. Perseus Books Group, New York.
6. Katz B, Milhedi R. 1965. The effect of the Temperature on the Synaptic Delay at the Neuromuscular Junction. *J Physiology*. (London) 181: 656.
7. McNeil D. 2005. *Gesture and Thought*. University of Chicago Press. Chicago and London.
8. Amir-Jahed AK. 2002. Seeing the Thought in the Mirror of Speech. (text in Persian). *Rahavard, A Persian Journal of Iranian Studies*. 63: 50-61.
9. Ivanoff J. 2005. Sea Gypsies of Myanmar. *National Geographic*. 207(4):36-56.
10. Cook IJ, Newson M. 2001. *Chomsky's Universal Grammar. An Introduction*. Blackwell Publishers.

CHAPTER ELEVEN

THE IMMATERIAL BIOLOGY

NAVAND IN THE GENESIS OF SPEECH

The Navand theory implies that our mind energy in its neurodynamic mechanism directed by the expressivity trait uses an objectivable energy amount by Navand activating the triggering concept formation and sustaining expression. This action of Navand, aided by preparatory plans as discussed in the last chapter, ultimately constructs our ideas that, in order to be communicable, they need to be formulated in an all comprehensive form. Thus, interpreting the syntax in its general broad sense of what it really is, i.e., combination of meanings directing dynamism to give an understandable expressive idea would allow us to date its origin as far back as to auto-hetero-gnosis with biognostic recognition of autos and nonautos and the inception of memory and developing self as seen explained in our past discussions.

At that archaic time, the visual-visional mental interpretation, no matter how primitive with the primitive memory fixation, served the mental organizer as an inner tacit expressive capacity as precursor for the language to be. This nucleus of the tacit language evolved to serve psychognosis ultimately with the refined holistic semions and moveons. So in short, the early biognostic senses, faced with time related dynamic change in the sequentially appearing actual events, was using syntax. This primitive syntax can be said to have been based on a binary principle of presence or absence of a single perceived figure in the present time, as yes or no for any meaning it could induce to the biopsychon's principle of SR, assimilated for yes and rejected for no. The foundation, purely biological, served as the building for all subsequent steps in the expanding expressivity trait. The passages through evolution allowed all necessary adaptive changes to ultimately reach the nonphysical contact expressivity, the wireless thought transfer through language, but still on the initial biological foundation. Thus, creation of the literal world, making our inseparable symbiont, has nothing artificial even though figurative single identity (psychognostic iconic) and double identity (idiognostic iconic and worded) concepts, sense or nonsense, are profusely present in the language as biologically originated but nonmaterial items in essence as meanings in our minds only. The binary principle is as dominant in the present expressivity as it was at the initiation, but now, with biognosis reaching full auto-hetero-gnosis, the nonbeing is by itself a being.

Starting with concrete elements in biognosis, if two figures were related, the effect of their relation in a synthetic singled meaning was again interpretable in the same way as holistic yes or no, but to actualize the synthetic meaning in this binary *yes-versus-no* mechanism, the process had to go through the steps of reviewing the sequential dynamism considering the matter and the material change. In short, the visual figure had to be integrated into time in the revision, or simply said, the frame for the matter had to be assembled with the frame for time. This basic binary logic evidently expanded beyond the exigence of the principle of SR to include the life commandment of logic and happiness and pleasure-pain principle setup on the same binary foundation. The elaborated outcome could be said to be responsible for the refined holistic semions and moveons in the psychognostic speechless mind and later transposed to the worded mind of idiognosis. The setup naturally operated in the present time, based on the binary axiom of concrete existence of visual or other sensed stimuli meaning presence or absence of *being* for the sensed entity. This realization plausibly originated the psychognostic infinitive *to be* in the mental understanding of the psychognostic speechless mind. In fact, we can presume that probably the very first verb in languages was exhibiting the *sense of concrete being* in the present time titan by the autos for self and for nonautos entities, a sense that was crucial in the mental formation of the holistic semions and moveons.

This original mental realization of the binary principle for *being* versus *nonbeing* must be accepted as the nucleus for the logic of *cause-effect* versus *no cause—no effect*, the very basis of the life commandment and all two-frame models that make real modular matter-time metaphoric sets with inseparable matter-time meanings. Evidently, the frame for the matter including the iconic visual percept is the one operating in the visual speechless psychognostic mind, and the frame for the time belonging to the inner time titan is the tacitly operating frame without specific iconic representation. The time becomes represented in idiognosis with words expressing the *being* carried in the time frame. Thus syntactic thought does in reality exist tacitly in the psychognostic speechless mind. It operates with the two-frame module and the iconic dynamistic meaning in the monodimensional time titan, unrealized to autos and becoming realized to self in idiognosis for the first time with worded confirmation in the monodimensional timed syntax.

THE TIME FOR "BEING"

It is of interest to find possible worded presentation of this present time sensing and expressing in the later appearing idiognosis. In the well-preserved Ghatic texts of Avesta and Vedic Sanskrit and in Old Persian, all estimated to represent a period of usage for these dialects of the Indo-European origin extended between 4000 to 8000 years ago, the infinitive for *to be* or *being* is *ah* in Avestan and Old Persian and *as* in Sanskrit. In both, the vowel is the solidly invariable phoneme /*a*/ and is also ubiquitously found as suffix at the end of most names and adjectives, inciting the idea that probably it contributed to these words' formation as a natural indicator for their mental confirmation of "*being*" in the tacit present time.

At the same time, coincident with this suffix phoneme /*a*/ that is intrinsic in the words generally giving the meaning of the static *being* in the present time, the same phoneme and other suffix phonemes and morphemes are found added on to names and to adjectives, more as additional than as natural formative ones,[1] which seem to induce some dynamic functional character to the words. These facts appear compatible with the presumed transitional period of the monodimensional era of the present time before the archaic syntax incorporated the finer subsequent specific verbs, tenses, modes, and function words. Indeed chronology of linguistic changes, from the archaic times to the present, evidences this course of events, initiating with suffix vowel additions exemplified in the Indo-European languages to indicate states of nouns and adjectives in relation to time and dynamic functions that gradually became replaced by additional refined elements. But relics of the old mechanism of inflections in words are still notable in most Indo-European branch languages with varying refinements as in others, specifically the Semitic languages.

To schematize the archaic syntax structure, the two-frame model will be used here in single concepts initially and will be expanded to two concepts, keeping with the inherent logic in the chain of meaning-governed expressivity. The processes will be shown assuming the hypothetical monodimensional time era that covered the transitional period of change from iconic psychognostic to full worded idiognostic mind with extended abstractive power. This situation is schematically shown in Table IV. (See over). The table includes the hypothetical archaic syntactic frames, which would match the two-frame models of matter and time with possible changes in phonetic expression for single concepts through various stages of biognosis with mental development. The matter is represented by the Namaa and the time by the Navand. The matter from the original, mainly visual stimuli of iconic nature in auto-hetero-gnosis goes on to be fully visual and visional in psychognosis and to acquire the additional worded symbolic figures in idiognosis. The time is shown as Navand (abridged in the NET trilogy) that includes energy and time in rendering the neurodynamic processes for perception to initial memory registration and fixation of the conceptual matter at earlier stage of auto-hetero-gnosis and for perception-fixation-recall at later stages. In all this deroulement of progression, the outcome will be the present time cognition of the meaning finalized as the concept tacitly understood to be in the present time, but not formally expressed.

With idiognosis, the matter becomes additionally symbolized in words over and above the iconic symbols formed and jointly with the worded time symbol that is inherent with the cognition of the concept in the examples of the archaic Avestan and Old Persian languages. This time symbol is represented in the phoneme */a/* preceding the later formed infinitive *ah*. In these examples, the time frame demonstrating the action of Navand in time must indeed show the Navand effect, which it does by the autonomously including the innate suffix phoneme */a/* in the natural way of confirming its cognitional weight for the *being* in the permanent present time. It is clear that the infinitive form *ah* only appears later in semantically advanced language with formal abstractive expression facilitating extension and generalized application of data in ideation.

In the originating step however, the simple phonemic */a/* is used as matrix indicating *being* as an innate constituent of the worded name. These two examples of Avestan and Sanskrit as archaic languages are chosen to demonstrate the reality of natural phonemic suffix */a/* that appears in examples of typical Indo-European words conferring the natural permanency status to them. It can be seen in Table IV that the outcome frame for the model of concepts starts from the general tacit cognition of the concept in the present time, to reach cognition-expression for words, still in tacit understanding in the present time. This is expressed clearly representing the time in the word giving the meaning of being for the concept in the simplest form of the phonemic */a/*. In terms of the Navand theory, it is the Namaa (the word) that carries the meaning, and it is the Navand (*/a/*) that shows the dynamism through time reflecting the energy in concept realization.

Progress of expressivity, to demonstrate emotion, made the dynamic states for the names forming complex symbolic concepts of noun plus adjective with meaning that changed from the original single concept with this added refinement and dynamism. In this change, the additional novelty was incorporated with the old cognized item invariably through an intermediate connection, meaning *being in the present time*, symbolized again by the same vowel suffix */a/*.

This connective element was initially reflecting the unconscious time titan and remained functionally the same, i.e., still the same old time expressed in */a/* that underwent minor diachronic linguistic change over millennia that followed. In the archaic syntax models for two items combining to make one concept, stages of changing examples are summarized in Table V. (See over).

TABLE IV

ARCHAIC SYNTACTIC FRAMES

FOR ADYNAMIC SINGLE CONCEPTS

Stages of Biognostic Expressions	**Frame for Matter**		**Frame for Time**		**Outcome**
Adynamic Single Concepts	**Namaa**	+	**Navand**	=	**Concept**
Fundamental Syntactic Stages	***Neurodynamic Realization***				
Auto-Hetero-Gnosis	Iconic + Perception-Fixation				= concept in present time
Psychognosis	Iconic + Fixation-Recall				= concept in present time
Idiognosis	Icon and/or Word + Fixation-Recall				= concept in present time
Avestan Language, Ghatic	Ahura (God) + Intrinsic Suffix /a/ = Ahura (/*a*/)				
Same Source	Hva (He) + Intrinsic Suffix /*a*/ = Hva (/*a*/)				
Old Persian	Baga (God) + Intrinsic Suffix /*a*/ = Baga (/*a*/)				
Same Source	Hya (He) + Intrinsic Suffix /*a*/ = Hya (/*a*/)				

Table IV. The basic formula II for the concept formation based on the matter-energy constitution, Namaa and Navand, given in chapter 10 is shown with the matter initiating in archaic times with the iconic visual figure of static or dynamic nature initiating the neuroperceptive process with the time effect for the Navand action resulting in concept realization. In subsequent archaic languages, this effect is shown phonetically with the phoneme */a/* as suffix attached to the worded icon. Examples show the intrinsic character of the inherent time representative assertion of the suffix that is inseparable from the name.

TABLE V

ARCHAIC SYNTACTIC FRAMES

FOR DYNAMIC TWO-WORD CONCEPTS

Stages of Biognostic Expressions	Frame for Matter		Frame for Time		Outcome
Dynamic two-word Concepts	**Namaa**	**+**	**Navand**	**=**	**Concept**
	Fundamental Syntactic Models				
Avestan/Sanskrit Source	Seta (/a/)	+	Vaesa (/a/)	=	Setavaesa[2] (Hundred Homes)
Old Persian	Baga (/a/)	+	Vazraka (/a/)	=	Baga Vazraka[3] (God is Great)
Same Source	Mana (/a/)	+	Kaara (/a/)	=	Mana Kara[4] (My Army)
Same Source	Kaara (/a/)	+	Hya (/a/) Mana (/a/)	=	Kara Hya Mana[4] (Army of Mine)
Modern Persian	Rooz (/e/)	+	Roshan	=	Rooze Roshan (Day is Bright)
Same Source	Khodaa (/ye/)	+	Jahan	=	Khodaye Jahan (God of World)

Table V. Archaic dynamic two-word concepts are shown emphasizing the intrinsic phoneme /a/ as suffix in examples shown from sources of Avestan and Sanskrit languages, Old Persian and new Persian, extending from about appproximately seven thousand to three thousand years ago to the present time. The principle has evidently continued in the present Persian as can be seen. The only change that has occurred in phonation is the normal diachronic change over millennia, but the fundamental principle of the two-frame model has not changed.

THE TIME FOR "BEING-DOING"

Starting with the first example of Table V, an Avestan and Sanskrit mixture, the two components both have the phoneme /a/, which they keep in the combined word in this example with the literary meaning of hundred homes and metaphorically meaning a star.[2] The numeral adjective *seta*, meaning hundred, and the name *vaesa*, meaning home, agree as adjective and noun in sharing the suffix phoneme /a/. The phoneme stands for the tacit confirmation of the state of *being in the present time*. The reinforcement of double phonemic suffix, one in each term, is rather the rule and is witnessed in almost all forms of the archaic dialects of Avestan and Old Persian. The second example showing a clearer instance of noun and adjective agreement with *Vazraka*, meaning great, which follows *baga*, meaning God, demonstrates the qualificative mode of the old adjective-name agreement.[3] Here again, the combination indicates the dynamic change taking place in the present time by */a/* that in reality signifies *is*. The third and the fourth examples are more interesting in showing the genitive mode, which, now in a more developed vocabulary, allows both forms of possessive adjective meanings to be used. The form with the phoneme */a/* alone, using the subject and the object both with */a/* successively, as in the previous example, obviously the older type, evidences the dynamic agreement of subject and object in possessing the state of *being* expressed by */a/*. The second form, shown in example 4, is the more refined expression. In this form, the object being the first and the subject the last, the pronoun *hya*, which literally means *it is*, after the object and before the subject indicates possessive relation between object as property and subject as proprietor, akin to *of*, confirming the possessive status by the subject. This pronoun *hya* is the origin of subsequent */i/* in Middle Persian and finally */e/* in the modern Persian language.[4] It plays the same syntactic role confirming the *being* or the *being of* in the genitive modes as in examples 5 and 6.

Other examples can be cited, but the only additional example that I believe is indicated to be shown, both for its authenticity of dating as close as circa 480 BC[3] and value of confirming the indicator role of the phoneme */a/* as naturally innate sense of *being* in the present time, is the well-known inscription in Old Persian in cuneiform writing by Xerxes that states

Baga	vazraka	Auramazdaa,
God is	great	Auramazdaa,

Hya	*imam*	*boomim*	*adaa*		
He (is)	this	earth	gave		
Hya	*avam*	*asmanam*	*adaa*		
He (is)	that	sky	gave		
Hya	*martyam*		*adaa*		
He (is)	man		gave		
Hya	*shyatim*		*adaa*	*martya*	*hya*
He (is)	joy		gave	man He	(is)[5]

This example turned into English, word by word, shows the repeated *hya* that could be translated as *He is*, *He who*, *He that is*, *He that* or *that*, but in reality the meaning of "*being*" or "*is*" is hidden in /*a*/ that is incorporated in the word "*hya*" from times immemorial. The refined translated version according to our present understanding should read ***"The great God is Auramazdaa who gave this earth, and that sky, and man, and happiness for man."***

These examples from well-preserved texts of Indo-European origin of archaic times reveal syntactic initiation based mainly on present time being tacitly confirmed for the subjects, objects, and adjectives in inflectional forms. Thus, the monodimensional present time syntax seems to have been real in the earlier initiating idiognostic language. These examples firmly support the Navand theory, which, in fact, must have been fundamental in the expressivity trait constructing the monodimensional speechless psychognosis, which formed semions and moveons on the two-frame models. From this basic speechless foundation to the fully structuring idiognostic languages, the idiognostic archaic syntax evolved in two directions both guided by the meaning expression. In one, the inflectional dynamism dominated to further perfection as can be witnessed in Semitic languages and, in the other, the evolution progressed with function words being added, replacing inflectional mechanism to a great extent as in branches of Indo-European languages. But both directions preserved each mechanism in accordance with the perfected dynamism and logic in the evolved syntactic forms.

The constancy of inflection in both evolutive directions remained with verbs that essentially meant *being* or *being engaged in doing* in the long period of the monodimensial time. The situation conserved the essence of the monodimensional time constant as *being* and exteriorized *doing* states of every *being* by inflectional suffixes in the examples we are studying. This is certainly the easiest and the most economical mental way for expressing ideas as long as repeatable actions do not obligate the speaker to express either present or past tense for both *being* and *doing*. When this obligation occurs, there is only one of two possibilities for the expressivity to work—either both *being* and *doing* are to be changed by inflection to satisfy expression or only one to change to indicate either the present or the past for both. In practice, in most contemporary languages, obviously it is the change for the time of the *being* that does the work, but the *doing* still remaining in the present tense witnesses the continued monodimensional pattern extended in the present time. For instance, in the examples of the most common current languages like in English, we can see *is sleeping* or *was sleeping*, or in French *est dormant* or *etait dormant*, or in Spanish *esta dormiendo* or *estaba dormiendo*. This stage appears to have beeen, and continues to be, the most likely lasting transitional phase from mono to pluridimensional time expression. It seems to have given origin to verb concepts from adjectives and nouns and or vice versa and have facilitated expression of various verbs and tenses including the differentiation of the earlier past to more remote past tenses using auxiliary verb combinations. This refinement seems to have followed the biological simplification based on conservation of energy, achieved in a most economically acceptable linguistic mechanism avoiding the burden of added frames. This was made possible by combining the two methods, inflection and new word formation (including function words), for making adequate lingual expression.

CHRONOPHORIC DYNAMIC IDIOMS

Based on the Navand theory, verbs can be viewed as chronophoric dynamic idioms and classified in three distinct categories—idiostatic concept verbs, idioconnective syntax verbs, and idiodynamic action verbs. This taxonomy is based on the relationship between Navand and Namaa in the mind and governs the expressive dynamism in that relation in speech presentations. The first two groups mainly represent structural binders saving and/or enhancing the meaning-related dynamisms. The third category is an idiodynamic type, which, per se, is inherently dynamic as singly expressing an action expandable in time to include actors, subjects and objects.

The first group makes the fundamental original concept forming idiostatic verbs that preserve the basic two-frame models of action effect type, best represented by *being*, in the form of inapparent tacit *being* and/or *doing*. The function then expands in few other worded forms in the developed language. Examples of their inapparent archaic suffix */a/* in Tables IV and V and in *hya* and */e/*, show the two applied forms this verb can take, which, in single or double worded concepts (name and name or name and adjective), clearly indicate the *being* of the single one Namaa concept in the present time, confirmed by its Navand, or the compound dual Namaas bound with the action of Navand in an inseparable whole forming compound concepts (genitive), hence the preferred name of concept verbs. The concept verbs in its simplest form is shown in the static archaic names shown singly in Table V with the suffix */a/* that confirms the actual state of being for the concept as a name. This concept verb and its newer form */e/*, derived from the Avestan infinitive *ah*, stands for *to be* that seems to form the very original prototype root for all verbs to form in Persian and, in principle, in various languages for inducing the meaning of *to be,* extended as *to do,* to show the state of being with variable meanings of *to be* as *being, making, doing, have been done, and to be done*, in short some state of *to be* or *to do*, either with the same root or with different roots. In Persian, it counts seven such different roots with meanings of "*will, can, ought, deserve, become, do,* *done*" for doing. These verbs can also serve as auxiliary verbs when indicated, similar to their very original root *to be*, according to semantically conditioned expressions. In reality, these verbs work as binders in a monodimensional chronophoric capacity binding the present time *being* and *doing* in static and dynamic word concepts.

Syntax verbs (idioconnective) make the group of function words that have no grammatical forms of verbs but have invariably correlative functional actions in forming and securing the syntax or expressing Navand to indicate the state of being. They bind the modularly formed two-model concepts in syntax. These verbs accomplish two essential functions in terms of adjusting the Navand-Namaa units, single or combined, in the syntactic chain by adding correct direction in two parallel lines: logic and continuity. Their actions are not only chronophoric in bearing the time, but in bearing the logical time for causality. These syntactic verbs form the group of all function words like *at, by, for, then, to, with,* etc., in English, for example, and equivalent words in other languages. This is made in the premotor stage of speech realization in the mind starting with the first Navand action in the attention-intention chain for sequences needed for sentence construction. These syntactic verbs are sine qua non for securing the final correct expressive idea without interfering directly with individual concept meanings or altering the constancy of the two-frame models. Thus, formation of a grammatical sentence, made on the basis of these verbs that are usually short phonemic-morphemic function words, allows the exact core meaning of Namaas to appear in the expression and all metaphorical possibilities too. Meaning ambiguities that can surface in syntactically correct sentences by the effect of these syntax binding words can occur, but speech errors other than word shifts in these syntaxes are rare.[6]

The third group of verbs, named the group of idiodynamic verbs, allows expressing dynamic actions and states to include actors and times in one worded expression. They include the classically known grammatical verbs for which, the Navand theory understanding matches the common description given for verbs in textbooks. These verbs secure a bond for three elements and their meaning in the actions they describe—the agent, the action, and the time. The issue of *subjects and objects* assembled in *agent* is in line with the fundamental holistic psychognostic concepts emphasizing the basic psychognostic comprehension, primarily of action and not of subject or of object. For the same reason, these idiodynamic verbs are not replicating identical presentations in the order of subject, verb, and object in different languages

In Navand theory meaning, the concept verbs and the syntax verbs more generally present the expressivity trait as a whole, acting as structural binders just the same as used in the archaic monodimensional time, and the idiodynamic verbs present expressivity specifically in terms of agents, actions, and verb tenses in the real actual time. These verbs are also chronophoric idioms, but additionally called idiodynamic to show that they can singly set the structure and the meaning specifically by securing both time and logic in the action transfer between agents—subject and object (transitive), or action condensed in the subject (intransitive), or in the object (passive) modal types.

NAVAND EXPRESSIVITY: THE POWER OF WORDS

Expressivity trait is emotionally based and by motivating lingual expressions naturally expands the domain of dynamic meaning transfer, which uses the two mechanisms of inflectional and noninflectional types in morphology and the broad metaphoric scale in meaning variations in languages.

The mechanism of the archaic inflectional type, which essentially opens the way for the effect of Navand, extending the meaning of Namaa as a concept through time and allowing its dynamic meaning change when combined with another concept, is the simplest way of expressivity expansion at the early stage of meaningful phonations. Thus, expressivity dynamism develops primarily on inflectional mechanism by the Navand function, allowing change in the Namaa meaning. However, using the same Navand mechanism, expressivity also initiates new Namaa formation with inherent new meaning. In short, lingual expressivity expands with new meaning introduced through new words from either new roots or old ones with expansion possibilities. The first category by new roots is of de novo type, concrete or abstract new mental Namaas created through Navand mechanism. The second category is by restructuring new meaning Namaas from old ones again through Navand mechanism. This second category is solely dependent on the predominant types of inflectional or noninflectional languages, being incomparably more expandable in noninflectional categories that are not subjected to rigid rules of inflections controlling the entire structure-meaning complexes. The inflectional mechanism has been satisfactory initially and, to some extent subsequently, beyond the presumed monodimensional present time period in a relatively restricted form, but the noninflectional progress has allowed astronomical expansion to the language and powerful dynamism to the syntax in realizing the intended meaning by the extending growth of number of new words.

The principle ruling the new word formation in this category of extension in essence is on the same archaic basis of keeping the two-frame modular system unchanged. The task is accomplished either by inflectional change of the root words through restricted rules of inflections in the same roots changing sequencing of letters and intercalating limited number of fixed additional letters or by using the root words in their original forms but adding affixes. The first modality, classically named inflectional, can only expand within the limits of the dynamic meaning modules, but the second named noninflectional has the liberty of unlimited affixes to be added to the root word. The potential expansion of noninflectional languages is remarkable. A comparative study of Persian as noninflectional with Arabic as inflectional has shown the calculated potential of the Persian for new words to exceed 226,275,000 as against Arabic with 1,750,000.[7] New independent specific meaning extensions, developed by Navand through mechanisms that can be abridged as perception-interpretation-expression, contribute extensive meaning variation conferred to Namaa of specific concrete or abstract material, still strictly observing the two-frame model originally established by the life commandment.

Elaborating on this last point, the most perfect syntax, if thought of in terms of meaning transfer by words is that which secures the truest perfect expression with the most perfect comprehension. The interhuman expressions of either emotional affective type or factual noetic nature, using motto, stato, or mixed words, therefore would best use the purest mutually comprehensible precision in the worded meaning in communication in the simplest natural two-frame models. If we wanted to study this aspect of the syntax, we really should go back in time and try to find the very initial step, the nucleus of such expressions that could be showing the purest form of the formula **II** given in the previous chapter.

This regression in time is possible, thanks to historical anthropology. In fact, historical anthropology gives us examples of social bonding in primatology, based on pair grooming, that occupies a good 30% of the living time of primates and some 23% in the *Homo habilis*[8-10] and also occasions phonetic expressions of oohs, aahs, and ouches as signs of feeling and emotional elations accompanying grooming.[11] These interjections reflect the purest affective meaning in similar situations today. Should they be interpreted as the clearest expressions in the purest syntaxes? The answer is of course affirmative and cannot be otherwise in terms of expressivity by meaning transfer. This is the clearest example of noniconic worded expression of motto type with Navand at its maximum in meaning transfer versus Namaa in its minimum, but dynamism in this type of one-word primitive syntax is fully operating and clearly evident as it would be with greater or lesser extent by all function words playing practically the same role in developed languages.

The difference between this aspect of Namaa and Navand (Namaa at minimal material/meaning content and Navand at maximal emotional/meaning effect) concerning the meaning transfer is clearly evident in the study of conduction aphasia where the conduction between the sensory and motor centers of the language comprehension and production (through arcuate fasciculus) in the left hemisphere is damaged.[12] It is therefore quite clear that the Navand mechanism in transferring the meaning without a material Namaa in chance phonation, or with nonsense Namaa in nonsense words as in conductive aphasics, is restricted to emotionally meaning senses of archaic type, mottos of limbic origin, needing no intact conduction between the two speech centers (sensory and motor), which developed subsequent to the limbic phonetic function of emotions. This development occurred gradually in psychognosis with correlations between iconic visual and hearing auditory stimuli establishing together meaningful single identifiable Namaas by words in idiognosis. The synthetic effect initiated the era of worded meaning with further associated development in neural connections among visual occipital cortex, sensorial temporal cortex, and motor frontal cortex. With word meaning in Namaa added to Navand already in operation in the iconic nonworded but meaningful holistic semions and moveons, the expressivity trait entered its unlimited scope of abstraction into the world of metaphysics.

From this sketching, supported by evidence, mainly from conduction aphasia that essentially separates word meaning from word production, a conclusion and an avenue of further considerations naturally ensue in the structuring of expression-based syntax. The conclusion is that Navand definitely preceded Namaa in initiating and conducting the expressivity in the archaic times when a neuroperceptive stimulus without a precise Namaa (iconic or else) could cause expressive reactions. I have so far discussed this point in psychognosis that led eventually to semions and moveons. The avenue that opens up is the relationship shown between the Navand effect and expressivity in idiognosis with function words that may not have a single precise content value as Namaa and instead stand as forceful motional indicator as Navands in syntax. Probably, the clearest handy example would be the one I used in the sentence 2, in the elements of expressivity section in chapter 9, which shows confirmation for the present time bond, forming the syntax without any word indicating the *being* in relation to the present time. That sentence was *what a beautiful spring day*!

In this example, there are two categories of Namaa in terms of precise content and meaning versus the imprecise types: in *beautiful spring day*, every word is one Namaa with precise strict content; and, in the other, *what a*, both words acting as function words are Namaas that concur to express rather imprecision by lack of definite material content in their isolated forms, but express the greatest effect together in the syntax by their included Navand expressing meaningful emotion by structuring the tacit bond mixing the components and confirming their *being* (in the present time) with their carried dynamic overwhelming meaning. In terms of autonomous limbic emotional expressivity, the expression of *what a* in this example and all expressions of aahs, oohs, and ouches enumerated for the archaic limbic-made phonation are equivalent. The only difference there can be is the unintended tacit present time, concommittently incorporated in the latter expressive interjections from the intended but yet unidentified present time in syntactic binding, which can be intuitively detected. This effect, which is obvious and accountable in the archaic language examples seen earlier in Tables IV and V, becomes syntactically detectable in the fully developed idiognosis depending somewhat on the dominant meaning of the function words being of logical order preference (sequencing and cause-effect type) or time order continuity predominating (connection and continuation type). In fact, the basic archaic phonetic /a/ for the fundamental present time of the monodimensional era has been the pattern for many originated and differently oriented function words with variable meaning implications in languages. The ultimate change can be seen in the reversing name-adjective relation in the genitive mode (discussed briefly below), completely eliminating the archaic connective element between the two words in several languages once the two words are united in meaning as unified sameness or defined belonging. Thus, function words, though originating on the archaic principle of replacing inflections, not only carry the archaic time conduction effect with tacit present time confirmation, but also develop conditional logical implications in meanings as well, which may subordinate their time connecting role.

For a clearer understanding of this transformation in function words implications, I will take the syntactic genitive mode for an example in the contemporary Persian. As it was pointed out earlier, the phoneme /e/ in the modern Persian is the diachronically simplified form of the /hya/ of the Old Persian.[4] This phonemic /e/, preceded by a consonant ending word, or /ye/ preceded by a vowel ending word, establishes both meanings of belonging to (implying logic of possession) and being connected with (implying logic of continuity, closeness, or sameness) in the genitive mode. It obviously shows the pattern of carrying a transformed meaning that initiated in the Old Persian and became evident in the middle and modern Persian in approximately 2,500 years, showing relation of two-word concepts together. In short, it is an idiostatic concept verb. In that form, it can be preceded by a name and succeeded by another name or by a pronoun and implies possession of the first-named item by the second-named one, similar to the function of English /of/ in the same position. For instance, the example 6 in Table V was such a case with the meaning of *God of world*. Or the phoneme /e/ or /ye/ could be of the same value between a preceding name or pronoun and a succeeding adjective like the fifth example in table V, meaning *bright day*. In both situations, the implicit *being* in the present time is naturally included in the syntactic meaning by the concept verb as the ground dynamic effect binding the two conceptual meanings together.

Studying further Persian examples, the natural duality of the phoneme /*e*/ in the genitive case showing both logical and temporal meaning predominances in syntax structuring can be clearly observed. The following Persian-English-French-German matching examples are used.

1. Maadar /*e*/ Shirin, Mother of Shirin, La mere de Shirin, Mutter von Schirin[13]
2. Shirin /*e*/ ziba, Shirin is beautiful, Shirin est belle, Schirin ist schoen

In the first instance, the genitive case indicates possessive relation by /*e*/ in Persian, /of/ in English, /de/ in French, and /von/ in German. The equivalent grammatical function of these connective items between the two Namaas is beyond doubt and confirms their similar roots from one origin, each serving the purpose as did the archaic /*a*/ or /hya/ of the Old Persian. In the second example, confirmation of being beautiful by the adjective following the name is by /*e*/ in Persian, /is/ in English, /est/ in French, and /ist/ in German. These examples leave no doubt for the dual meaning implication of /*e*/ in inducing predominantly a logically oriented sense or a time oriented state of *being* according to the mind's logic in directing or interpreting the expression. If then we consider the reversed genitive mode, we can see that the simplified forms in all four languages omit the function of the added equivalent of /*e*/, like ziba Shirin, beautiful Shirin, belle Shirin, and schoene Shirin, respectively. This similarity further confirms the equivalent value of /*e*/, /of/, /de/, and /von/ in analogical functionality when the two Namaas are united by the Navand in the reversed combination, entering the adjective meaning in the names and revealing the dual function of the phonemes that act as idioconnective syntax verbs. This state is meaning dependent and shows the lack of necessity for the function words when sameness of the two connected elements is expressed. These verbs, in fact, exhibit the dual functions of possessive dynamism either as possessive pronouns or as adjectives. It can be noted also that following the initial phase of preparatory mental combination that used the simple inflectional mechanism in the archaic time with /*a*/ (vowel suffix) intrinsically attached to the word, the use of the separate phoneme of /*a*/, or its changed equivalents came into practice, often reflecting more than a single meaning.

Evidence of dynamism making changes in the concept formation and syntactic structuring is at the origin of dissimilarities notable today in various languages from the same protolanguage root. In fact, as noted in discussing the

Navand theory earlier in this chapter, the universality is solidly explainable on the basis of the Navand theory and differences noted in various languages are defendable on the same basis. Several points of Chomsky's universal grammar as discussed by Cook and Newson[14] could be reexamined here to further defend Chomsky's views. But for the sake of brevity, I would recall the origin of the psychognostic mental mechanism constructing semions and moveons, the original neuroperceptive facilitation that is essential to help understand the mechanisms of these variations.

As already discussed, the foundation pillars for movement coordination of the detected sequential action as stimulus and elaborated reaction as response by reflection of both activities by the mirror neurons at the site of the Broca's to be the future speech production center in the left frontal lobe, support a unified synthetic reflection of the entire action in the mind in a holistic view. Therefore, the semions and moveons form and act as unitary perceptive and responsive actions mimicking and replicating repeatedly what is necessary to deal with the environmental imposing effects in refined rehearsals most apt to face any given situation. When this mechanism is using the two-frame model in idiognosis, what is important for elaborating the meaning of the action observed is essentially the action itself, but not the subject or the object even if detected. Just the actual action is reflected by mirror neurons forming the related specific meaning, and the iconic subject and object, included or not, are contingency items. Thus, there is no mandatory line of conduct to orient an action interpretation in terms of active verb or passive verb type, or head first or head last for verbal phrases, or transitive versus intransitive verb representation discussed in Chomsky's universal grammar as examples.[14] These voiced syntacticly exteriorized examples in idiognosis naturally are not the unchanged holistic ones in psychognosis.

Thus, all conceptual meanings in terms of actions as *change through time* given for gnosis (c.f. auto-hetero-gnosis in chapter 3) reflect the governing foundation that was realized by Navand in psychognosis but show idiognostic words coming in the field of expression with their accompanying changes for assuring meaning by Namaas. In final analysis, psychologically interpreted, both anterograde and retrograde reviews of an action in the mind lead to the same final conceptual result, which is invariably constant, indicating the fundamental universality principle for what would be the nucleus of the *internalized language* of Chomsky.[15] This psychological interpretation certainly follows the dominance of egocentrist versus allocentrist mind's tendencies to judge introjected or extrojected result to self as subject or object and the related wording in expressing the respective meanings. Perhaps one should even think in terms of deep structure concept to admit an ***abysmal structure*** as the deepest, corresponding to the psychognostic speechless meaning in the speechless holistic syntax of semions and moveons as fundamental primordial model of deep structure in worded idiognosis. The fact that all languages so far appear to be structure dependent in syntax, though not using a strictly similar structure, is a solid evidence for the universality obeying the archaic semion-moveon models prior to idiognostic worded languages.

Studies on inseparable simultaneous gesticulated expressions and worded expressions[16] also point in the direction of one fundamental pattern of expression, which can use motor demonstration of the expression in gesticulation and phonation equally. The pattern is the same basic immutable two-frame model, and with the meaning in mind, both proactive and retroactive dynamism interpretation of the action result can be shown in the guise of expressive movements, or words, or both.

THE LIFE OF LANGUAGES

Languages live with people and change as the human life changes. They reflect the old symbiont, the nonmaterial biological being, discussed in the initiating chapter of this book, and show the human life's mental dynamism in its immateriality expressed in words. Any language survives many generations of humans speaking it and is estimated to have a life of about six thousand years, an estimate not universally accepted by all linguists.[17] In the Navand theory understanding, what dies are Namaas made by words that may completely change and become unrecognizable after thousands of years, but Navand at the root of the expressivity trait lives with the human phenotypes in the form of a constant biological mental autonomy of expressivity. This fact is noticeable in words of languages with varying emotional charges. The greater the emotional meaning charge of a word, the greater its longevity and its universality character, as in words for mother, father, for examples, that are making motto idioms; Namaas with emotionally significant concrete material contents; or in *aah*, *oh*, *o*, *ouch*, Namaas of pure motto idiom types with all dynamic Navand and no concrete material content. Evidently the groups of Namaas with strongly significant emotionally oriented meaning or affective induction effect are more uniform in the human races and survive linguistic changes longer. This fact has helped to search such common words for classifications of earlier archaic languages based on common word cognates from contemporary word classes. As I have expressed earlier, our visual-visional life is our primordial mental life forming the quintessence of our biological immaterial life, which goes on in our inner permanent time titan whether or not complemented with words. But idiognosis brought words into being, giving our visual-visional life a replicator or a duplicator register in reality, which used memory to register events and, with them, frames of our permanent time titan exteriorized from its hidden continuum in the form of actual real present time with words, exposing our mental life. In this metamorphosis, our biological mental life left traces in spoken and written words. If we study these traces, we can analyze our mental evolution in a retrograde direction and possibly find the very initiation of word formation, languages, and their evolution with time. This type of study is, in fact, concerned with the genealogy of both the species of *Homo sapiens* and the life of their languages in the world as these two entities are inseparable. Thus, one decisive point to settle is whether we are dealing with one root for both humans and their languages, or with multiple origins or as anthropologists word it, with monogenesis as against polygenesis.

As concerns the Navand theory of languages, both mono and polygenesis are compatible in principle with language origination and development and the included similarities and differences between various language classes. In Navand theory, the expressivity trait is the determinant evolutionary force that leads to psychognostic speechless concept formation facilitating the further worded idiognostic language construction. This evolutionary scenario could not be specific to just one species if the anatomical brain functionality would be the same in other candidate species as well. The speechless psychognostic mind could theoretically have been shared reasonably equally by all hominids. Neanderthals, for example, indeed seem to have had it with even phonetic expressions judging their brain sizes that seem to have exceeded that of *Homo sapiens*. On the other hands, monogenesis to be chosen as the solo possibility needs to be confirmed genetically in order to allow restricting the choice of the application of the evolutionary scenario to only one species that is *Homo sapiens*. But even with this understanding, borrowing of words by new species from an older one, say *Homo sapiens* from Neanderthals for example, cannot be excluded. Therefore, confirming *Homo sapiens* monogenesis does not tacitly enforce absolute lingual monogenesis and does not exclude possible borrowing by *Homo sapiens* that were present before extinction of Neanderthals (with possible intermixing) for whom no solid control studies seems feasible.

Linguistic research looking for the origin of languages started in reality over 220 years ago with deciphering studies of the cuneiform writing[18] and the hieroglyphic inscriptions[19] and kept a progressively rewarding continuous pace. With the discovery of the Indo-European languages by Sir William Jones[20] and their status proved as being a largely related language family, an era of unifying sameness in the biological human soma and psyche seemed sumptuously inaugurated. However, the puzzling inability to group all extant languages in one Indo-European large class caused perplexities. In fact, some distinct differences in words and syntax of what had been accepted as Indo-European languages compared to other languages stayed in the way for such global generalization. This status quo, though with occasional allusions for a possible global unique origin from sparse sources, persisted more or less to around 1960s when the major group of Indo-European family was still regarded distinctly separate from Semitic and Turkish languages, not to mention several hundreds of other less well-known dialects.

Further sporadic discoveries in comparative linguistics with more and more evidence in retrograde analysis were suggestive of possible older relationship of disparate classes when protolanguages were tried to be reconstructed. Thus, the idea of a possible shared one single protolanguage class was gaining some support. However, debates continued as it was difficult to classify languages on the basis of both historical phonetic changes explaining differences in the cognates being examined to reach the presumed protolanguage classes for detecting similarities on the one hand and to correlate these findings with genetic similarities spread over the world to effectively find solid evidence for a possible common ancestry on the other. Such expected and hoped findings came about in 1980s and 1990s with the work of Greenberg[21] and Cavalli-Sforza.[22] An elaborate discussion concerning these authors' findings can be found in the monogram of Merritt Ruhlen.[17] According to Ruhlen's reviewing and summarizing the research exposed, a classification that can be specified as an anthropogenetic genealogic classification of the human species' and their languages resulted. This classification establishes the unity of origin for all human languages. Ruhlen's resultant classification starts with proto-*sapiens* that divides into African and non-African, keeping with the concept of out of Africa spread. Each division then subdivides in the respective human groups with their local languages; the African group gives Khoisan, Niger-Kordofanian, and Nilo-Saharan, and the non-African group contributes to form the two main Southeast Asia/Pacific and Eurasian groups as roots and subsequent subdivisions including the Indo-European languages.

This extremely abridged historical account was given here for two main reasons—one being the solidly important conclusion it includes, which is in accordance with the general evolutionary evidence, paleontologic and anthropologic linguistics genetically checked and firmly establishing the unified origin of *Homo sapiens,* and the other being its value in confirming the fundamental soundness of the mental evolutionary course through psychognosis and idiognosis described in this book, leading to speech and the mental symbiont. So it seems that we do not have the problem of monogenesis or polygenesis any longer, believing the scenario given in this brief historical account, but we still need an understanding of how the initial worded concepts were mentally constructed and phonetically expressed. How words came ultimately out of the proto-*Homo sapiens*' mouth or some other hominids (as the question of the extinct Neanderthals is not clearly eliminated) and started to form languages? Suggestion of the gene labeled FOXP2, suspected to play a role in the development of language, if proved in the recently advocated Neanderthals genetic study attempts, may shine distinctive light on the confusing question of monogenesis versus polygenesis and on possible interbreeding of Neanderthals and *Homo sapiens* species.[23]

The phonetic expression, matched with known visual iconic material as psychognostic concepts taking double identities with words, worked in archaic times initially unconsciously on emotional and affective phonation basis, and then consciously and intentionally, using the invariable module of two-frame expressions. We can say we notice this emotional basis every day in our pets' voices with affective tones and meanings expressing needs, joy, or pain, often also accompanied with distinctly meaning looks and gesticulations of a formidable prelingual expressivity. Similar observations are recorded by observers in the wild.[24] At this stage, phonation may carry some meaning and can be said to concurrently progress with the psychognostic mind in primates and hominids, but only extendable in hominids of upright positions with the laryngeal descent and propensity for further voice refinement to produce words. Birds' singing goes through a similar progressive change, but on a shorter scale, matching their younger biological expression to their adult forms.

This archaic period that starts from simple primitive biological expressivity, to gradually include refined semantic content of significance, can be emotionally spontaneous or intentional in both psychognostic speechless mind and idiognostic worded language. The speechless mind handles both in using holistic concepts on the mental two-frame models in hominids, but the intentional expression also succeeds to code the iconic concept with words representing solidly established meaning. The psychognostic representation of the iconic content of the first frames in the two-frame models with words probably initiated by *Homo erectus* who also used fire around 1.6 million years ago, made stone tools of shapes suggesting simple two-concept reasoning, and traveled out of Africa.[25] A protolanguage, a very crude form of one iconic-worded material concept and one motion expressive type (yes-no, attracting-repulsing) could have existed at around two million years ago. Some phonetic utterances indicating concrete objects of immediate surroundings and intimate family members' names or pronouns also could have been the earliest meaningful utterances at that time.

FROM VISIONS TO WORDS

Discussing the foundation of idiognosis in chapter 9, I made it clear that iconic holistic concepts in psychognosis that are all visual are the models for the future worded concepts in idiognosis. Either phonation or gesticulation, or both, must come to function in order to make coded and public the inner psychognostic concepts that are purely visual or visional and private. Regardless of what abstract aspect the visional concepts may include that could not be clearly definable, visual concepts on the contrary are all invariably concrete and well-defined. These fundamental two-frame psychognostic concepts as described, in becoming idiognostic by words, will gain double identity, one basic original identity by the self-recognizable icon, already innate, and another by the word specifically added. In this process, the Navand that originates the neuroperceptive concept with a tacitly recognized iconic figure, which is only known to self and represents an innate meaning will gain a conventionally recognizable Namaa by word, serving the ultimate expressivity purpose for communication. Similarly, the possible abstract visional concept that could not be defined by a distinct icon alone now finds a potential ground to be defined sometimes in some worded form, even if only on an approximate scale.

Wording of visional concepts like unreal monsters or deities (or states of mind with no defined limits like happiness, sorrow, greed, etc.) also appears to follow this line of progress, but the order prioritizes concrete visual iconic psychognostic concepts before expanding to abstract thoughts. In reality, a visual iconodynamic or iconokinetic state in psychognosisis expected to be replaced by another that is an auditory iconostatic state in idiognosis is yet supposed to offer significant potential and real dynamism to eventually complete the replacement. How could such change be realized? The answer is that total replacement is impossible except with pure Navand in extreme emotional phonation with intonation or with motto idioms in general, and partial replacement with word supplementing to be significant can be expected according to motto-stato proportion of words content. Idiognosis must copy psychognosis with the new items being only attached to the old ones as identifying tags, and a partial replacement may be expected after a long supplementation and with the help of emotionality in the phonation used. Complete replacement will be assuredly impossible, either quantitatively or qualitatively alone, and will suffer incompleteness with lack of emotionality. Thus, the mechanism of wording for the iconic concepts must follow the normally established psychognostic rules and must logically include two steps in transition—spontaneous-natural and conventional-natural

In an original initiating step, words will be normally created with important emotional charge and intimacy to autos. This step uses spontaneous biological means in as natural a way as possible. In this step, triggering of the mechanism is by conditioned autos' interpretation of the emotional charge that such iconic concepts may induce normally; it is autonomous, and the word that is formed may be called autonymic. A second step of additional conditioning by secondary factors possibly influencing the wording mechanism conventionally, difficult to clearly define, and possibly transitional from psychognosis to already-begun idiognosis, must follow the first step for all other iconic concepts but conventionally and nuanced metaphorically-metonymically. This wording mechanism must include a combination of psychoemotive to psychonoetic range, providing motto, stato, and mixed idioms in established idiognosis based on facts of concrete nature and significance.

In line with the spontaneous natural word formation, the study of the contemporary linguistic cognates leading to possible original word forms in protolanguage classes may provide understanding of how words initiated, identifying the psychognostic iconic concepts. The order of such concept categories including time, orientation, quantity, and quality as related to the egocentric position of autos can be further checked in the light of such linguistic studies. In the following related examples, I aim to explain circumstantial evidences to assure understanding the variations associated with word cognates to support the sound foundation of this type of linguistic study.

For the cognate category of words related to time, examples outlined and cited in connection with the Navand theory of languages clarified the time concept as contingent of being. This is the biological root of the time concept that led to its instrumental applicability in later abstraction. The worded archaic phoneme */a/* in the Avestan and Sanskrit and analogous examples of it in the old and the contemporary Persian languages in their cognates, indicated the time in the meaning of *being* by filling up the second frame of the two-frame psychognostic models. This fundamental origination gave all refined verb forms for *being* in Indo-European languages. The example symbolizes the affective meaning implication of time inherent in being, and not the abstract meaning, which took millennia later to be formed. In general, the actual task is clearly an undertaking that starts by autos and is conditional to autos' environmental impressions. In the normal human language acquisition and development this imbedded time in being becomes tacitly known during the passive-reactive earlier time in infancy and its conceptual form only in the active-reactive later maturity period over several years. The philosophical abstract concept of time from the mythological era had yet to wait for its refined spacetime concept until the twentieth century for being firmly accepted in its new physical definition.

Other examples in the study of cognate words to possibly reach the original roots are of concrete type and with affective charge to the egocentrist autos; they are motto idioms. These examples cover instances to be shown in the other three classes—orientation, quantity, and quality. These are the iconic structural concepts already formed on the two-frame psychognostic models that are based only on the benefit from the connective role of the second frame content, i.e., the time used by Navand mechanisms. This basic understanding explains the scope of immense natural changes of essentially metaphoric types that fill the domain of the developed languages. Also, egocentricity in defining primitive concept formation appears not to be an inflexible rule, and some early allocentric abstractive capacity may actually work according to conditional factors.

These considerations concern autos' self-preferences primarily that follow closely the life commandment. These strictly personal preferences based on interpersonal bonds are dominant in the autos' immediate surrounding, which is the family. In this immediate environment, human contacts bear the greatest emotional charges that may face autos and originate the most private conceptual definitions to be developed. The same analogy with the contemporary human infant's psychology is to be considered for the archaic human, and the roots of words, directly or indirectly related to one or more of the four spacetime classes of conditioning factors for worded concepts to be formed, can be constructed by autos-self mainly emotionally and egocentrically oriented. Plausibly, the paramount priority objects, inducing such private emotional concepts, are the caregiver and family members.

Regarding the initial word application to these priority concepts to be literally identified, the mechanism involved is clearly psychoperceptive-psychoreactive. I would consider this twofold mechanism that is naturally biological and representing the natural combined work of the expressivity Navand and the physiologically available functioning of mirror cells to reduplicate motor activity, as being fundamental in gesticulating, or talking, or both. Shortly named, this complex mechanism, which characterizes the first normal autonomic step for wording can be simply designated as psychodynamic. Psychologically, the emotional charge in the iconic visual concepts of the close relatives (father, mother, brother, and sister) conditions the expressivity trait, and the Navand coincident with the iconic Namaa may add psychomotor phonetic or gesticulated expression discharging the emotional pressure simultaneous with the mental realization of the visual concepts. The stereotyped repetition of whatever spontaneous utterance normally occurs may form a fixed idiom that would ultimately become the worded Namaa to the privately known self's mental icon. This double identity with icon-word Namaa then will be easier to recall in memory and use in communication rather than the single icon concept impossible to exteriorize without a chain connection or a psychoperceptive-psychoreactive semion-moveon mechanism. On the practical side, the spontaneous utterances for these emotion inducing iconic figures would be phonetic sounds that possibly related to these figures focused on autos with the easiest natural way, which would be through the autonomic mirror cells inducing action in phonation.

I believe, the mother among other possibilities, as the emotionally closest person to autos, is also defined closest in the autos' bodily spacetime (the intimate breast-feeding unification) and is psychologically interpreted as a bodily part by autos. Thus, at least this figure's emotional charge to autos (dependency and possessiveness) may be presumed to occasion an utterance showing its iconic concept and its connection to autos. This reasoned inference as a supposition seems to be born out in actual observation by the consonant *m* in initiating the words *mother, mama*, and *me* in the utterance by autos-self, all three reflecting the emotional charge of the bond between *autos-self, me, mother, and mother's breast*. In fact, it should not be by coincidence that the easiest labial consonant *m* is also the unique common consonant found initiating both the personal pronoun *me* for autos-self and for the closest sensed bodily contact part of the mother, *mama,* or *mamma* projected and personified with *mother.* This affective trilogy (*me, mama, mother)* as evidenced in basic Indo-European cognates is most emotionally bound as a unity in the early life and later in great many languages. This unique fundamentally emotional scenario, highly plausible in retrospect, seems to have legitimate psychological and linguistic support to form *mama* that demonstrates the combination of the emotional charge and the linguistic reflection of mirror cells' function reduplicating syllabic sounds *ma-ma* as credential of authenticity.

Along the same reasoning, another fact that corroborates my inferred conclusion above is the comparative incidence of consonant */m/*, */p/*, and */k/* for *mama* (mother, breast, or mine-mine?), *papa* (father), and *kaka* (brother) in the various languages studied by Ruhlen.[17] The first two words are widely known and have a ubiquitous distribution in languages; but the last one, *kaka*, meaning brother, older brother, older male sibling, or uncle, is not so well-known although it is also one concept with strong emotional meaning nearly equaling that of the other two words *mama* and *papa*. Ruhlen's study shows *kaka* or a variation of it invariably including */k/* to occur in forty-three language representatives that I have counted in Ruhlen's report or constructed roots (not counting Persian and its variations) with wide distribution throughout the old and the new world. This evidence confers ponderous credibility to the psychodynamic explanation assembling psychological and practical factors in setting up the mechanism of natural wording for the psychognostic iconic concepts.

A realistic argument in favor of the natural priority for the consonant /*m*/ in early phonation, in the significant complex affectivity just discussed, is the two-factor agonistic psychodynamic mechanism that appears to act as a principle: the principle for most emotion expressing with least energy consuming. The final word in normal autonomous word creation indeed seems to be the effective functionality of mirror cells action in the frontal cortex in relation with the Broca's motor lingual center on the one hand and, on the other, the linguistic principle of syllabic doubling with the easy rhythmic consonant-vowel combination as examples of consonant-vowel syllable timer pairs *ma-ma* given by Foley[26] or *mama*, *papa*, *kaka* of Ruhlen.[17] The duplicating presentation by mirror cells cannot be ruled out in these examples of double syllabic utterances and are indeed one solid basis for the normal psychodynamic word formation, precisely on the most emotion-charged and least energy-used principle.

Other than the emotional prioritizing of wording by autos-self in its predominant egocentric position for close items of emotional value, there are autos' body parts that also form concepts of emotionally significant value given priority for natural wording to complement their single iconic content. This is a second order priority, however, that can be realized when meanings for the word show variations that indicate possible two or more expressive ideas. One prototype example is the concrete single icon of finger, which indicates a number of variable concepts such as numeric quantity counting in units, oneness or more, or quality as part or whole with hand or arm, and orientation with sides to egocentric position by the words representing it. For example, the linguistically reconstructing operations for roots and the contemporary cognates in languages indicate the root TIK for finger proposed by Greenberg as reported by Ruhlen. This root and its cognates in many possible variations are based on consonant /*t*/ and /*k*/ and are shown in some 150 or so variations with vowels added and consonant changes mainly /*t*/ to /*d*/ and /*k*/ to /*g*/. The most frequent meaning was *one* as a single meaning for unity and the next most common was finger, with combination of these two, and further meanings of hand, arm, toe, and finger pointing, and number five and ten coming down the line in decreasing order.[17] At this time, one may think that the reason for variations noted may be a transition from the unique autos to initiation of selfs from the psychognostic to idiognostic mind, a transition no longer allowing the solo emotional power of autos to rule but also imposing added conventional constraints served through self or self(s). Notwithstanding this possibility, the spacetime orientation related to the example of finger as the unity and its derivative meanings into the number five or even ten and the expression of direction and motion by pointing with the finger are all served with one self's body part, finger, or hand, or arm. An observation that indicates that this great variability around the focus of unity (the finger) in this presumed transitional period is constraint related and is not caused by a different mechanism than the normal psychodynamic word formation by autos is the scarce numerals of one, two, and possibly three for example, noted in the case of Papuan languages of the New Guinea. The fact is that lack of need for a higher numerical expression in these aboriginals is the only cause for this scarcity in that particularly conditioned culture.[26]

In such instances of limited, poorly defined, or absent mental definition for concept expression, the psychodynamic mechanism to initiate adding word identity either totally fails or errs on substitution of bodily parts for ill-defined quantity, quality, or function with conceptual vagueness. Meaning changes can therefore appear and induce understanding in mental interpretation to physical contiguity of the iconic unity (finger, hand, arm) or may allow psychointerpretive possibility of expansion (orientation, direction, quantity) on metaphoric ground. This state of affairs still includes only natural-conventional word formations.

The meaning expansions extend to both physical concrete and mental abstract concepts. We can say that sometimes after the words giving the double identity to icons, a field opens for new combinations to appear in idiognosis, which would allow all sorts of connection between concepts according to generated ideas. The Namaa conntectivity of specific meanings, together with the basic more general expressivity by Navand connectivity, allows particularly the metaphoric meanings to be formed. The mechanism for metaphoric words and their meanings, briefly discussed in chapter 9—where semantic connectivity by any cause allowed a metaphoric meaning change with substitution of iconic figures for different meanings—could also permit single word's abstract meaning to be formed without any concretely identifiable icon. In other words, nonspecific meaning application to a worded icon is not a one way road, and nonspecific icon (icon never defined or impossible to be defined) also can be formed with words of abstract meaning with a theoretical range of iconic variability from zero to multiple shadowy ill-defined figures like time, God, happiness, etc. We can remember this possibility from the Navand theory definition with both Namaa and Navand potentiality of maximal versus minimal, reversed reciprocal values, forming motto to stato idioms in the order of decreasing Navand value with increasing Namaa value and vice-versa.

To explain mechanisms for words to give double identity to fuzzy iconic psychognostic abstract concepts, the example of finger is the best. In fact, the index finger, by virtue of its shape and function with relation to the whole hand, has played a great role in autos-self's expressivity either as a part singled out from hand or as a whole with it. From examples of meanings in cognates cited earlier, three domains of abstract thinking appear to be served with the icon of finger and its connections and functions with hand and self. These are the fundamental three categories of orientation, quantity, and quality concepts following the basic concept of "*being*" in the time titan. They are enumerated in increasing order of potential abstract ideation with corresponding words ultimately added to their psychognostic iconic concepts as will be seen below.

Orientation, as primordial necessity between self and nonself, naturally singles out nonself environmental items distinctly from self. In this function, both static concrete positional and abstract directional or motional intents of indicating an idea are included. Metaphoric possibilities that can be mentally coming in expressions by the root word TIK can occasion words indicating variable ideas with minimal changes by inflection. The English words for *fingering* and *fingered* are clearest such examples. In regarding *quantity*, abstracting thoughts from the unity to more than unity—one to ten (numerality) in examples cited by Ruhlen—indicate a further step of structuring chain connection using time direction and increasing or decreasing order. This effect is also evident in the cognates for TIK in practically the same way. It can be seen in YAK and TAK in Persian, for example, the first as numeral, meaning one, and the second as qualificative adjective meaning unique, with just a change of affix (*/y/* to */t/*) to impart variation from the numeral to qualificative meaning.

In addition, the index finger combined with the other fingers, not only has served for numbering forming other cognates as separate concepts in chain, but also as continuity measuring mass connection of whole quantities (size variation) using hand with the word handful as an example. But the most abstract word formation representing quality in this basic fundamental way is with both a change in the meaning of the shape of the index finger (iconic shape change) and the word meaning (Namaa semantic change) of far reaching ideas expressing the shape change as a visual-visional stimulus for the abstract message to be conveyed. An example of historical value is the expression of obedience and submission with presentation of bent index finger, in archaic bas-reliefs of anciant times shown by a subordinate to a superior as a sample, reported by Jaynes.[27]

Onomatopoeic origination of words, as another biologically conditioned formation of worded concepts on the auditory basis of sounds, is evident in words like hissing, buzzing, ticking, etc. A documented example of linguistic significance is the word *flying*, which shows more variability in cognates of different forms, but the oldest languages of Indo-European origin seem to show */p/* and */r/* as basic in their forms.[17,28] The present Persian and Greek words for *flying* in their shorter forms clearly remind *prrrrr* sound indicating bird flying. The word *par* in Persian as root for the verb "to fly" and as wings is a common example. There are certainly several other examples for word formation on onomatopoeic basis. But I mean to reiterate and emphasize that the fundamental basis originating words is biological and has governed the mechanisms by the principle of psychognostic to idiognostic transformation, which has taken a long evolutionary time not necessarily in sharp consecutive steps, but quite harmoniously and in part simultaneously, with initiation of deep psychognostic concepts of high emotional value. The word *mama* may be probably two million years old. We can expect significant clarification on the initiation of language through studies of cognates leading to identify protolanguages. In any consideration, however, idiognosis will never assuredly exhibit psychognostic affective expressions with words, but in exchange, it leads us out of psychognosis to metaphysical realms of abstracts extended Gnosis forming our immense immaterial biology.

Chimpanzees exhibit significant short term memory and recognition for numbers[29] and are said to have about three dozens distinctive calls considered analogous to a proto-protolanguage. But a definite protolanguage based on refinement of the million years old sensory-motor functions by psychomotor adjustability seems to have started only in *Homo habilis*, evidenced by right handedness, indicating left brain potentiality for psychomotor expressivity. Psychomotor operation in *Homo erectus*, initiated at some unknown times and probably using protophonemes for meaning transfers with better short-term memory, better tool making and their domestication of fire seem to be at the right time for the metamorphosis of psychognostic iconic to idiognostic worded concepts. The change has been gradual, probably in the last two million years, and has reached a stage of grossly comparable language ability nearly homologous to our present basic expressive ability in *Homo sapiens* around four hundred thousand years ago. It is assumed that by about two hundred thousand years ago, phrases restricted to two or three words, clauses rarely exceeding one argument, and none or rare embedding were current.[26] However, changes with extension of idiognosis, along with the unlimited possibilities of metaphorization increasing manifold the semantic horizons, occurred gradually with expansion of descriptive expressivity and abstract thinking.

At around forty thousand years ago, abstract thinking, forming an extended metalinguistic domain with well-established linguistic self and nonselfs, had already allowed artistic extroverted production, propositional descriptive language, and established introverted communication with self. This state of affairs led the seed of philosophy in the abstraction sol to grow from the initial ego figures of mythology and god kings to monotheistic religions and beyond, with the self remaining in oblivion as did autos and its reality for long time. The philosophical thinking, initiated by the innate puzzle of life, and extended in every direction as if the central question of life was naturally explained on its own by the figure of God, leaving nothing more to be resolved. The sense of eternal permanency of self with the time titan, going on naturally with the commandment of logic and happiness incorporating self and nonself positions in that permanency, submitted also naturally to the principle of pleasure-pain conflicts to be controlled. This trend invaded the domain of idiognostic philosophy. Thus, the obvious puzzle of autos and its nature never really came into account except in Cartesian thinking. If I had to redirect the idiognostic expression to redefine the biocracy with autos, I would simply say that autos in us is our inner origin that our idiognostic abstractive mind has wrongly projected out as a unique personified superego in God. Or said more metaphysically, perhaps our immaterial biology as a principle of perpetual immortality assimilates the projected figure, akin to the anthropic principle,[30] incorporated in what can be called the holognostic creation that will be discussed in chapter 13.

REFERENCES FOR CHAPTER ELEVEN

1. Razi H. 1989. *Avesta Grammar*. Part II. (Text in Persian). Fravahar Publication. Tehran.
2. Bahrami E, Joneidi F. 2000. *Dictionary of the Avesta.* (Text in Persian and Avestan). III part. Pp1406. Balkh Publication of Nishapur Foundation. Tehran.
3. Razi H. 1989. *Old Persian. Grammar, Texts, Lexicon.* (Text in Persian, Avestan, German, and English). Fravahar Publication, Tehran.
4. Mo'in M. 1984. *Izafa. The Genitive Case.* (Text in Persian). Fourth Edition. Amir Kabir Publications. Tehran.
5. Razi H. 1989. *Old Persian. Grammar, Texts, Lexicon.* (Text in Persian). Pp199. Fravahar Publication. Tehran.
6. Garrett MF. 1990. Sentence Processing. In: *An Invitation to cognitive Science. Language.* Daniel N Osherson, Howard Lasnik. Eds. Pp 133-175. A Bradford Book.The MIT Press, Cambridge, Massachusetts, London, England.
7. Mithen S. 1999. *The Prehistory of the Mind. The cognitive Origins of Art and Science*. Pp 111. Thames and Hudson. New York, NY.
8. Dunbar RIM. 1991. Functional significance of Social Grooming in Primates. *Folia Primatologica*. 57:121-131.
9. Dunbar RIM. 1993. Coevolution of Neocortex Size, Group Size, and Language in Humans. *Behavioral and Brain Science.* 16:681-735.
10. Aiello L, Dunbar RIM. 1993. Neocortex Size, Group Size, and the Evolution of Language. *Current Anthropology*.34: 184-193.
11. Carlson NR. 1998. *Physiology of Behavior*. 6th Edition. Boston. Allyn and Bacon.

12. O'Grady W, Dobrovolsky M, Aronoff M. 1989. *Contemprary Linguistics. An Introduction.* Pp.257-263. St. Martin's Press, New York.
13. Rastan H. 2002. Personal Communication on current German.
14. Cook IJ, Newson M. 2001. *Chomsky's Universal Grammar. An Introduction.* Blackwell Publishers.
15. Chomsky N. 1991. Linguistics and Adjacent Fields: A Personal View. In: Kasher A. Ed. *The Chomskian Turn.* Pp 5-23. Oxford, Blackwell.
16. McNeill D. 2005. *Gesture and Thought.* University of Chicago Press. Chicago and London.
17. Ruhlen M. 1994. *The Origin of Language.* John Wiley and Sons, Inc. New York. Chichester. Brisbane. Toronto. Singapore.
18. Shmitt R. 1993. Cuneiform Script. In: *Encyclopedia Iranica.* Edited by Ehsan Yarshater. VI: 456-462. Mazda Publisher, Costa Mesa, California.
19. Robinson A. 2003. Champolion Breaks the Egyptian Code. In: *The Story of Writing.* Thames and Hudson, Inc. New York, N
20. Lehman WP.1993. *Theoretical Bases of Indo-European Linguistics.* P 289. Routledge, London and New York.
21. Greenberg JH. Cited by Rhulen M. Reference 19.
22. Cavalli-Sforza LL, Piazza A, Menozzi P, Mountain J. 1988. Reconstruction of Human Evolution: Bringing together Genetic, Archeological, and Linguistic Data. *Proceeding of The Natural Academy of Sciences.* 85: 6002-06.
23. Dumiac M. 2006. The Neanderthal Code. *Archeology.* 59(6): 22-25.
24. Griffin DR. 2001. *Animal Minds. Beyond cognition to Consciousness.* University of Chicago Press.
25. Lewin R. In The Age of Mankind. *Smithsonian Book of Human Evolution.* P 91. Smithsonian Books, Washington, D.C.
26. Foley WL. 2001. *Anthropological Linguistics. An Introduction.* Pp 67-72. Blackwell Publishers, Malden, Massachusetts.
27. Jaynes J. 1990. *The Origin of Consciousness in the Breakdown of the Bicameral Mind.* Houghton Mifflin Company. Boston.
28. Buck CD. 1998. *A Dictionary of Selected Synonyms in the Principal Indo-European Languages.* The University of Chicago Press. Chicago and London.
29. Springer M. 2006. Champ Chimp. *Scientific American Mind.* 17(4): 12-14.
30. Phipps C. 2006. A New Dawn for Cosmology. *What Is Enlightenment?* 33:42-48.

CHAPTER TWELVE

IN SEARCH OF TRUTH

THE ABSTRACT ONENESS

The mind's imaginative wondering indeed started earlier than the speech and extended unlimitedly into the literary abstract ultimately, searching for the unseen unknowns beyond life in the literary coded abstracts—infinity, eternity, and God. And man has wondered and has searched for that firmament of truth to find the answer, which originated the metaphysics that continues the search unendingly in an endeavor beyond concrete facts, in the abstractive domain transcending from the material to the immaterial biology.

We must realize that our human capacities are limited to what we are, soma and psyche, or matter and mind, but a matter that is intrinsically mindful and therefore is biocratic in essence. However, this matter sets our limits of imagination and logic. What makes out our system of the biological matter and psychological states are set for us the way we understand them and accept them as norms. We also must realize that this human-made nature and norms for logicality is our only means of reasoning for a possible metaphysical truth for which a proof of being right or wrong is nonexistent. Yet in our logical trust and fairness, we still must adhere to tested rules of our own available logic to validate our steps in searching the truth, not a mathematical logical truth but a psychological one in line with our life commandment compatibilities of logic and happiness.

In objective facts, revealed in the domain of medicine and psychology, effects of strong belief such as firm sincere religiousness or mysticism have proved undeniably beneficial to physical and mental health. Particularly, mystical meditation has proved to affect both soma and psyche favorably, especially if focused on the concept of God and salvage beyond physical life, in short, if the hope for continuation of the soul to reach the ultimate unification with the expected origin for a secured ultimate sustaining is repeatedly rehearsed, reiterated, and revived by meditation. This metaphysical thinking with its principle and practice, giving observable results in enhancing well-being has opened windows to scientific minds. Scientific research to delve the fathomable depths of mind for reasons of observed benefits has been developed and is expanding, and some openness is presently observed in scientific thinking to examine metaphysical ideas and observations for finding out about possible truths beyond objective facts known to science.

A clear evidence for this openness in scientific thinking was witnessed with the inclusion of an invited discourse to be given by Dalai Lama at the American Society for Neurosciences' annual meeting in November 2005 in Washington DC.[1] This event indicates that scientific thinking indeed is as logically open to ideas of interest, old or new, as to allow any idea of nonscientifically proved nature with beneficial result for the mankind, to be validated and used rationally. Thus, metaphysics as a product of the human abstraction with all it may offer also deserves respectable scientific evaluation.

The scope of metaphysical ideas is vast. The entire philosophy of life and the being, as real or as virtual, and all that relates to it, some having been cleared and some still remaining untested in metaphysics start entering in scientific fields to be tested. These can be condensed in basic fundamental questions. The essential questions have been and are, Does indeed life hides any metaphysical reality? Does that reality include any Gnostic immortality independent of our imagination in a holistic definable unity as God? Is there an origin to be traced and discovered for that holistic unity? And the last question of practical value is: Why should even all these questions be posed, which in their last inquisitional wondering raise biological, philosophical, metaphysical, and theological issues that, even if resolved, probably will not change physical and biological laws. But human mind has its curiosity reasons inherent precisely in the fundamental life laws: the life commandment of logic and happiness with the biological expressivity trait.

It seems that in spite of what we are sensed to have grasped through the content of this book so far, you and I are coming to a point of decision to ponder more on some of the enumerated issues, namely, the philosophical, metaphysical, and theological ones. I will therefore direct my discussion in this chapter to follow along these three fundamental lines.

I will focus on the logicality versus possible illogicality of our arguments for an all inclusive superpower creator to be. My inquiry then focuses on the metaphysical aspects of the question of our minded abstraction of the concept of God, respecting our intuitive sensing with our unexplained motives, as well as our evidences to support acceptance, rejection, or modification of the concept. Finally, regardless of our eventual conclusion, we will have to realize its predictable potential psychological effect on the human mind. Indeed, the ultimate aim for choosing a proper line of conduct in our approach to this abstractive endeavor should not neglect to secure logic and happiness in line with the life commandment and in accordance with the medical adage of *primum non nocere*.

THE CHAINS OF INFINITIES

Physics and biology, unified by the bioforce in forming biocracy, fully explain that life must be the phenomenon made by both of these realities joining and making ultimately ***the matter that sees its own self and thinks***. This is the evidence we face and are involved with in this book. On this basis, we can clearly see that the mind is the biological abstracting organ, and its power of abstract thinking and syllogism to reach conclusions of literary precision is entirely a natural biological fact, realized by language, in the form of concepts that are manipulated by the machinery of conscious logical operation at will. On this basis, we can conclude that biocracy could have theoretically included deterministic direction ***to make the matter that sees and thinks*** or even ***to make the matter to see and to think.*** This determinism, as unscientific as it may appear, gains an aura of credence when the question of the superpower is also envisaged with it, even if the superpower is only hypothetical and unknown. The reason is the logic of causality that is inherent in our abstractive mind and is ipso facto the reason justifying our first basic question on questioning the metaphysical reasons at all. This fundamental logic also considers the hypothetical unknown superpower as unreal, but not categorically deniable as it can be sensed without needing explanation and keeps it as an abstract potential for lack of better alternative, which would be an observable scientific fact that is unavailable. Thus, metaphysics is born with the quest for a unitary, all-inclusive real totipotent power: the almighty God, not on logic alone but also on intuitive reasons searching for happiness. Consequently, ***the matter to be made to see and to think***, constituting a truly appearing deterministic setup (in potential) and suggesting dualism with an innate direction in the bioforce, which must have been given by the creator (but who else!), becomes plausibly real except that, in a second thought by the same logical argument, a vicious circle of redundant causalities imposing unobservable creators in succession of interminably no end, makes the deterministic creation thought to evaporate. Evidently, this mental annihilation of determinism is because of lacking scientifically observable affirmation for the figure of God, an iconic representation of the causality process, but not if the causality is figureless. In fact, such figureless causality as pure energy is most evident in physical immortality principle in the photon and as perpetual immortality transfer in the gene. Genetics, establishing basic structural unchangeable patterns but allowing environmental conditional morphism and functionalism according to natural selection, reduced determinism to only the initiation of the biological force, which would maintain itself through biological carriers to continue permanently in phenotypic steps.

If we accept the theoretical possibility of pure energy as the cause providing the force of creation, we can regard the irresolvable chain of successive creators ending with our inability to proceed any further. Thus, the least in determinism that must be accepted would be for the ***matter to be made that could develop into seeing and thinking***. Then the second question that may appear apparently irresolvable is "***If the created matter indeed sees and thinks, it should see both self and nonself and comprehend entities in more than one dimension and one time.***" This, as an established truth in the biological material life for the self(s) and nonselfs, psychologically speaking, is a truth defined in auto-hetero-gnosis in chapter 3. However, the unreal abstract metaphysical entities including the unknown God as nonself remain unproved, making intruding nonselfs to the logical mind to be denied even if perceived by the idiognostic literary imagination. Indeed, the logical impossibilities to entertain the reality of one metaphysical abstract God appear overwhelming; the real phantasmagoric figure can never be found. But against reasons, this impossibility remains with the psychological self as a permanent quest to search for that being of hypothetical nonself. If therefore the abstract idea of God as nonself, based on the described attributes of God in religious terms against scientific logic is not categorically rejected, the reason is either because our psychological self is incapable to recognize that nonself as such (first reason) which is impossible, or perhaps because it has no power per se to reject this unreal nonself even if recognized as such (second reason), which is possible, or it is simply because that religiously sketched entity has no other figure by which to be identified (third reason) that is also possible or finally because it is not a nonself (last reason) that is also entirely possible. And in final analysis, as we will see further, the second, third, and the last reasons may give the final answer.

The debatability raised with the second reason of lack of power for rejection of the illogism adds significant clarification. This seems to be the major reason that is based on the fact that our literary self is faced with a superego of God, which has been modeled initially after the autogenetic ego models in archaic times as an undefined superpower. This superpower takes shape with our literary self in figurative description with its old roots in abstraction that could not be as easily rejected as was later hardly elaborated and perfected in imagination of Godhood in idiognostic descriptions.

As we shall remember, in the auto-hetero-gnosis in chapter 3, we understood that the pristine autos of the bioforce origin, in transition from bion to biopsychon and ultimately to literary biopsychon, is represented by psychologically formed self(s) with abstract heterognosis that includes all nonselfs. Thus, God is a heterognostic abstract entity of possibly amorphous psychognostic or literary morphic idiognostic nature of nonself type. Consequently, this nonself hypothetical entity is naturally illogical to our conscious self that is primordially based on logical essence of the perceptive detection-deduction type, and it must be rejected if not found, but perhaps it is not illogical in the logic of our pristine autos, the logic of subatomic quantum, and the logic of life commandment conditioned by pleasure predominance. And we well know that both our literary logic of our representative self and the bioforce quantum logic of our autos are real fundamental axiomatic systems, only at different time scales of time definite and time indefinite realities, respectively. By the same token, the literary idiognostic concept of God originated from autogenetic ego models of external environmental powers and refined into one religiously embellished figure through idiognosis cannot be rejected as the figure is abstractively real as a whole and as pleasing in terms of protector in line with the life commandment. In a sense, our idiognostic literary logic has no authority to categorically reject the figure on suspicion of unreality. So the gate of logic must be expanded, so to speak, to include not only the logic of time definite (macrocosmic classic physics) of our *self*, but also the time indefinite logic (microcosmic quantum physics) of our ***autos*** allowing probabilities to accommodate the hypothetical God with more inclination to happiness principle than to logic of the life commandment. Therefore, efforts for proving God's reality by sense only and by the pristine autos reflecting pure bioforce keeps continuing; abstract Gnostic extension with Gnosticism and mysticism then walk on more solid grounds in metaphysics. Thus, our consideration to settle the issue of logical impossibilities finds reason to continue searching alternative ways that would also consider the reason of the human heart for the sake of the likable figure in this overall venture. The third and the last reasons will need some elaborate explanatory discussion to be examined as we go on.

THE SHINING LIGHT FROM WITHIN

The reason of the heart conquered the human mind, I should explain, on the basis of the life commandment of logic and happiness detailed in chapter 6. It prevailed when its logical cofactor failed and the happy one did not. In fact, the literary concept of God included significantly attractive humanly character to please the human mind. It offered the lovable picture of both father and mother to the ever-maturing human mind from its very early beginning. The story to be recalled is that of the legitimate question of the unknown cause with known effects that initially shaked that absolute determinism inclusive in the causality logic yet plausible with theoretical repeatable big bangs and scientific facts by evolutionary evidences that made the absolute determinism to further change from its totipotent theoretical fatalism to its practical rationalism accepting evolutionary norms. But all these could not eliminate the attractive attributes of God in its humanized ego figure. Thus, this unfathomable blindly accepted nonself puzzle persistently remained the same unknown. To the inquisitive minds, it may remind the puzzle of the intrinsic unknown force bound to and activating the matter-wave appearance in quantum physics called bioforce in this book.

Thus, based on human logic, itself based on the power of abstraction realized with linguistic precision, the imagining of a unique totipotent superior force akin to an idealized human personality with will and determination to have created the whole world and the human kind, now becomes untenable unless considered as abstraction of a mirror image of self, or a humanoid ego figure constructed abstractively by self. This man's Gnostic phantom has originated from the biologically established auto-hetero-gnosis and extended into metaphysical thinking; it has created a sort of gnosis of unknowables in suspense, based on the same validity for Socrates at the dawn of serious logical questioning with the question—what is the meaning of virtue, if there is no virtue to have a meaning?—without rejecting the concept of virtue in essence. So whatever the meaning may be that has justified the theoretical possibility of metaphysical theosophy for our system of logic to be entertained, is fundamentally touching the heart and not the mind. It has a meaning, which is sensed but not explained. The definition of the superpower, the unique almighty God, if any, will necessarily heavily lean on mystical sensing devoid of any literary conscious time definite logical rule, but on time indefinite quantum logic of the bioforce in the pristine autos assuming nonrejectable probabilities. In reality, the subject is the same as we have seen in chapter 3 discussing holognosis in the service of metaphysics. There, I endeavored through reasoning in hypothetical terms and proposing a formula for holognosis as ***energy through time*** without material carrier to reach the philosophically abstract immaterial being as an outcome that can be linguistically condensed in God.

We can clearly see that the legitimate question based on human logic, itself based on the power of abstraction for imagining a unique totipotent superpower with will and determination to have created the whole world and the human kind, in fact, has no other originality than suggested by the autobiographical memory reflecting the analogy of such humanoid God in the mind. If this God concept has continued, substantially remaining in the human mind through philosophical debates, the reason is rooted in the human psychognosis, not on logic but on feeling. In fact, exact logic to solve the problem and tell us the truth between accepting versus rejecting the humanly described God on the basis of our present scientific knowledge, irrespective of our inner feeling, our biocratic autos, and our reasons of the heart would categorically disappoint us. Rationally, unfailing objectivity based on a pure binary foundation assumingly in line with the life commandment would take the logic plus happiness for yes, against the logic minus happiness for no, as final. So, the reasons of the heart would prevail, God would not disappear, and mysticism would endeavor in metaphysical philosophy to see the expected light. Let us take a quick look at mysticism.

Mysticism has psychognostic roots in the deep quest for protection. It is innate in the biocracy, axed on survival of soma and psyche, forming the ultimately hoped autogenetic superpower ego figure to identify with for eternal living once the earliest abstract thinking of the unified figure of God is realized. All mythology, multiple Gods, philosophical and metaphysical assumptions, and arguments for the unique superpower to exist originated from the old root of mysticism on the ancient oriental lands, mainly conceived of Mithraic origin that remained in the assembled old Persian theology from the ancient pre-Zoroastrian Aryan mythology.[2,3] Probably that oldest mysticism was documented as mystic doctrine in its present form, notably defined and recorded much later by Baayazid of Bistaam, a Persian mystic of Zoroastrian ancestry in the ninth century AD.[4,5] The essence of this mystic's doctrine and his followers held the belief that if man could transcend his physical being in pure spirit, devoid of any and all material interests, he could unite in pure truth with God. In reality, mysticism overshadowed theology by making the comprehension of God possible in sense feeling alone, the comprehension that had failed to be desirably gained only by indoctrinated religious beliefs. Mysticism in reality revealed the innate natural force for a Gnostic quest of the living spirit for continuing beyond the physical being, a force that was available to the living brain even before logic would concretely organize the abstractive mind.

It is true that all we can take for evidence to theorize a possible superpower for the phenomenal causality of life in the cosmos, with our present knowledge, is essentially limited to only observable analogy of physical matter-energy relation and biological events powered by the same energy. This analogy strangely continues in every step of matter-energy interchange without failure, as exemplified in the genotypes and phenotypes invariably following the path for reduplicating the same analogies and strangely in ontogenesis recapitulating phylogenesis. If viewed through psychological eyes, as there seems to be a quest for extension of sensing into refined estimation, interpretation, concept formation, and abstract ideation, all manifestations are in line with a progress to self-replication ending in memes.[6] The pattern replicates chains of thoughts endlessly further and further in cultures to come. From the ontogenetic phase to birth and growth, the humanoids and then *Homo* species have matured their psychognostic states culminating in holistic semions and moveons on the unfailing drive of saving and prolonging life. Consequently, the human mind must have been prepared for realizing this self-endowed quest for continuing to live and identifying with the autogenetic ego models fulfilling that quest, mainly with ergosynthetic and chronosynthetic ones discussed in chapter 4. Thus, mysticism comprehending God as a nonmaterial being representing the unique creator, opened the way to both infinite power and infinite time concepts to be personified by autos-self as that nonmaterial being to reach and unite with. This unification would be only realized if transcending the physical limits by purification of mind. This mystical doctrine reigned through the time and reached us in plain scientifically flourishing knowledge and rigid logic contrasting more adamantly with the elaborated concept of God because this deeply subjective mystical feeling followed the reason of heart. Now, should mysticism continue to value the reason of the heart over scientific evidences negating it? The answer may seem unattainable without personally experiencing the mystical state of mind in which the heartly feeling with the subjective experience is real and the mystical doctrine rooted in psychognosis only needs objective proof for gaining rational value.

Such experiential proof, though only subjective, seems real in the transcending state by mystics being solely interpretable through the mystic's self. Indeed, considering the soma and the psyche of our biological unit, if our psychological general sense of self that represents our organic soma with its idiognostic literary semantic wholeness that controls the reason, can be theoretically made to disappear with all its material attachments but leaving our sensing detector to remain, our biological unit will only sense our pristine autos engendered by the bioforce, which essentially forms our psychognostic wholeness. A genuine mystic transcendence also seems to use the same pattern to achieve the same result intentionally allowing contact with the deepest own self: the autos. In fact, mystics talk about the sense of complete selfless feeling, timeless, limitless, and wholeness when they reach the level of transcendence out of body. This state of transcendence, which can be interpreted as reminiscent of *Anaghra Raochanghaam* (infinite lights) in Avestan terms and its contrast with *Fana'l fanaa* (nothing of nothingness) in Arabic from its subsequent mystical change and renditions by Persian mystics, bears witness to complete separation of the soul in mystics' terms from the physical body, necessarily annihilating psychological self. In contrast to this, hypsoconsciousness, a conscious intentional concentration elevating concentrated awareness without annihilation of self, advocated for the benefits it would give in practical living[7] is beneficial but not in essence mystically oriented. So in principle, the mystical unification with God, theoretically sought and experienced if all soma and psyche are purified to the point that the mystic only senses the superior being, the expected God, is bona fide a personal experience objectively interpreted as evidence.

These considerations reinforce the idea of the mental force discussed in chapters one and two. We have to accept this reality that our mind's force is real with all its observable manifestations. It acts as a nonmaterial energetic effect being directed by attention focusing it in an intentional direction to enhance the conscious awareness in the form of hypsoconsciousness or to dominating controlled consciousness on uncontrolled obsessive compulsive disorders[8] or in variable modalities used in autofeedback exercises for enhancing mental concentration and sustaining control,[9] or finally for reaching selfless, limitless, and timeless state of mind in the holistic meditation gained in mysticism.[10] The reality to be interpreted properly is that the nature of the force sensed in the mind's fathom in the mystic state of transcendence can be either an enlightenment of wholeness shining from within or possibly from without. In my logically inclined interpretation, a source from within is more certain than one from without in this essentially subjective experience.

SKETCHING THE PHANTOM

In an objective undertaking considering human mind tendencies and its logical abilities, we can safely say that the affective penchant to wish for a superpower to identify with its sacred figure is regarded generally as a welcome comforting sense, a valued gift for the average human. This consideration implies that the average person's receptive tendency and educational influences for or against such abstract sublime figure, plus the level of her logical control, contribute to the final belief in her mind of how seriously she looks to metaphysics, theology, and theosophy. But certainly all affective tendencies will have to satisfy logical norms in order to be grounded solidly. This norm seems to have been ultimately fully respected and finally scientifically tested in the work of Newberg, D'Aquili, and Rause.[10] These investigators used single photon emission computed tomography (SPECT) to check the brain function in deep mystical meditation. They found that the parietal lobe areas of the left hemisphere responsible for securing self-orientation and indicating the physical bodily position, keeping the self-identity separate from the environment, showed significantly decreased activity in the deepest meditation state. Their result coincided in time perfectly with the highest state of the mind described by the meditating subject giving a typical account as provided by meditating mystics in general: a feeling of being detached from himself and united with a serene encompassing uniform wholeness with no boundary, infinite, and with an indescribable sensing of elation and joy in a timeless infinity. Similar experiments of other subjects including Franciscan nuns at prayers gave identical results, with noting a sense of unifying with God.

This scientifically undeniable evidence, in my interpretation, indicates that the concentrating meditation with an aimed focused intention reaches the ultimate neurodynamic stage of effacing the psychological self and its egocentric spacetime limits discussed in chapters 9 and 11, leaving only the autos activated by the bioforce, which feels not alien to the general spacetime of the cosmic wholeness. The brain synthetic association areas of the right hemisphere, not being suppressed or even showing more activation simultaneously in the tomographic brain picture, in reality reveal the archaic autos of the limbic origin in whole reinforcement by the synthetic right hemispheric liberated power. In reality, the actual psychognostic present time realization detects the holistic autos of biognostic or even paleognostic origin. The phenomenon likely reveals a retrograde passage from idiognostic to psychognostic comprehension and even deeper biognostic sensing of wholeness in wordless, limitless, timeless form of the creation spacetime. In fact, *the inner God, the autos*, sees its own light coming out, shining, and being united with the ultimate whole of the origin. The same vision was described as self, the feeling that lent interpretation of self Godhood in frank declaration by two great Persian mystics, al-Hallaj and Suhrawardi, both martyred for their revelations by fervent extremist Moslems.[11]

This mystical inner sense of sublime purification to seeing nothing but immense infinity of light and wholeness, of being one and the same with all surrounding, without boundary, interpreted as God or Godhood state has been times and again described by sincere religious minds. The target in transcending exercises has generally been the unknown hoped origin that in psychological interpretation represents the concept of **TRUTH** (**T**o **R**each **U**ltimate **T**rue **H**ope) related in chapter 2, sometimes regarded as a personified phantom figure named God. It is not possible for the idiognostic mind dealing with literary-defined codes to concretize a definition for that phantom figure other than copying a human model. In fact, the logically confined idiognosis to words cannot look back to psychognosis except with retrograde metaphorization if feasible, through the infinity of time in theoretical imagination regressing possibly to the origin of life, but hardly beyond.

Descriptions of visional telepathies out of body experiences and near death and return visions, in general, are unexplained perceptions in the same category of mystical impressions indicating elements of light vision and strange senses. In mystical meditating states, these are mainly of three general types described in decreasing order of importance noted as: (1) light sensed surrounding self, (2) happiness sensed within self (with lost boundary revealing autos), and (3) time sensed limitless reminding time titan. These are the primordial elements of mental realization inseparable from the bioforce indicating defined concepts of the biological life perpetuated in phenotypes in holistic semions, moveons, and time titan in psychognosis. The same is true for individualized concepts and segmented time titan into real time in idiognosis that are separated in the individualized literary concepts in parceled forms recounted in words.

GODHOOD IN DISGUISE

God in fact can exist for mystics and sincere religious individuals in both forms of abstract concept and detectable neurobiological phenomenon comprehended in a specified state in the mind but as interpreted by the subject feeling it or the examiner interpreting the SPECT or a similar document. For others, God exists in concept alone, or not at all, depending on culture, knowledge, and feeling. But in essence, God's entity concept to be formed in the human mind initiates by the psychological literary self from its psychognostic nucleus to join its abstract concept satisfying psychological affective need. Evidence so far makes this psychological sensing scientifically proved by showing the concentrated state of mind with its original force that is documented neurobiologically by the SPECT examination. The transcending experience that reveals the conscious sensation may vary according to religious tendencies and personal inclinations, but ultimately may end in one and invariably the same constant conclusion that the experienced state of mind is real, is personal, is whole, but overall undefinable except with the elements of light, happiness, and infinite time, and of course, we should reiterate, not without the force behind it, the innate power of autos. In fact, this experienced state of the mind, described by those attaining it, is invariably in unassuming variable terms of literary description except for these three unchangeable elements of *light*, *happiness*, and *infinite time* that are invariably experienced. The reason may be that this state of mind is very basically psychognostic and reflects the pristine auto-hetero-gnosis of the biopsychon when all is still both self and nonself in practical terms, self being autos at that time and nonself representing environment as a whole, defined through autogenetic egos and only as nonautos without further precision. At this archaic biopsychon stage, there is no formal concept of God, but that of some external superior force.

In later psychognostic visual holistic concept formation giving semions and moveons, there are concepts of concrete visual or visionary imaginable, but somewhat fuzzy entities including external superpowers or an ultimate single one. In idiognosis, literary coding applies worded identity to the iconic concepts and allows undefined Navand-made concepts of nonvisually sensed and elaborated initiating abstract forms of feeling to be shown by Namaas with words as discussed in the Navand theory of languages. All abstract concepts and metaphoric meanings then will be formed in the mind with their literary identifying codes that show the range of some undefined being and not a single precisely defined one per se, like with all adjective words of opposite meaning—cold/warm, dark/light, small/great, happy/sad, etc. The revelation by the intense meditating state of the mind, demonstrated in SPECT, seems also revealing an abstract unreal object as result of focused attention-intention process. In reality, the imaginary Godhood or Godness is vaguely personified, but not the literary God, which has never been defined in reality in any comprehensive way but with good human attributes plus immortality as an abstract entity in the human mind. This state of mind reported by true spiritual meditating monks, even if absolutely interpreted as unification with God, has never been associated with epiphanies of an external being and never anything but light, happiness, and timelessness. This is quite different from hallucinatory apparition or auditions claimed by prophetic statements and strongly questioned by some serious writers.[12]

The history of theology regarded in scientific scrutiny can only confirm that the notion of God, as generally elaborated in religions, has been in fact the real main tendency for the inquisitive logical minds to reject the concept of God rather than to accept it. We are therefore faced with an overwhelming reality in our lives that this neuropsychological transcendence phenomenon that reveals mind force sensed spiritually, though not per se an external objective reality for confirming any external super entity, may be still of significance showing a force (the bioforce) within reach and with potential advantage[13] if controlled, regardless of its origin that presently refutes identification. Such controlling would mean total mastery of our mind using our innate mental force to our advantage; it would assure stable mental integrity extendable in action throughout our life. Further, all mental insecurities with disorders ranging from instability to neurosis and to psychosis could be expected to become more amenable to effective control. This control that appears to be centered on the abstract concept of God being enforced in deeper beliefs is in reality centered on the axis of accessibility from our idiognostic literary mind to our psychognostic holistic mind and is also attainable without using the concept of God, like in atheistic spiritual Buddhist monks.

The concept of God in the human mind should be regarded in reality as that of an exteriorized psychological ego, as old as autogenetic egos described in chapter 4, that can be recalled in a way from regressing in time by transcending exercises. This autogenetic ego is naturally created for permitting identification with it for the benefit of the psychological self and by the reason of the heart. The authentic autogenetic ego figure presumably present in the bion without organized brain and self, as discussed originally, can be regarded hypothetically as a model to affect autos directly. This autogenetic figure being purely theoretical, imaginary, securing the purpose of discussion in line with descriptive models in chapter four, of course does not transfer unchanged to the abstractive mind where self(s) representing autos can fabricate investitures of semantic nature on it. In such a situation, this ego figure justifiable on autogenetic ego models for the self, is that of an anthropomorphic entity with unlimited power and infinite permanency able to be communicated with in the same language, understanding the same system of reasoning, and sharing the same emotional ability of loving and being loved and always available to help but never there to disturb. In reality, this concept of God is a mirror image of self but a super self, or a friendly superpower to talk to with imagined attributes not possessed by self. All religions have in fact presented, more or less, essentially this figure in a virtual personified form with an all inclusive infinite power for eternity in one indivisible all inclusive unity. The only exception is Zoroastrianism to which we will have reason to return.

Regardless of psychological and theological debatable considerations on this concept of God, the fact is that using this concept has revealed one of several ways to reach and exteriorize the force of mind. Thus, in broad medical sense, psychodynamic mechanisms to be used to reach and explore the mind's force are not limited. As shown in chapter 1, the force of mind can be accessed and used to effectively treat the compulsive obsessive disorders through simple adamant autosuggestions. Thus, regardless of the mechanism used to reach one's force of mind and to use it in directed ways, which can be left to the individual's preference and feasibility, the medical understanding on this subject is that enforcing the accessibility to the force of mind through more controllable mental exercises can open horizons of broad psychosomatic therapies.

THE CONDITIONED ONENESS

The all inclusive unity of God in religious descriptions incites reasoning to understand this particular oneness concept that seems to impose perfect sameness and absolute uniqueness in one defined global unity. The very first reasoning is on the basic logic in the definition of the concept itself. Analyzing this special oneness of unity, totality, and sameness can become exhaustively confusing as it can represent only a mere impossibility in the human logic based on the fundamental simple binary principle of thesis not allowing antithesis to exist simultaneously. In simpler words, a unity by definition must have dimensions and boundaries and cannot be an infinity without limits except as an abstract whole but necessarily integral in function, exhibiting functional unity. Or for such infinite unitary entity to make some sense, we can argue that this unity can be regarded an abstract unitary holism in function only, excluding dimensionality, to include abstract integrity. This line of reasoning leads us to opt for one limitless unique power functionality, which would mean energy without matter, the only theoretical entity with simultaneous opposing holisms of unity-sameness and infinity-sameness. So in essence, the hypothetical God must be theorized as being unlimited energy detectable through its functions only, an entity most perfectly represented by light philosophically and physically and by the immortality principle securing material and immaterial biology.

Similarly, letting aside all other presumed additional attributes for its personified image of unity and sameness but keeping one single symbolic function for it, the concept of God would represent one singularity of totipotency as function in one unitary being. Again, assuming uniqueness for its inherent function, but presuming limits for this unitary being would exclude sameness with any thing beyond that limit imposing duality. Thus, inclusiveness of integral uniqueness for being and functioning as a whole singularity dictates infinity to exclude binary state that would force accepting thesis and antithesis simultaneously as true. Indeed, if God is all *beings* as a hypothetical infinity and any and all *beings* are God's, therefore *nonbeing* would not exist and no binary system can be imagined to be true, a conclusion that is opposing time definite logic. The uniqueness inclusive capacity of God imposes that if God exists, everything exists and nothing could be nonexistent, even nonbeing. This logical contraposition causing perplexity, based on being versus nonbeing, is only compatible with subatomic uncertainty quantum principle. It excludes the entire foundation of the concept of God based on unity with any dimensionality in our time definite logic that forces limits. The reasonable conclusion is therefore sameness and infinitness of function-only for the God concept, meaning a paradoxical massless and timeless spacetime of energy alone, a theoretical possibility only at the subatomic level.

Thus, the abstract holism of function-only concept is the only metaphysically compatible understanding for the God concept to be defendable. The truly such possible unity is the binding force of feedback between matter and energy in subatomic quantum principle, the force that is existing conditional to both energy and matter, as ***both either and or times*** in the indefinite subatomic time scale, but may appear nonexisting or existing with energy-only detected as dynamism or matter-only adynamism in the human scale time definite logic. And we can still name this force our God for our reasons of the heart if we so desire, but with the understanding of its uniqueness, limitless, and timeless functional entity in quantum terms that I have abridged in the name of **bioforce** as a more appropriate identifying name. This is the basic ideological connection we can recognize between the humanly God concept and the subatomic energy in the theoretical massless, timeless spacetime. This basic ideological connection leads us to a more elaborate, more inspiring one, which is a discernible analogy between the three elements of the quantum principle, mass, energy, and their relation, a trilogy remotely reminiscent of theological type trilogies that I will discuss later.

To understand the interpretation of this subject of particular oneness more clearly, let us consider again the overwhelming scientific evidence with SPECT demonstrated neuropsychologically for the state of mind presuming unification with the ultimate unity, the almighty God. The question here is, what concept of God is this state of mind uniting with? If your answer is that the accepted concept is naturally *preformed* in the mind according to culture, knowledge, and feeling, you are on the right tract by using the word *preformed,* but you may end evidentally wrongly if assuming that the concept's precedence is limited to the religiously indoctrinated beliefs in contrast to scientifically evident counter beliefs. The concept has been naturally *preformed* in its essence, not by culture, but in reality by autos at the initiation of biognosis reflecting the bioforce and in the form of a superior external source of energy, an autogenous ego as stimulus causing a response carried through cultures. The concept became holistically comprehensible in psychognosis in terms of inclusive infinite power and infinite presence, mimicking autos' archaic ego figures, and was given form and significance in idiognosis with literary coding somewhat confusingly mimicking a human figure with supreme characters. Even the unitary indivisibility is a feature indoctrinated by self in idiognosis reflecting out the figure of the God ego with the evolved natural adaptability using abstractive expressiveness. The literary concept of God, in fact, is an item of the abstract repertoire of the mind not needing transcendent meditation to be recalled. But the transcendent elation can be interpreted as reflecting the effect of directed mental focusing as can be seen in autofeedback exercises and shows the archaic imprints of autogenetic egos on autos revealing its own reality.

The effect of the transcending meditation can be explained more comprehensibly to be due to removing the boundary of the idiognostic self, focusing in the direction of the desired target that could be the concept of

God with that originally modeled autogenetic ego figure by autos' transfigured literality in self, separated from its forgotten origin. Removing this self can show up in pure form the bioforce in autos, a force of the same nature that unites macro and microcosm. The actualization here is not from the semantic memory, but through the semantic memory as the trigger to delve the fathom of the unrecognized archaic memory theoretically reaching the very origin of the biological creation: the bioforce. Thus, the state of mind sensing unification with God is in reality sensing the inner bioforce in unified enhanced form. This sensing is produced by maximally realized potentiation of the right hemispheric synthetic power, liberated from the limiting self's egocentric unitary quantitation with positional environmental connections exerted by the left parietal lobe. So the interpretable sense should be rather the sense of feeling the intense unified and harmonized bioforce, a Godhood entity, a wholeness of the autos with the autocratic mind force.

So we can only say that in all probabilistic reasoning, our God exists in us as energy, like in every biological material, in the form of bioforce in the primordium of soma: the genes. That same energy forming the force of mind in us is in our deepest autocratic being, our autos, and can be revealed in our senses as a whole if our force of mind can orient our senses to that focused revelation without allowing any other sensation to interfere with that orientation. The point that now becomes appropriately clear is the significance imbedded in the last reason brought up in the Chain of Infinities earlier, rejecting the recognition of the figure of God by the self, because indeed it is not a nonself—it is the self's deepest origin.

This explanation can be understood vividly if made visual by a metaphorical visional example. Let us imagine that you are in the center of a planetarium and experience the sumptuous effect of all stellar lights suddenly coming on around you automatically and in increasing intensity to extreme shining. You would feel overwhelmed by the splendid infinity and intensity of your surrounding lights, giving you the happy splendor and the moment limitless time and beauty. In this experience, you sense you are observing a beautiful phenomenon that is made outside of you totally independent from your will as observer. You could be thinking that you are perhaps sensing one transcendence thrill by chance as you have read earlier with SPECT description. But if the planetarium's multitudes of light bulbs, when extinct, also act as mirrors naturally reflecting light from a central source, and if there is such central source of adjustable light that you can switch on by a remote control and increase brightness of reflections from the surrounding mirror effects to give you the same surrounding scintillation, you would realize that you are observing the same effect you have made your self. In this second instance, you are sure the cause and the effect of your experience is none but your own self, by using a device capable to realize your will, and your surrounding light is just reflected from your central light that was there but you had to activate it to get the experience. You would sense the identical experience, though you know that it is only the result of reflexion from your central light. In this imagined setup, you can logically conclude that: (1) the experience is primarily the result of your own perception, either as detector alone (peripheral light source and central detection by you) or your will as both producer and detector (central light source with peripheral reflectors), (2) the experience is dependent and proportional to the intensity of the central light source in controlled production by your will through reflectivity by outside mirrors, (3) the evident existing connection between your will and the outer reflexion mechanism is only by light energy, and (4) if you or your central light did not exist, there would be no experience. The very initial demonstration reflects the assumption of an independent outside source revealed by your detecting senses, but your logical results enumerated from one to four above, only confirm your person that could be the only origin of the experience with possibility of outer reflection mechanisms capable to use the same medium of communication.

In line with these conclusions, more explanation must be sought for more clarification. As a matter of fact, the neurobiologically demonstrated SPECT indicates our own action, observing our own production, strictly in our own mind, and more precisely, our very particular vision and our own interpretation of the scene that is absolutely different from the materially recorded SPECT picture. This experience, even repeated in scientifically controlled designing by independent observers, could not have been carried out still under categorical double blind restrictions due to its psychological nature necessarily uncontrollable both because of variability in the subjective results and in the unblended interpretations. As a matter of fact, an electroencephalogram (EEG) evidence of epileptic seizure can be argued to also form a scientific evidence for the neurobiological change in the brain showing intense uninhibited concentrated stimulation resulting in brain-made uncontrolled motor manifestations. In this example, the cause of the epilepsy and the means of treatment are scientifically clear, and all ghostly spirit causing the seizure no longer exist as control of the focal cause of the stimulation medically or surgically cures the seizure and eliminates all conjectures. However, it should be remarked that epilepsy is not a state of mind, not conscious, and not willfully triggered, and above all is an essentially material biological phenomenon. No doubt results of studies similar to SPECT, examining other spontaneous emotional enhancements, elation or sublimation by musical effect, or intense fear, rage, or pleasure, if observed, are not excluded to reveal some similar documentary evidence with nuances or frank differences according to each type. But all these experiences from our deepest sense of helplessness to greatest ecstatic elation are brain-internal, *states of mind,* confined in our brain neurodynamic processes making our immaterial world of emotion in abstractive expressivity. The essential difference between the mystical meditation ecstasy and these spontaneous experiences is the willfully focused internal intention in the directed mystical meditation. Of course, any and all spiritual events in these situations are also internal, and the state of mind they produce is entirely an internal affair, endogenic in origin by the intentionally focused will and without independent external intervention.

Although one may feel personally inclined to accept the possibility of unification with an external source of force or forces acting on the brain through the mystical experience to give the SPECT effect, with the planetarium example mentioned above one must recognize that the mechanism of the experience in these alternatives is internal and is in essence realization of an expressivity initiated within and possibly exteriorized and sensed or even reflected further out and sensed as from beyond one's own limits. Telepathy, for example, evidences connection between minds that correlates with conclusions 2 and 3 of the planetarium experience. When vividly experienced emotions or intense states of minds are shared between related individuals causing unexplained revelations, the only evidence is often the common psychognostic reciprocal value existing between them that suggests evidence of unquestionable relays. But even in this situation when undeniably documented communication is revealed between minds, it establishes the truth for the mind force and for the possible medium for its transfer. Sharing neuroanatomic structural similarity together with the mind force in some identical twin examples has also been invoked for telepathy explanation. *But is this state of affair possibly under the control of a unified superpower outside of one self's mind force, a religiously described unique God?* ***The answer seems to be a categorical no, but could be a conditional yes for a potential universal connective network of relays for the bioforces and in a comprehensive holognostic creation***, and further clarification must be sought.

We can see that the entire processing controlling the system in meditation or telepathy is intrinsic in our minds as indicate the references to the mind power to engender observable material effects at distance. One such example, occasionally reported, is twisting of metal objects by the mind force at distance.[14] All triggering external and internal factors, we have to accept, act as inciters only in terms of stimuli to and from our minds that could be willful with Libet constant time or Navand in conceptual idiognosis to be exteriorized in act or declaration, as we have seen in chapter eight, or could be inner emotional charges without being expressed. In either instance, some degree of mind force enhancement could be actualized that could bring about the meditating experience or could trigger telepathy to other affectively related minds. This enhancing effect in the mind force seems to work neurodynamically to unify a greater number of synaptic connections in a harmonized way. Interestingly, this harmonizing effect can be assimilated to dynamos working in harmony as said by Thomas Edison in his speech in 1880s on colossal electrical power illuminating the New York City. If Edison was alive today, he could say the same for the harmonization of the neurons in the right hemispheric association areas of the brain, exhibiting focused power acting as dynamos concurring to give the enlightening experience in the mystical state of mind. What is enhanced and sensed seems in fact to be the exponential increase in the mind force in harmonized unification at height of the mystical experience that reveals our bioforce in our autos.

Here, we stand at the moment of decision; we have evidence of seeing our own creator inside us through our mind's bioforce harmonized and intensified, and we can accept this inner force to possibly affect outer similar forces in a two way communication, but we have no clear evidence that any unique single outer force exists except as the totality of all such forces making a unified functional principle. This situation is that of *all-or-none* in terms of sameness, and we must accept *all* in that sense of functional sameness as an integral unity. This idea would suggest a universal mind, a theoretical metaphysical *hologgnosticism* that I have ventured to call psychothymic, remarked previously, whose physical unitary element is the light and its spacetime is defined by energy, the immaterial biological energy and indefinite time as inherent and inseparable from energy. The psychothymic symbolic existence can be seen in the light as the uniform sameness for the function of the bioforce. We can say that there are infinite unities of these forces in biological material with or without evident auto-hetero-gnosis, and that for the moment we are the only such unities with evident auto-hetero-gnosis living two lives, somatic and spiritual. The spiritual life elaborates the ***conditional yes*** remarked earlier, which still seems more definite only for that unified field and unified function of possible relay rather than a unique single body. If that theoretical unified field through which our telepathies seem communicated is scientifically proved, it can be regarded as part of the psychothymic universal whole. This consideration invokes the transfer from material to immaterial biology to be interpreted as deterministic, reminiscing Asha of Zoroastrianism in its immaterial purity of mind as sine qua non for reaching the sublime purity of Ahura Mazda, the absolute pure wisdom. The functional principle that is innate in creation and in biology, the bioforce, thus appears to lead the deterministic way in philosophical terms from the pristine autonomy to the abstractive metaphysics to close the circle.

An interesting broadly matching idea to biocracy already mentioned is the anthropic principle, also a science-philosophy concept in a braod sense, well-explained by James Gardner.[15] This principle is based mainly on the deductive reasoning that in the infinite possibilities (googol or googolplexes, one googol being ten to the power of hundred) of the cosmological creations with subatomic variabilities in accord with M-theory and with the unlimited number of big bangs allowing such astronomically extended possibilities, the occurrence of a life friendly universe allowing the evolution of man must be regarded as an anthropic principle inherent to that creation. In fact, the strongest assertion of this anthropic principle by John Barrow and Frank Tipler, referred to as final anthropic principle by Gardner, claims that "once life has arisen any where in this or any other universe, its sophistication and pervasiveness will expand inexorably and exponentially until life domain is conterminous with the boundaries of the cosmos itself". This hypothetical conclusion, though reached through pure reasoning and is not strictly biologically based, is in agreement with the biocratic principles based on bioforce that assures sustained forward progress of the material biology to ultimately reach the immaterial limitless biology with the abstractive mind.

THE TIME OF TRUTH

Considerations ensuing from our discussions seem to enforce that the imagined Godhood-Godness, sensed in the intense meditation, is nothing but the harmonized intense energy in the form that is more accurately identifiable as the bioforce existing in our soma and psyche. Further, that this force is reachable by appropriate measures, controllable, and can be enhanced, directed, and focused by will and could possibly relay minds together. How this level of mental achievement was eventually developed, as evidenced in mysticism, is an interesting question to answer precisely for finding out the reasoning behind the mystic belief and the means that could enhance the realization of this intense state of the mental force. In the discussion to follow, I will try to trace the Persian mysticism based definitely on God and the Buddhist mysticism of no definite God and will try to interpret these two apparently opposing doctrines that astonishingly can end in one outcome with similar meditation ecstasy. Evidently, the God figure must not be essential to achieve the mystic experience, and in fact, it is not expected to play the decisive role in Buddhism with no defined God to be focused in reaching the meditation ecstasy but is expected to be significant in the Persian mysticism believing in God. When pondering on the similarity of the result in either way, the reason indicates that the meditating mind, in fact, may be likely triggered by any adequate directive reason as the cause for focusing in that similar direction to reach the same effect.

Psychologically, this God value in the general metaphysical idea is bound to the literary concept of God based on the reason of the heart which stands on anticipation of security in unification with God and gaining the guaranty for the old SR in the material biology and in ***TRUTH*** discussed in chapter 2 for the immaterial biology. Therefore, the whole life commandment is affected by the concept of God. Both logic and pleasure principles are therefore involved by the reason of the heart, and evidently, the logic for the God's existence by no means is flatly rejected, even if completely failing, because the inseparable pleasure component in the life commandment of logic and happiness is psychologically dominant and adds to the reason of the heart by the very anticipating ***TRUTH*** (see both chapters 2 and 6 for this eponymic ***TRUTH***). Furthermore, the civilized human mind has come a long way from the biopsychon and the pure autos life commandment of the solo SR fundamentality and has developed modular self(s), psychological value definitions including human rights and even the coevolutional altruism that regards good doing to others as real virtue, animals and plants included. In this status, except for those minds still in the religiously bound medieval restrictive beliefs allowing reprimand and harsh punishment, the reason of the heart hoping for God has gained even more, not because of the hypothetical human figure God, never revealed by science, but because of the realized happiness with peace and progress under the auspices of that theoretical figure as the ultimate hope in life. This new valuation respecting the solemn rights of all creatures and the human species has influenced both logic and happiness principle in a new way. This forcing trend in the life commandment principle of logic and happiness implies that the reason of the heart may even find a principle of higher psychological value than God to adhere to, ultimately replacing the God figure and promoting the reason of the heart even more and gaining a rewarding sense of fulfillment and salvation not attainable with the confusingly hypothetical God concept.

In such an ideal system, the reason of the heart that has persisted with the mankind needs validation not only by mere belief, but also by sublime reality that could replace any other validation and secure absolute logical validity for what the sublime quest of happiness would find, gaining an ultimate sense of practical fulfillment in life. Such system of purposefully oriented thinking is by no means new, and the oldest nucleus of such an idea with an elaborated early ideology can be extracted from the millennia old Zoroastrianism of remote antiquity, plus a partially modified aspect of it in the subsequent mysticism from the hidden meaning in Zoroaster's original teaching used in a modified way. The original precept aimed for happiness during life on a sublime principle of a tangible truth above everything (equivalent to SR **TRUTH** in the life commandment) and for the subsequent mystical view in a similar way, but mainly for the acronymic metaphysical ***TRUTH*** beyond life. This modified application was targeted later to the mystically created God figure. This difference represented the Zoroastrian doctrine as the unquestionable principle preserving nature and the physical world of creation for happy living, contrasting paradoxically with its subsequent transformation into Persian mysticism as neglecting the life and everything with it for what is beyond, showing a completely opposing views. In reality, the original Zoroastrian belief adhered to **TRUTH** of the life commandment, keeping ***TRUTH*** of the metaphysical world above human level, but its derivative mysticism in disfigured belief completely deviated to promote the metaphysical ***TRUTH*** and neglect the life commandment.

It is ironical that the oldest monotheistic Zoroastrian ideology is also the most scientifically oriented and psychologically sound for happy living when examined and interpreted in line with facts and practicalities analyzing functionalities rather than forms. Indeed, the philosophy of Zoroastrianism[15-18] that in my belief has not been inquisitively scrutinized through its authentic sources in scientific terms reveals a doctrine on the reason of the heart validated on objective facts in practical living psychology and supported scientifically, physically, and metaphysically. To discuss this subject of interest to our purpose, I should start with a summarized understanding of the quantum principles implying bioforce in terms closely connecting analogically to this old monotheistic ideology that strangely gives us what I feel may fit in today's philosophical application for purposeful happy living without binding our minds to hypothetical unproved entities.

The cause of the phenomenon of the bioforce, believing the quantum physics norms, is intrinsic in the phenomenon itself and is in the unique mechanism binding the material and the nonmaterial components of the bioforce, making the phenomenon physically real. Philosophically, the very fundamental state in this situation is that of two real entities (electron and positron) supporting one another reciprocally through a third one (the wave function of light) that is unapparent but acts at once both as independent cause and as dependent result of the other two apparent entities in their reciprocal relation. More clearly said, the two real entities (matter and energy of potentially observable nature) are being made by a third real entity itself humanly undetectable (the unobservable interchanging feedback related to the uncertainty principle in the indefinite time) whose existence seems to be both the inciting cause of and the dependent effect from the other two entities, humanly observable as light. Thus, the third entity mandates time simultaneity for both either and or (rather than single either/or for each of the inclusive two entities) that will not exist separately in any relation without the third entity. This seems to be in short the understandable principle of the quantum string theory or its modified M-theory in nontechnical terms. This condition is intrinsically time indefinite. In the time definite logic of the human mind, either the photonic mass or the energy wave of light can be detected each at the exclusion of the other, though in their time indefinite being, they are always together. Therefore, in the macrocosm, we grossly detect the matter, the energy, or the motion in which the force itself is hidden to our time definite logic. Our realization of this hidden force, however, can be facilitated by the bioforce concept in autos and in the mind's force in meditation through attention-intention mental focusing triggered by will.

In logical conclusion, we come down to focus on the essence of this relational intermediary phantom between the matter and the energy as the possible creator. However, this element in fact is functionally both conditioning and conditioned. Its conditioning function seems to be beyond comprehension, mysterious and unsolvable as it appears, but its conditioned aspect depending on the two entities of matter and energy is comprehensible by being observable in the matter or the energy, particle or wave presentation in light.

This state of affairs that appears enigmatic may be focused more clearly considering the initial course of events in the big bang briefly outlined previously. Shortly said, the first 380,000 years after the big bang were spent in stupendous high temperature and matter-radiation interacting in the so called quark-gluon plasma until photons containing one electron and one positron in equilibrium were produced in the stabilized final stage.[19,20] The spinning of the two constituent elements (positron, electron) making together the material and the speeding wavy propagation of the light explaining the inseparable togetherness of the material and the wave presentation make the force governing this togetherness somewhat less mysterious. However, this unexplained force seems the only essential puzzle remaining, the third element, the unknown one that can be regarded as Godhood by its known function manifested with energy and matter showing no form but action, an action that is theoretically continuous, perpetual, time indefinite, limitless, and unidirectionally progressive by expansion. Evidently, this enigmatic creator can be defined both as a hypothetical Godhood in form and as a practical one in function. The practical functional one that seems real in the material world for human purposes according to its characteristics shown by matter and energy can be incorporated in a trilogy of functions cited earlier. This analogically elaborated trilogy, in my opinion, applies to Zoroastrianism by three names: Asha for the unknown reciprocity function together with Spenta Mainyu for energy and Angara Mainyu for matter. This trilogy leaves out of the material complex, the energy-mass, the hypothetical figurative God with its humanly attributed characters of ultimate wisdom, power, and love named the God Ahura Mazda as a purely abstract concept in the ultimate sublime Asha state, the state of infinite lights (Anaghra Raochanghaam in Avestan tradition), theoretically attainable through Spenta Mainyu.

This trilogy of Zoroastrianism (not classically outlined as trilogy) extractable from the *Gathas* of Zarathustra, as I understand, and not the divine triad of

God, Good Mind, and Truth described in somewhat different interpretation by other scholars, as reported by Professor Mehr,[3] is not easy to comprehend and interpret correctly because it is not a real trilogy of material, iconic, or anthropomorphic type like in Christian Trinity but is definable through terms explaining functionality meanings in *Gathas*'s belief system. The God Ahura Mazda, which is not described in any form except as an all-inclusive eternal entity of infinite knowledge, power, and love of creating and expanding the world of being is conceptualized through the logic of causality, the law of Asha, by Zarathustra's deductive reasoning. The law of Asha is the truth governing the creation as it covers cosmos, observed and comprehended by reason of the hierarchical truth of supremacies in energies. In Zarathustra's logic, Asha is the unchangeable truth dominating untruth, reality superseding unreality, and in physical terms, the greater ruling the lesser, the bigger dominating the smaller, all in a relativity scale culminating in pure truth and absolute power in the imaginable extreme unity of the all inclusive Ahura Mazda, the almighty God. This relative scale of Asha is the path open to mankind. The man can choose his way in that path either through Spenta Mainyu (the expanding spirit) or Angara Mainyu (the depressing spirit)—the first one leading to ultimate truth with happiness, success, and salvation and the second to the opposite, madness, failure, and loss. In fact, these three functionalities—Asha, Spenta Mainyu, and Angara Mainyu—making the Zarathustrian trilogy as I see it, are realities for the human mind with functional values. Zarathustra's mind, in fact, must have grasped the idea of the abstract supreme God through sensing, understanding, and elaborating on Asha before conceptualizing the ultimate hypothetical Godhood. This fundamental trilogy became more concrete and understandable in later Avestan writings in Humata (good thoughts), Hukhta (good words), and Huwarshta (good deeds), which clearly brought out the desirable noble human functionalities reflecting God's characters implied in the old abstractive trilogy. We can automatically sense the analogical functionality between these three Zoroastrian principles and the three subatomic quantum elements of light.

In this analogy, Asha may be considered as the third entity, the originating energy, the unapparent subatomic binding force between positron and electron maintaining their spinning. Its humanly perception is the light of no definite time in continuous all-directional progression of expansion, never stationary and never regressing. Spenta Mainyu is the energy continually expanding that could be regarded standing as antigravity, ultimately represented as light, and Angara Mainyu is the energy condensing along the gravity ultimately representing darkness as black hole. In other words, one functional spacetime seems to be made of gravity and antigravity as opposing poles harboring prototypes of pure ultimate energy and pure ultimate mass as extremes, and Asha is the subatomic force of the system originating and continuing the expansion phase or reversing and forming the compression state. The expanding energy shown by Spanta Mainyu is constantly overcoming its opposite Angara Mainyu. Although this explanatory elaboration cannot be interpreted to reflect such possible depth of knowledge in physics at that archaic time; interestingly, Zarathustra's logic seems to have solved the quantum puzzle at inception in the quantum way proper by accepting the God Ahura Mazda as the nonhuman ultimate of pure truth of wisdom, power, love, recognized as the unreachable ultimate goodness, but approachable in choosing the Spenta Mainyu, the good spirit, or avoided in selecting the Angara Mainyu, the bad spirit. In this philosophy, the hypothetical God remains rightfully the abstract, outside the realm of the human deeds, not in any way influencing the human affairs. It is a targeted superego in the human mind (a theoretical time definite ultimate in time indefiniteness) that is impersonal and never targets the human mind that has full liberty of action with absolute freedom of choice (time definite entities). I vividly remember being questioned, during a speech I gave on the philosophy of relativity versus absolutism in religious thinking—what I would consider the Spenta Mainyu and Angara Mainyu of Zoroastrianism to represent in the actual physical world today?—I answered, light and black hole.

This scientifically matching analogical interpretation that provides exceptional logical weight to Zoroastrian philosophy dominating its theological aspect, reinforces the happiness component of the life commandment not so much by its only analogical scientific weight but by the meaning of Asha with sublime truth that reveals itself as "truth above all," the only reality acceptable by the human mind. Indeed, the first maxim of Zarathustra saying, "***Truth is goodness, it is the best goodness, it is happiness per se; happy those whose best goodness is truth***," is a prophecy of "truth above all." Thus, Asha as a tangible reality for truth above all superseded the concept of God in the subsequent inquisitive minds originating post Zoroastrian mysticism. But later-Persian mysticism, in fact, originated on this precept and essentially believed in purifying the mind from all wicked tendencies, acting none but following truth and doing none but good deeds for the benefit of mankind before self. Without binding to religious tenets other than to the name of God as substrate for representing supreme truth, this mysticism continued to secure salvation of the soul beyond material life. This supersession of God concept by ultimate *truth* observable in good deeds but in neglecting religious norms by mystics in Islamic times was not professed openly and such mystic beliefs still rule in sects under various names for the original Asha. However, these believes are oblivious of their original roots, keeping with the original motto as ultimate token of love for the creator. This level of abstract thinking seems impossible with the religious God figure except in the analogical trilogy with modern physics. Zoroastrianism with the ***truth above all*** condensed in the love for the creator and the creation, and Buddhism where there is no God, but there is ***truth above all*** in the love for all creatures share, so to speak, one motto through two opposite views.

A historical gap seems to exist between the Zoroastrian and Buddhist origins for the fundamental ***truth above all*** principle that seems indicating the reasons of both philosophies in their similarities but also differences. A precise evidence for exact dating of the origin of Zoroastrian monotheism cannot be found in historical recordings that are all subsequent to Zarathustra's time and confusingly citing a range of six to two millennia BC. The more reliable means to use, though questioned by some, I believe, is the comparative antiquity of languages. Searching in antiquity languages for finding the trace of human thought on life, its reasons, and its connection with one unique superpower figure assumed to be the origin of the entire creation, only one such example in the remote antiquity can be found, which is in the *Gathas* of Zarathustra. The tongue of *Gathas* is as archaic as having nearly gone out of use at time of Cyrus the great, circa 600 BC. Subsequent cuneiform writings about the same time or shortly thereafter, written in the Old Persian, are more indicative of some diachronic change from the older Avestan, which itself was a younger dialect than *Gathas*. If the average life of a language of six thousand years is taken to have any meaning in spite of some controversies raised to that estimate,[21] the time of the dialect of *Gathas* of Zarathushtra could go back to the most remote estimate of about eight thousand years ago.[22] We can thus realize that the reason for the questions asked by the human mind regarding the metaphysical unique power to have created matter and mind has been with us from some very remote archaic time. During this long time, the metaphysics subjected to the scrutiny of the human logic could have been totally rejected, particularly after evolutional Darwinian principles were widely accepted. But this did not happen, and the concept of God remained in the human mind precisely for the reason of the heart. As said earlier, this reason that is psychognostic seems to be inherent in the human mind as a hardwired, meaningful, coveted, parental love, needed to be received and to be expressed. This same reason evidently finds exaltation with the axiomatic ***truth above all,*** in reality the metaphysical ***TRUTH*** discussed earlier, representing the sublimed ideal figure of God.

The time of Buddha in the sixth century BC followed the Zoroastrian time long enough for either transfer of the older Zoroastrian thinking changing into new forms or a completely independent formulation with new reasoning, an event that regardless of its roots did grant full liberty to ***truth above all*** principle with respect to all life forms in Buddhism. The human mind, unfailingly conditioned by the life commandment for logic and happiness with its reason of the heart supporting solidly the happiness principle, can liberate itself from the religious concept of God to adhere to the ***truth above all***, which is in reality targeting the same aim explained in concluding the thought experiment in chapter 2: ***T**o* ***R**each* ***U**ltimate* ***T**rue* ***H**ope* ***"TRUTH"*** transcending from the material onto the immaterial biology.

A final word seems in order on the three most philosophically minded religious thinking, Zoroastrianism, Buddhism, and Mysticism. All three have religious foundations but variable degrees of supporting science and philosophy as well as other mental conditioning factors of culture and belief affecting the life commandment principles with effects ultimately on the mind permissiveness for the transcendence ecstasy.

In Zoroastrianism, scientific and philosophical aspects are most prominent. The religious meaning in adequate harmony with both science and philosophy leaves no incompatibility with the principle of ***TRUTH*** explained in chapter 2, and again in chapter 6, in accord with the metaphysical meaning reckoned. Therefore, profitability in actual life and applicability in transcendental exercising by the concept of ***TRUTH*** (reflecting ***truth above all***) used as trigger are solid reasons for this philosophically and scientifically plausible doctrine to be advocated for securing the sense of fulfillment and happiness for religion lovers.

For Buddhism, in brief, there is no scientific ground supporting the significant hard-to-believe life renewal with transfiguration of creatures that is a significant belief in Buddhism. The ***TRUTH*** of the metaphysical abstract concept does not lead to any firm analogy with the contemporary scientific thinking. The SR based **TRUTH**, as foundation for the metaphysical one, leaves debatable reasons considering life values of all living creatures regarded as equal, which values are subject to cultural variations. The belief in respect or love for all creatures seems primary and general rather than contingent to love for the creator targeted as the unique aim for the metaphysical ***TRUTH***.

Mysticism in contrast to both Zoroastrianism and Buddhism is exceedingly inclined on the reason of the heart and forms a metaphysical atmosphere of absolute love for God to lose the material body mystically, become pure spirit, and unite with God. This basic theme is the dominant motto and many variations of mystic sects adhere to it regardless of their ego motif often being exemplary saints in various religions. This mystical thinking in real adherents indeed gives mind properties of extreme tolerance, impeccable peacefulness with reassuring composure, and variable degrees of clairvoyance. The value of **TRUTH** changes from that dominantly SR targeted logic of the material biology with applicability to living to that of the unlimited ***TRUTH*** of the immaterial biology, an imagined reality of beyond life, and with the hope for gaining the beyond life security by good deeds in life. This ultimate mystical belief leads to exaggerated altruism, minimized self-esteem and materialism, renouncing to possession, and maximizing immaterialism in the life ideology. It is incompatible with the scientifically progressive society life.

Quantum physics opens the horizons to the human mind for allegorical views and shows the possibility of logically impossible feats considering its uncertainty principle imposing the logic of indefinite time, the time titan of the biological permanent present time discussed before, the same time for both ***either and or*** choices at once. But the essence of change is in rendering the theoretical wholeness of unlimited mass, energy, and space of the single all inclusive unitary oneness with knowledge, power, love—the originally singularized illogical entity—to pluralized unities able to make up unlimited scope for unifying mass, energy, and time typically exemplified in the unitary subatomic dynamism: the hidden bioforce. This force renders possible for the ***matter-to-be,*** to be also immaterial, creating the immaterial matter, the functional capability, ***to see and to think*** as in the biological nerve cells ***to see*** their representation in the immaterial mind and for abstractive ***power to reach*** beyond limits of infinity. Our abstractive thinking seems to have reached the ability of controlling our mind, showing us our original creating force and allowing us to use it for having better human mind and happier purposeful life, ***using truth above all*** in a most natural conduct to reach ***TRUTH***. If we wonder philosophically about what makes indeed this abstractive thinking to search for, to and beyond the limits of infinity, we come to face our symbiont self reflected in the mirror of time by memory and searching its own origin that when seen, the symbiont would not recognize it in the tiny subatomic bioforce: its own God.

REFERENCES FOR CHAPTER TWELVE

1. Fields RD. 2006. Meditations on the Brain. *Scientific American Mind.* 17(1):42-43.
2. Razi H. 1992. *Mithraism; Cult, Myth, Cosmogony, and Cosmology* (Text in Persian). Behjat Publications. Tehran, Iran.
3. Mehr F. 2000. *The Zoroastrian Tradition. An introduction to the Ancient Wisdom of Zarathustra.* Zoroastrian Benevolent Publications. Zoroastrian Center. California, USA.
4. Naficy S. 1998. *Les Origines du Soufisme Iranian* (Text in Persian). 9th Edition. Librairie Foroughi, Tehran, Iran.
5. Vayghami KM. 2000. *Bayazid-e Bastami* (Text in Persian). Tarh-e No Publications. Tehran, Iran.

6. Blackmore S. 2004. *Consciousness. An Introduction.* Oxford University Press. Oxford. New York.
7. Baines J. 1995. *HypsoConsciousness. Techniques for Achieving Personal Success*. Published by John Baines Institute, Inc. USA.
8. Schwartz JM, Begley S. 2003. *The Mind and The Brain. Neuroplasticity and the Power of Mental Force.* Regan Books. Harper Collins Publishers Inc. New York, NY.
9. Kraft U. 2006. Train Your Brain. *Scientific American Mind.* 17(1): 58-63.
10. Haghighat A. 1999. *Sohrevardy* (Text in Persian). Behjat Publications, Koumesh Publications. Tehran, Iran.
11. Newberg A, D'Aquili E, Rause V. 2002. *Why God Wont Go Away.* Ballantine Books.
12. Jayne J. 1990. *The Origin of the Consciousness in the Break-Down of the Bicameral Mind.* Houghton Mifflin Company. Boston.
13. Talan J. 2006. Science Probe Spirituality. *Scientific American Mind.* 17(1): 38-41.
14. Guiley RE. 2001. *Encyclopedia of the Strange, Mystical, and Unexplained.* Gramercy Books, New York.
    Phipps C. 2006. A New Dawn for Cosmology. *What Is Enlightenment?* 33:42-48.
15. Boyce M. 1992. *Zoroastrianism, its Antiquity, and Constant Vigor.* Mazda Publisher in association with Bibliotheca Persica. California and New York.
16. Ashtiani J. 2002. *Zarathushtra.* Translated in English from Persian by M. Nourbaksh, and Edited by H. Pirnazar. Sherkat-e Sahami-e Enteshar. Tehran, Iran.
17. Jafarey A. 1994.*The Gathas. A Glance.* Zarathushtrian Assembly Publication. Anaheim, California.
18. Jafarey A. 1989. *The Gathas, Our Guide. The Thought Provoking Divine Songs of Zarathushtra.* Ushta Publications. Cypress, California.
19. Riordan M, Zajc WA. 2006. The First Few Microsecond. *Scientific American* 294(5):34-41.
20. Frank A. 2006. The First Billion Years. *Astronomy* 34(6): 30-35.
21. Ruhlen M. 1994. *The Origin of Languages. Tracing the Evolution of the Mother Tongue.* John Wiley and sons, Inc.
22. Katrak JK. 1948. *Zarathushtra's Time.* Translated in Persian by Keykhosro Keshavarzi. Ferdosi Publications, 1994. Tehran.

CHAPTER THIRTEEN

THE HOLOGNOSTIC CREATION

THE REALM OF THE UNKNOWNS

Physics has made an epoch making change in the last century particularly due to Einstein's influential new ideas brought to the understanding of the matter-energy values, reshaping the whole physics with the special and general relativity theories. The introduction of the speed of light in the Einstein's formula defining energy limit in relation to mass, and the concept of spacetime in physics, brought out the specificities needed for understanding the matter-energy-time complex but also the limits enforced on that complex in concrete and abstract ways for matter, energy, and time as a whole. The limits enforced were precisely the result of the special relativity and the significance of the speed of light representing the ultimate manifestation of kinetic energy in the balanced physical world. Thus, a hope dawned for an early significant further clarification searching for a unified theory of everything, and the idea was further revived in the physics' scientific milieu following Einstein's work.[1] However, subsequent profusion of theories mainly related to atomic structure and subatomic particle physics made the conclusive expectation for a global understanding unrealistic and reaching a unified theory of everything became simply impractical.

Regarding interactions of the fundamental physical forces of quantum chromodynamic, quantum electrodynamic, electroweak force, and gravity[2] at the present time, one is left with enthusiastic elaborate hypotheses, but not all proved and not supporting a sustained hope to unify the vast array of data that cause more confusion than clarification and eliminate all hope for a possible unifying theory. Not only there are still noticeable unknowns defying a solid conclusion as with the force of gravity for example, but there are also contradictions with some expressed new thinking forcing rethinking.[3,4]

In this dilemma, philosophical logic necessarily comes into the field, and the first point of debate precisely concerns the puzzle of the subatomic particle that could have initiated the matter-energy complex causing the pristine initiation of the physical world. In fact, there is no convincing evidence for the nature of the very original subatomic particle responsible for creating the initial matter-energy unitary entity before the stable photonic complex appeared and solidly showed the equilibration of the matter-energy bonding. Whether and what before big bang and in what exact form or order the pristine initiation could have occurred remains entirely hypothetical.[5] Although reconstructed findings reveal significant clarifying data for the first and the possible repeated big bangs,[6] nothing seems to be yet reconstructible for before big bang time. The present theories propose Quarks and Gluons in the proton and neutron of the atomic nucleus for the strong force, and bosons and electrons around the nucleus for the weak force. These subatomic particles known to have formed the atoms have dimensions and functions well-outlined and indicated that an earlier level of smaller size particles that could have been the very initial form(s) of the matter making these more elaborate ones is a fact. Interestingly, however, particles like neutrinos thought to be the smallest energy carriers of no mass of possible dark matter origin are not without theorized antineutrinos[7] and other hypothetical particles like bosons or tachyon (faster than light) of pure energy nature only add to the hypothetical complexities. Indeed, whether one can stop at this level, theorizing that the smallest size pure energy particle similar to photon is a reality or wait for some newer information the no-mass energy is the crux of the problem of the physical creation in which one is too far from a unified theory.

The solution to the puzzle of the creation concerning the knowledge of the starting particle of energy-matter complex, not definitely attainable in physics at the present time, is yet imaginable on the account of physical experimentations particularly simulating the big bang with the formation of the stable photonic particle and more complex atomic structures.[6] But still, physics is not any closer to the unified theory than before this scenario was known.

Hypothesizing on the origin of the creation, the human logic goes for simplicity and clarity and tends to opt for simultaneous mechanistic and idealistic causalities for the creation, meaning occurrence of a physical setup of matter-energy in which the mechanistic process also holds its idealistic meaning of perpetuation of reciprocal matter-energy conversion and imparts autonomy as well, though by unknown causality, to a particular type of matter able to both generate energy and perpetuate using it: the biological matter. This hypothetical interpretation option does not seem to be explainable by the known physical laws alone. Philosophically, the idealistic causality of creation could prevail, and if prevailing, it would mean deterministic metaphysics with contingency of physics. The difficulty is in the hypothetical causality to be taken for deterministic that, as evidences indicate through physics, only reveal laws of reciprocity between matter and energy. For logic to choose between matter and energy as being causal, considering the subatomic physical scale, the choice would plausibly go for energy in preference to matter for several reasons.

In the first place, energy and matter at subatomic level reflect the mechanistic and the idealistic meanings together as expected from what has been so far discussed, although it is yet subject to the subatomic Heisenberg's uncertainty principle.[8,9] And the subatomic physics as yet allows no possible absolute priority related causality to be detected for either energy or matter with the exception of hypothetical Higgs particles of force that assumably induce matter production.[10] This theoretical particle awaits to be revealed in the newly built Large Hadrons Collider in Geneva in the near future. However, a clearly relative energy to matter preponderance compatible with idealistic causality is observable in subatomic physics that seems deterministic, not convincing to physicists in an absolute way but perhaps to metaphysicians, and needs further clarification. The causality of deterministic nature initiating the creation that human logic can debate is based on the initial self-sustaining energy observed in the light between wave form and matter form manifesting as a form of autonomy. This autonomy will be philosophically and biologically supported with the progress of paleognosis to neognosis through material biology into immaterial biology with mind, abstraction, and metaphysics to holognosis. This reasoning establishes the perpetual cycle of phasic change from one immaterialism to another, but precisely through an intermediary biological materialism. The opposite reasoning believing in materialistic creation ends in physical nihilism without secured renovation and indeed ending in no renovation but total decay and change, as brought about randomly by physical laws. Therefore, the mechanistic exploration of the creation in physical terms by the same token is appropriate only for the physical world with physical laws in matter-energy exchange, physically autoregulated and practical, but not for explaining consciousness, abstraction, and metaphysics that transcend from material physics into immaterial metaphysics, or more simply said, from the material biology to the immaterial biology. This idealistic extension leads the abstraction to beyond physics and relativity.

In reality, the question of a deterministic causality is beyond physical truth and scientific proof and is the monopoly of philosophy, strongly supported on the biological ground namely in observing the ontology of biological evolutional phases through a matrix of material biology with autonomy, a matrix that itself has always been scientifically proved as both physical and metaphysical. This matrix has been supporting indeed the impact of time through the evolution along with the pristine autonomy to consciousness.

Debating in philosophical terms, the pristine causality aiming to eternal perpetuation of the phasic creation, though plausible to the human logic, must find a way out of the pure physical process causality to show itself in bare truth. The way is through material biology to immaterial biology. This way out is precisely the course of the evolution explaining phylogenesis shortened in ontogenesis, which evidences the metamorphosis of the material biology under a directing self-autonomy. This long evolutionary change, manifesting the directed progress as with an aim, appears passing through material biology to achieve the phasic cycles of material-immaterial biology into an ultimate holognosis.

I still remember the eminent zoologist professor Matthey in Lausanne medical school in early1950s when he was emphasizing in his teaching, with a noticeable philosophical air, the recapitulation of phylogeny by the observable ontogeny. My astonishment of realizing the millions of years of evolution of phylogenesis being condensed in one gestational time period in ontogenesis, so beautifully observable on the evolving embryo during gestation, never left my mind but gradually changed into a philosophically permissive acceptance as for my professor. It gave me the hypothetically attractable idea that time condensation in its immaterial form in the mind with the evidence of memory and abstraction is the exact analogue of what we observe in the course of embryology with material ontogenesis. My possible explanation, reshaped at the present, is the phase difference in time scale of the time indefinite bioforce in evolutionary progress, changing scale with time definite periods in biology, making time definite material changes from the conception to the birth. This reflects itself in the time definite condensation for genetic transfer at molecular scale before another time indefinite allowance of expansion with psychognosis and again a time definite one with idiognosis in linguistic scales, allowing metaphoric indefiniteness and metaphysical abstractions to occur. This nature's documentary evidence with ontogenesis for the material biology is clearly visible in the embryo with the evidence of the substrate differentiation, i.e., the brain anatomy and in the mind with the mental function in evolutionary phylogeny of the mind. This observable mental refinement from birth to adulthood serves as living evidence supporting parallelism of phylogeny and ontogeny of the mind. The phylogeny of the mind appears to be beyond doubt, implying that mind's naturally expected ontogeny is undetectable in its immateriality, hidden in the nucleus of the bioforce with phasic biological expansion in life, in an "amorphic morphogenesis," developing with life and terminating in appearance with it in the phenotypes, but continually perpetuating in the genotypes materially and immaterially into the infinity of holognosis.

SYLLOGISTIC SPECULATIONS

It is ironical that the story of philosophy and science after indicating their long separation caused by the exponential expansion of scientific knowledge, and shrinking of the field of philosophy, now seems to mark an epoch making union. Wittgenstein lamented that "the sole remaining task for philosophy is the analysis of language" as remarked by Hawking[1] in his theory of everything, but now philosophy is showing a revival after the uncertainty principle in quantum physics has taken a solid ground. The change is to the point that philosophy seems imposing the final word. In fact, the theory of everything, as has been long searched by Einstein and other scientists, cannot be a real theory of everything without considering the mind, consciousness, and metaphysics that seem indispensable for complementing the conclusions reached in physics, harmonizing their philosophical meanings. What metaphysics means can indeed complement scientific evidences particularly when the global understanding of the physical and biological puzzles of creation is being explored.

With this in mind, the overall impression from what this book has exposed seems to show conditional determinism in the creation of physics and metaphysics together and with harmony. The triggering for this thought is the impression of a principle of immortality common to the autonomy seen in the stabilized photonic particle in physics and in the replicating live matter in biology that appears to work through bioforce. The impression of determinism is induced by observing the repeated physical and biological phasic changes in the macrocosm and microcosm, respectively. Physics, biology, and metaphysics are observed to be governed by one deterministic algorhythmic phasic change. This change covers from the initiation to the end of one phasic creation for the material world through the energy-matter-energy reciprocity, ultimately big bang to big bang repeatedly, and for the metaphysical world through the phasic immaterial-material-immaterial biology in the form of the unapparent ontogenetic course of the pristine biological autonomy to conscious abstraction and to holognosis. In this course of events, the sustained time used by the directed change seems having been consciously self-realized in phasic periodicity, gradually, from pristine biological autonomy to holognosis through consciousness repeatedly. This sketching is analogically compatible with a pattern of phasic repetitions in physics and metaphysics and seems to follow a fundamental deterministic pattern that appears immutable and logically hard to reject. The facts may look reminiscent of Nietzsche's philosophy with the eternal hourglass of existence if turning repeatedly upside down, but can differ from it substantially in close scrutiny.

Considered on pure philosophical ground, this conceptualized determinism, a priori compatible with the human logic, seems to find a fortiori standing if we assume one common cause acting as one factor but revealing two analogically related but not similar manifestations of the deterministic phasic change, for physics and for material and immaterial biology (the metaphysics). Indeed, in philosophical thinking, the unknown unique causal order for initiation of *being* (a time consuming process of independent matter-energy complex) out of any and all possible hypothetical choices, will be reduced to one with theoretical reciprocity between two potential poles, one for the state of absolute being and the other for the state of absolute nonbeing, both relatively conditioned according to the reciprocity principle between the two poles in a way that would prevent absolutism of either state alone. This continual hypothetical back and forth motion would exclude ending in any static state of either absolute existence or absolute nihilism and would assure endlessness in continuous phasic change.

This philosophical foundation leads to ultimate specification for the possible unknown unique cause to be at the subatomic level with the simultaneity of the being and nonbeing by the Heisenberg's uncertainty principle and Schrödinger's exemplary cat. This anecdotal cat, as said before, exemplifies being alive and dead at the same precise moment according to the uncertainty principle. Thus, this principle is acting in the infinitesimally short time (in the range of or *theoretically* shorter than the reciprocal of the speed of light) for the shift between the near absolute potential state of either hypothetical pole as discussed above.

This concluding logic imposes two contingent facts as realities. First is the deterministic principle for phasic change, or better for vacillations or waves, that can be regarded to vary from the level of subatomic energy-matter-energy-and so on to the level of cosmological changes, and the second is the principle of gnosis in biology apparent as autonomy initially in the material biology and as conscious abstraction with auto and heterognosis to hologndosis in the immaterial biology, both manifesting phasic renovation. To enunciate the final philosophical conclusion as a general axiom, the deterministic phasic renovation reveals immortalization in an eternal continuation as an ***immortality principle*** innate in the subatomic unique cause for all subsequent manifestations, material and immaterial. This immortality principle, inculcated in the pristine autonomy, is the unexplained essence for the subsequent abstractive refinement to reach comprehensive Gnostic ability assuring holognosis as part of the ***global holognostic creation.***

To give a tangible representative example for this philosophical determinism, the clearest is the eternal hourglass example of Nietzsche but one which would cover the biological world with a nuanced explanation in function. It would have an intrinsic force of turning upside down automatically and endlessly, would continue with time forever recapitulating its phasic changes of filling and emptying, and would renovate the whole content (total material decay and renovation) in each phase for biology but saving the autonomy in its gametogenetic material vehicle.

The force turning the hourglass could have been assumed to be extrinsic to the hourglass as in the classical Nietzsche's, indicating bare determinism and suggesting matching evidence for a nonphysical spacetime beyond the physical one imposing hypothetical pure energy. Although such energy is not substantiated, it can be hypothesized to possibly result in such hourglass for biology containing biological matter and biological energy behavior indicating determinism, which, if true, shows also the relation with the external deterministic principle. Such presumed relation could solely be maintained through the energy, but not through the mass, i.e., through the immaterial abstractive biology. Thus, we can assume a theoretical spacetime of infinite energy, infinite time, infinite space in which the spacetimes of material biology would be included as the metaphoric hourglass. The force, now, can be only the autonomous energy inside the hourglass, the energy in the occupant matter manifested as *energy-matter* in physical essence in the physics' spacetime and as *energy-time* of biological essence that admittedly includes *biological material spacetime* and the abstractive *biological immaterial spacetime*. Such presumed situation dictates distinct phasic changes for the biological material, transmitting also a hypothetical tacit unapparent biological immaterial spacetime. Nietzsche's hourglass, therefore, could be regarded as a metaphor for phase change in the classical spacetime of energy-matter-time with energy conservation in physical material terms and in biological terms but through the immaterial biology in both the living state and after decay of the material biology. Thus, the metaphysical world in this version of Nietzsche's hourglass would, in theory, work as the relay between the biological energy (released in the living brain matter) and the energy of the infinite spacetime of infinite energy-infinite time-infinite space.

Thus, we can say that the hourglass analogy raises in reality the possibility of three hypothetical spacetimes—one purely physical of matter-energy-time, one material biological type with autonomous matter-autonomous energy-time, and one of immaterial biological type of autonomous energy-time-space, the last two being combined in one carrier. The time is a common stable contingency in all, and the difference of the three spacetimes is reflected through the autonomy separating the second and the third from the first and the space singularizing the third with the autonomy in the energy with time but without mass is the infinite space that includes finite phenotypes. The autonomy seems to be the only causal factor securing perpetuation in the material physics in photonic permanency and in the material and the immaterial biology in phasic perpetuation and finally in the world of metaphysics closing the circle of immortality connection between the inner and the outer energies.

This theorization allows one to conclude that in reality the classical spacetime for physics can have its original analogue in metaphysics made up of infinite energy-infinite time-infinite space that could be regarded as the spacetime for the creation. In such theoretical spacetime, the autonomy principle must be carried by one or more of the constituents and most plausibly the choice would appear to be the energy or the energy-time as one principle, as the space represents equivalence for mass leaving the choice limited to the other two components.

The autonomy principle in physics seems to be exemplified best by photon according to the quantum electrodynamic principle and in material biology by bioforce engendering autonomy as the unit of consciousness energy through the time dimension and by consciousness to holognosis in the immaterial biology. So in terms of autonomous being versus nonbeing, all the rest of the material may undergo the metamorphosis in the phasic change to and from one pole to the other with the exception of the unique transmutable coded parcel that keeps perpetuating in the biological models carrying the unit of consciousness of the pristine autonomy. Thus, at the microcosmic and macrocosmic level, energy and matter at variable scales would represent the content of the hypothetical hourglass, which would interchange alternating emptying and filling theoretically indefinitely, following only physical laws for one, and by autonomy transcending to holognosis for the other.

In the biological system, the autonomy in the living material secures and conducts the energy and the steps to perpetuate phasic changes. Progressive expansion of the pristine autonomy to consciousness and autognosis-heterognosis to holognosis appear as hidden power inherent in the transmutant model deterministically transferred in phases. To simplify, this analogical exemplary autonomy should function in a graded consciousness ultimately reaching holognosis. The analogy assuredly connects an undeterminable time of phasic change from incalculable infinitesimal subatomic scale to any scale in the material biology, assumably also suggesting total all-inclusive state of fundamental holognosis in immaterial biology.

THE INDIVISIBLE CREATION

The link provided by consciousness between physical and biological world started in reality by the bioforce disguised as the biological autonomy in the early life forms. The story started with the smallest independent biological entity and its bioforce that, contrary to the basic elementary force in the photon in the physical realm, which seems theoretically eternal considering the billions of light years of its cosmic continuation, is limited in the biological world to the biological life span, though both forces seem to be from one origin and one essence. These biological units, bions and biopsychons, although different from the photon by the fact that they are mortal in comparison in terms of life span, yet they are similar to it by the fact that they carry the essence of the ***same immortality*** but through larger scale steps and phases. So in extended analogy, the similarity can be, in fact, a true identity of principle with phasic time differences in time scale durations, related to time indefinite subatomic world in fundamental physics and time definite atomic-molecular world in biology. In fact, physics is biology minus autonomy in simple metaphoric terms, and both physics and biology manifest their stories in phasic steps. The phase is shortest at subatomic level and in photon, being between the wave and the matter states (@ 1/C) longer in biological life spans between phenotypes breaking the continuity with genotypes, longer yet in macrocosmic matter-energy interplays, and still longer in succeeding big bangs. Thus, the basic principle is the same, but presentation follows the size theoretically from neutrinos or even smaller yet possible unknown particles to giant stars. This consideration imposes the foundation of an essentially time indefinite logic with a universal uncertainty principle that could accommodate infinite number of time definite logics with infinite different time scales. The time definite scale in the material biology differentiates from the time titan, handling biognostic processes from paleognosis to neognosis using memory scales and creating the ultimate immaterial biological mind with its world of metaphysics.

Therefore, the fundamental general conclusion can be that the basic principle, which is subatomic and time indefinite is the immortality essence, universal and constant, which can extend into large scale time definite physical forms with extendable variable phasic duration with or without autonomy, detectable by, but not limited to human comprehension. The time indefinite subatomic norm governs fundamentally, universally, and invariably throughout the creation irrespective of changes included in it in the form of time definite presentations, both in physics and biology. This understanding that I believe must be regarded as an axiomatic theorem should be considered basic in defining the conclusion that creation equally concerns physics and metaphysics in a holognostic representation.

My hypothesis of biocracy elected the bioforce as being the most naturally permanent expanding energy for the organic substrate to go on as an independent spacetime harboring the peculiarities of the organic material allowing adaptation through biognostic stages to reach the end result transcending from dimensional to nondimensional immaterial biology. If any cosmological origin should be invoked for the bioforce, as the subject has been entertained earlier, the choice would be restricted to the most expressive expansibility power that is in the light, seemingly as an expression of the ***hologn*****o*****stic creation***. By the same token, the unlimited infinite time for that expressivity can be regarded founding hologn*o*sis as we will see in further discussions, reflecting the ***Biothymic meaning of Gnostic permanency*** based on the pristine biological autonomy extended continually through time transfer, so to speak, by gametogenesis. This time extension of the biological autonomy secures the philosophical immortality of the hypothetical idealistic pure energy.

The reality of the indivisible holistic basis of the creation thus imposes a conclusion that this truth must be explainable in a unified way, in clear understanding, irrespective of its complicated multifactorial components of both physics and biology. This thought has haunted the scientific minds according to their variably scientific orientations and has caused extensive search for a unifying "theory of everything," mainly in the field of physics since Einstein's time, but also in philosophy. However, generally, the regrettable omission of either field's proponents has been the restrictive focusing on the informational contents of each single field of physics or philosophy, not providing a synthetic explainable view englobing both fields for the indivisible creation. Perhaps such comprehensive synthetic unification counting all details may appear too pretentious to be achieved and somewhat unrealistic to be expected. The reason is precisely the unpredictability of physical and biological modifications according to the related time-controlled variations between extremes of matter-energy changes and time uncertainties, but these unpredictabilities can be tested and cleared.

Let us now see some discrepancies of major significance in the field of physics that, a priori, seem to clearly hamper any attempt to reach a unified explanation in the world of physics alone and, a fortiori, support such unified explanation for the indivisible creation as a whole for both physics and metaphysics.

SCIENTIFIC CHAOS

The history of science is marked by trial and error before a reasonably acceptable norm is acclaimed for a solid theory proved by its absolute applicability. This is clearly seen in physics, astronomy, and biology. What concerns physics and biology, however, is the classical three-dimensional physics changing into the four-dimensional spacetime and multidimensional quantal physics theories on the one hand and the evolutional adaptation in natural selection theory tending to clarify reasons for SR and genetics on the other. These are directly related with the crucial question of creation, both for the worlds of physics and of biology.

Therefore, incongruity in the fundamental theories of physics can easily cause deviation from the right path to approach the question of creation, especially when the very fundamental original force at the subatomic level is to be accepted for both physical and biological creations, as proposed here.

An important discordance between the Einstein's general relativity law and quantum subatomic physics is one such major incongruity. This is aggravated by the inability to reconcile fundamental forces explained in physics with the force of gravity remaining unexplained, to reach a synthetic unified theory, albeit only in physics. The multitudes of the essentially hypothetical subatomic particles and their forces without fully knowing their limits of possible interchanges or modifications and the question of further unknown possible particles below neutrino size—if they exist and if they are from dark matter or are different—also add to the problem. This state of affairs is the present status, with some confusion without considering a real chaos, should the so called expansion theory[3] or De Sitter space implication for example,[4] also come into serious consideration. Indeed, such theoretical novelties eliminate a solid uniform understanding to be gained on firm basis by perplexities they cause. Thus, changeability in the present physical beliefs, particularly as concerning the nature of the gravity in subatomic particles' characteristics and other novelties in established physical norms render a solidly needed foundation impossible for a unified theory in physics to be based on acceptable ground.

The most significant presently persisting discordance in physics concerns the accepted general relativity explaining gravity through gravitons as waves for gravity facing opposing views trying to explain it as the result of space geometry variation by Maldacena,[4] interpreting the phenomenon in a holographic viewing from outside the sphere minimizing distances gradually from the center to the periphery as gravity effect. This theory considers the analogy that can exist between the quantum particle theory on the positive surface and the quantum gravity theory on the negative surface of the spherical spacetime, keeping with relative size proportions of the particles.

Irrespective of such novelties and their untested veracities and, in fact, in spite of all physical theories that could be imagined, no theoretical elaboration can reach a unified understanding explanation for the creation, which is not purely physical but is physical and metaphysical. Thus, any physical-only theory must be regarded incomplete that should be ultimately replaced by an integral theory of creation for physics and metaphysics.

In this unsettled situation of theoretical physics' world, even if one succeeds to find solid realities that could be included in the logical searching and reasoning concerning the creation puzzle, the task would be only half completed.

In reality, the creation puzzle originating the material world of physics, the material world of biology and the immaterial world of metaphysics can hardly be explained for anyone of these fundamental manifestations of creation alone without being defective, as all three manifestations are facets of one firm complex energy entity. The task is really a complex trifactorial problem that could be reduced to a bifactorial one of physics and biology at the first glance, though philosophy, holding the world of abstract metaphysics, may prevail in providing a unique solution but yet trifactorial for including physics, material biology and immaterial biology. For the moment, notwithstanding the unknowns and the chaotic knowns in the present scientific physical beliefs, the crucial philosophical valuable gauge that is time related—memory and consciousness representing the recapitulated autonomy unit in time—seems primordial in concurring to whatever conclusion must be reached by also using whatever solid trustable belief of physics can be included in the final explanation.

Fortunately, contrary to discordant theories reaching such incongruity in scientific thinking, the fundamental constituents of energy-mass-time as a whole remain valid as solid physical beliefs. These basic entities functioning in the still generally acceptable spacetime notion are reflected in the famous Einstein's formula of $E = mc^2$ keeping time inherent in the M-C as a symbolic reciprocity in physics, and this expressive formulary presentation can be regarded as a code that also represents the pristine energy of the creation at least in an abstract meaning. However, there is no evident expression of time infinite reality in this formulated presentation to implicate the possibility of the phasic continuation of unlimited permanent nature, not only of the physical world alone, but also of the metaphysical abstractive entities of immaterial biology. Indeed, instantaneity that is a bona fide characteristic of the physical world in macrocosmic moment changeability or in the time indefinite microcosmic quantal scales at the very tiniest time intervals has its counterpart in the pristine autonomy and its continually time reconstructed permanency with memory and the extension it takes with the abstractive thinking. Therefore, to attain a valid conclusion in a unified way on the physical and metaphysical origin and nature, the synthesis must include the essence of the metaphysical root, that is, the theoretical infinite time with the theoretical infinite energy in the conclusive expression.

IN SEARCH OF THE METAPHYSICAL PHANTOM

The abstract notion of time, as partially discussed in "timelessness in time" in chapter 1, is in reality that of an unrecognized carrier in our minds for the power of reconstruction of mental images, freely and repeatedly, to revive single or multiple frames of events in one uninterrupted continual virtual dimension from indefinite past to indefinite future. The background of this time power is the time concept of an immaterial entity, inseparable from the material biology, the time titan as named in this book that unassumingly operates in the live matter prior to any gnosis and without self-awareness, inherent in and balancing all internal biological feedbacks. This time matrix, already in psychognosis and clearly in idiognosis, manifests features of progressively more conscious type developing the explicit memory power from the earlier implicit one. The evolutional history of this time progress reaching to conscious abstraction is phasic, with infinite alternations of phenotypes and genotypes. In a way, there seems to be time effect condensation evidencing the progress from the pristine autonomy of the early replicating cell to the stage of the fully developed material brain with the immaterial realm of mind in spite of the phasic gametogenetic interruptions, or more correctly, due to it with the ensuing permanent biological continuity. So in essence, this continued replicating time-based power, realizing abstraction with self-awareness, is the energy liberated through time vehicle for consciousness, allowing ideation processes. The abstract conscious mental presentations are time-defined and/or subject-defined, that is, they are basically time definite with acceptable relativity that can suggest time alone as carrier for the energy without mass. This thinking allows one to postulate a principle of theoretical pure energy-time to exist. This presumed principle could be symbolically presented by the speed of light C and the infinite time T as CT for all that could be immaterial energy including metaphysics.

Along this line of thinking, the fundamental time indefiniteness of the quantum subatomic scale could hypothetically continue, carrying energy in biology in the form of time titan. In the material biology, it is accompanied by an obligatory associated morphic transfiguration into time definite forms, and in the immaterial biology, its transfiguration is similar but pseudomorphic to amorphic of abstract type. In the material biology, the obvious morphological changes providing organogenesis with refined functional adaptations are gross observations. Mental activities of the nonmaterial biology in consciousness, on the other hand, are pseudomorphic to amorphic of time definite or indefinite concepts and ideas. These are contingencies of the time indefiniteness of the fundamental time titan. In fact, the course of events through the transition time from psychognosis into idiognosis in evolution, reaching time definiteness, also left a permanent document showing the ground probabilistic relative indefinite time world with examples of dream visions in every life. In short, the story of time holds that repetition of phase difference does not stop with the time definite creation of idiognostic concepts, and even Namaa and time definition limitations still do not affect the unlimitedness of the mental creative phase with its inherent metaphoric changeability recapitulating the trend in the nonmaterial biology of the mind and in the no frontier time. All these scenario variations are in fact showing time as the fundamental carrier for energy with or without mass in the hypothetical symbolic CT.

In essence, more clearly said, the story of time shows that the creation of the metaphysical world by the mind indicates an obligatory outcome through NET principle, with Navand presenting the autonomy directing the energy through time in the immaterial biology and without dependence on mass. That assures intervention of both Namaa and Navand in time definite expressions and leads to the abstract metaphysics as part of the infinite ***holognostic creation*** of infinite limitless time. Along the same line, the process invariably confirms the time vehicle for the energy, a reality that suggests the veracity of a unique principle to exist as energy-time in the metaphysics as hypothesized symbolically with infinite CT.

The pristine biological autonomy, expanding into the complexity of the mind by the help of evolution, further suggests the legitimacy of consciousness creation for the complex abstract thinking, securing holognostic principle and, by the same token, seems to indicate an obligatory outcome. This is supported by the objective evidence of ontogeny of the gradual progressive elaboration in organogenesis of the brain. Similarly, the ontogeny of the mind, shown by the gradual changing of the pristine autonomy reaching consciousness, supports logical acceptance of the link between the material and the nonmaterial biology traced back to the very initiation of life. This confirms the accumulated time effects in biology from the material to the immaterial type that needs to be included in the unifying explanation of creation. Therefore, the nucleus of the consciousness, the autonomy principle carried by the bioforce through biocracy, should be considered as a constant inherent constituent of the subatomic particle-energy surviving or renovating through big bangs. This primordial energy particulate, seemingly matches the ***biothymic characteristic*** elaborated in chapter 3, evidenced as analogues in light in the physical world of subatomic level, and in the pristine autonomy reaching consciousness in the biological world evidencing tacitly the energy-time principle. The fundamental general understanding that issues suggests that the basic principle, which is subatomic and time indefinite, is the immortality essence, universal and constant, which extends into large scale time definite physical forms detectable to human comprehension. This immortality principle personified in time permanency carrying energy as energy-time principle (CT) is the legitimate foundation for the indivisible creation to be represented by one grand unified theory of creation of both physics and metaphysics.

THE SPACETIME FOR METAPHYSICS AND PHYSICS

Could we indeed look for a unified spacetime of the indivisible unique creation for both physics and metaphysics?

Trying to transcend from physical to metaphysical would mean traveling from the subatomic level to supraatomic and further to cosmological level from time indefiniteness to time definiteness and further again to time indefiniteness in our capacity of transiting through our abstraction. We may then reach a theoretical spacetime that can comprehensively include physics and metaphysics as entertained earlier with the hourglass examples. Such spacetime must be regarded hypothetical as it would have to include both the material world and the immaterial world, which is abstract and nondimensional. Even as hypothetical as it can be, surely it cannot be the same spacetime proposed for the mass-energy-time in physics but can be an abstract limitless space, a quasi emptiness, a void, a hypothetical dimensionless infinity but one in which mass-energy entities could also exist along nonmass energy abstract space and time. Accordingly, such idea of common spacetime for metaphysics and physics incites the thought of the precreation spacetime of infinite energy, infinite time, infinite space as noted before, in which creation of matter would initiate the spacetime for physics. Logically, the infinite energy of the precreation must be regarded theoretically as pure energy representing the ultimate no mass energy as exemplified with light or shown in quantum physics with negligible subatomic mass counterpart to the lowest limits of measurability. The principle of the infinite energy-time then can be regarded as the hypothetical carrier for the autonomy principle in the precreation spacetime as the causal factor for the creation. In fact, the other element of the precreation spacetime, the infinite space could only stand for the space occupying matter, the mass, engendered by the creation, a by-product of the carrier function for the autonomy principle.

Hypothetically, one may face the argument that metaphysics having no physical dimension needs no spacetime. But this argument could be valid if no pristine autonomy and no pure energy would exist, an assumption not validated in the face of biological material and immaterial evidences. According to these evidences substantially covered in this book, the immaterial biology allowing holognosis forms the *primum movens* for the material creation giving the substrate with the biological material to carry the pristine autonomy to the immateriality of abstraction. In further scrutiny, if time is regarded as a physical dimension only, then it should not exist with nonphysical immaterial entities of abstract type, but conscious realization with neuroperceptive time scale excludes illusory absence of time. Human logic, in fact, accepts time as an abstract entity revealed by material physical bodies and by immaterial metaphysical entities. The simple logical reasoning on priority of the inclusiveness logic also agrees more easily with the unlimited to include limited measurables than the other way around, and the unlimited metaphysical spacetime can theoretically be defined as infinite energy-infinite time-infinite space containing finite energy-mass-time components.

Studies in astronomy have revealed surprising facts that characterize physical spacetimes in the absolute unlimited space, or more exactly emptiness in degrees, or intergalactic medium using astrophysicists'expressions.[11] In essence, the light year chronology surpasses imagination with the revelation of clusters of galaxies that have been formed in the estimated age of fourteen billion years of the universe.[12] The multitudes of forms and light intensities and other characteristics of these galaxies have permitted to recognize various shapes, motions, and other behaviors that permit grouping them in three main groups of spiral, elliptical, and irregular galaxies. The gigantic unknown dimensions of the spacing containing these galaxies, hypothesized to have been originally spherical, is estimated to be such as to present these forms on extended curved surfaces as plane flattened spaces occupied by the galaxies including our Milky Way, for example, with our sun as one of the stars in it. The overall impression of this cosmological picture allows three main facts that seem to be generally accepted today.

The first is galactic formation from protogalactic clouds and dusts and multiple individual stars that necessarily must have started at smaller material dimension and grown to larger size by accretion of lesser masses.[12] This cosmological observation solidly agrees with the pristine atomic formation as extrapolated from the experimental big bang duplication in particle accelerator studies.[6] The second fact theoretically agreed upon by recent studies is the expansion of the universe that is admittedly progressing. This expansion of the universe, as is the expression customarily used, is in reality by the hypothetical infinite endless space expanding but not by the material universe expanding into any known space with possibility of limitation. This situation means that distances between stars and galaxies are supposed to progressively increase irrespective of any change that may occur in the content of the expanding universe. The third and the most significant fact is the unavoidable impression of gravitational force exerted by dark matter as halo surrounding the central luminosity of the galaxies. This halo of the dark matter may have a radius ten times that of the galaxy's central luminosity.[12] Also, the notion of dark matter meaning gravitational force by an imaginary matter substrate is the essential point facing a counter point in physical theories in general, the counter point being the dark energy exerting expansion. However, dark matter and dark energy have not been clearly substantiated any further. Theoretically, dark matter is controlling gravitation (contraction of the material content within the space) and dark energy is engendering expansion of the universe (extending the infinite space) as presently viewed. This picture reminds one of the fundamental duality in the subatomic reciprocal energy-matter interaction, which seems to represent the essence of the autonomy principle. By the same token, the cosmological interplay of dark matter-dark energy in reality denotes the same autonomy, an autonomy of constant reciprocity principle assuring continual integrity of the creation energy at all time.

Notwithstanding the theoreticl basis of this arrangement, dark matter and dark energy imply three theoretically possible events: the first is predominating gravitation and collapse (theoretical pre-bigbang black hole state of maximal infinite energy compressed in infinitely minimal space), the second is overriding expansion (infinite expansion to ultimate annihilation of energies into dead stars), and the third is possible equilibrium in a stable universe. This scenario reconstructs the cosmological evidences according to my earlier logicality of the two poles of absolute being versus absolute nonbeing and their intermediary phasic status. And for the universe to continue at equilibrium, the material density needs to be about a critical level called the critical density. Present scientific ideas seem to predict greater probability for the progressive expansion based on the argument that gravitational effect of the dark matter does not match expansibility of dark energy as judged by the total mass of stars in the universe (0.5%), which would need about two hundred times more dark matter to counter the expansion.[12] This fact interestingly enforces the supremacy of energy over the matter here used in the hypothesis describing the precreation spacetime and supporting the CT as pure energy-time principle.

This present understanding, however, is not totally without flaw due to present shortage of accurate observation of the farthest galaxies not completed for estimating the total dark matter. So in essence, we can still hold on to the phasic big bang repetition for the manifesting life creations with physical and metaphysical presentations deduced by our philosophical reasoning. Therefore, the spacetime for the metaphysical and physical creation can be regarded the total perspective we gain, understanding the physical data, which leaves the major unknown that must exist without being explicable except by implicating the energy-time as an immortality principle symbolized in CT.

Viewed through present norms of physics, the time is said to start with the big bang, but is not shown in the best presentation of the mass-energy expression that is summarized in Einstein's formula, which, as characteristically showing matter and energy, is specifically and adequately limited to mass and to an absolute speed presentation as symbol of energy inherently time inclusive. Therefore, infinite time with its subatomic indefinite and supraatomic definite aspects in physics, material biology, and immaterial biology can be ultimately regarded unlimited, in abstractive scale permissibility, as logically part of the metaphysical spacetime. This infinite time is a must to be defined in connection with energy in the hypothetical precreation spacetime expected to hold the autonomy principle within the infinite creational energy. This metaphysically understood time should be incorporated in an expression of spacetime that would complete the physical presentation together with the infinite metaphysical time to express the indivisible creation of both physics and metaphysics.

An interesting debate may ensue if the infinite metaphysical time and the finite physical time logically interrelated are considered in relation to the hypothetical beginning of time with the event of big bang. The physical time of the mass-energy-time expression that is theoretically subordinate to the metaphysical time appears once the matter forms. This could mean that dark matter and dark energy in functional presentation, as a unified energy principle of continuity between two poles of consecutive contraction and expansion, could have been present in the metaphysical infinite time before big bang as no-mass energy-time in one uniform field and could have appeared with big bang as a duality with mass-energy created in the already present no-mass energy metaphysics. This unknown state of hypothetical energy without mass, metaphysically suggestible, could be presumed a potential energy (Higgs or other force carrying particle) of the infinite precreational spacetime in the infinite space in which the appearance of matter-energy causes differential gradient and polarization of the uniform potential energy into kinetic to and from the focus of the matter formation. Thus, the finite physical time would appear in reality to be only the relativity-bound time with appearance of matter sharing the same continuity of the infinite metaphysical time. The same continuing analogy is noticeable for the explicitly defined memory developing from the general implicit one, or the sophisticated mind from the pristine autonomy through the course of the evolution. In this theorization, the movements and shaping expansions of the energy-matter is conditioned by the preexisting potential energy made to change in two oppositely vectored forces: the gravity and the antigravity. This scenario would better fit the hypothetical before big bang time as it could be imagined to have occurred from the pristine matter-energy formation at subatomic scale in contrast to all presumed big bangs to occur subsequently that would result from one focus of conditioned preexisting energy-matter to expand in explosion. Thus, one initial holognostic creation for nonbiological and biological matters would suffice to trigger the phasic repetitions. In the present beliefs of accepted big bangs era, the short time of 10^{-43} seconds (Plank Era) following the beginning explosion (zero time) of the infinitely compressed energy-matter of the pre-existing state leaves a fragment of time that remains unknown, and all other hypothetical knowledge of related facts to big bang do not reach in retrospect beyond the known beginning. Thus, no evidence can be ever found to negate the metaphysical hypothesis of ***holognostic creation*** that theoretically precedes the initial big bang. Accordingly, the initiation of time could be regarded to be with the indefinite time of the subatomic character, i.e., simultaneous with manifesting energy suggestive of the hypothetical energy-time principle, the symbolic CT entertained earlier. That symbolic figure can be taken as the primordial constant modus of energy-time subatomic presentation extending beyond relativity.

The infinity issue of the time for the metaphysical realm with its hypothetical infinite space and imaginable infinite energy and no-mass may appear a pure theoretical possibility. The puzzle would be for the creation, a priori, being simultaneously integral and infinite, meaning that the infinite space would have to be ultimately occupied totally by mass, a situation incompatible with the actual cosmological evidences requiring a hypothetical range of unending $-\infty$ to $+\infty$ for space (as mass) and for energy and similarly for time. This thought represents a theorized notion of illogical infinity-integrity once creation occurs. Therefore, the plausible alternative would be an absolute infinity for the metaphysical spacetime with infinities of energy-time-space with intercalated partialities of physical creations never changing the metaphysical infinity. Such infinity would mean to represent a constant unalterable connectivity principle holding matter-energy in any scale of minimal to maximal space, allowing infinite number of such created foci within itself. This connectivity principle, as a primordial constant, is in fact another facet of the same reciprocity autonomy in the matter-wave relation remarked earlier and can ultimately explain the duality of dark matter-dark energy governing the entire phasic repetitions from big bang to big bang. In fact, believing physics' weak nuclear force question of symmetry and asymmetry,[10] if we interpret symmetry as independent single particle's and asymmetry as dependent integrated condition for groups of such particles, the former state would represent the hypothetical infinite precreational space where the electroweak force particles are considered massless energy carriers, and the latter shows the physical state where the same particles would be imbued to have mass by the Higgs' field effect. So the physical explanation for creation of mass in essence matches the metaphysical postulate of energy initiating matter, and with it inducing polarization in the nonpolarized energy field of the precreation infinite space into opposing energies (dark matter and dark energy), the same effect being explained in physics' terms with symmetry broken into asymmetry.

With physical foci of creation, mathematical infinities point to existing relationship of fundamental nature with geometrical forms that ultimately show integrity in every physical spacetime. The most significant proof for this relationship is in the circle and the sphere, the two outstanding forms of the matter also governing the physical dynamism at micro and macrocosmic scales, manifestly demonstrated in cosmology. Thus, the hypothetical connectivity principle of the precreation state seems to repeat itself for every focus of creation after the initial big bang, explaining the subsequent gravity and antigravity.

In fact, time indefiniteness as a form of infinity at geometrical multidimensional subatomic levels and curvature effects of basically spherical form presenting plane extensions at cosmological levels are well-recognized to rule the quantum and the astronomic dynamisms, respectively. The essence of the geometrical rules of the circle is the well-known theorem of antiquity ascribed to Thales for triangles based on circle's diameter that constantly preserve the integrity of squares of two sides equaling diameter square. The significance of this phenomenon confirms compatibility of energy-matter expansion to infinity of time and in line with conservation of the pristine infinity-integrity secured with the principle of connectivity that remains unalterable. This unalterable constancy is evident in the observable physical laws governed by relativity.

Two infinity concepts related to time that are in reality nothing but integrities in line with the above examples are first the infinity of changes in the length of the two sides of the triangles, conserving constantly the 90-degree angle with the infinite number of triangles possibly formed, and second, the numerical infinity in π for the circumference relation to the diameter, which does not exist in the geometrical integrity of holistic relation of the triangles thus formed. From the circle to the sphere, these geometrical integrities can be imagined to multiply an infinite number of times so that the spherical physical spacetime for the expanding energy-matter-space would keep inherently infinite, replicating the pristine integral infinity. And by the same token, the hypothetical energy-time principle, in any number of expansions still leaves the integral infinity intact.

Thus, the unsolved infinite time of the creation, if considered as a principle of integral energy-time (symbolized in CT) can be regarded forming a precreation metaphysical spacetime as energy-time-space in analogy with the physical spacetime of energy-time-mass. This principle of energy-time, regardless of space or mass, although appearing hypothetical, is not without solid basis when the foundation of the abstract metaphysical concept of the creation is considered, the foundation that is based on the reality of memory as time personification in all abstractive mental work. Indeed, the very basic Gnostic abstraction is not attainable without abstract concept formation for which neurodynamic energy through the memory time is essential to permit abstract concept realization. This realization for gnosis based on the formula 2 given in chapter 3 shows

GNOSIS = NEURODYNAMIC ENERGY × TIME

This formulaic presentation can be taken as the nucleus for the foundation of gnosis in metaphysical terms and can be expressed more generally in:

1. **GNOSIS = ENERGY × TIME**

This summarized expression shows that in abstract terms both of this metaphysical and the physical formula ($E = mc^2$ of Einstein) share one common term, which is the energy, the symbolized C governing relativity equivalent to c in Einstein's formula or shown as infinity in the symbolized CT principle beyond relativity. Therefore, the mathematical solution for reaching a final unified formula is within the logical reach.

THE HOLOGNOSTIC CREATION

As is obvious, if biological life as part of the whole creation does consist of both physical and metaphysical components, the holognosis must include not only the immaterial metaphysics but also the material physics as represented by the global energy concept related to mass, as explained by Einstein. This all applicable ideal general relationship of energy and mass in physics is undeniably clear and is not only restricted to the physical world but is also expandable to the metaphysical world if considering the hypothetical metaphysical spacetime of infinite energy-infinite time-infinite space and remembering the equivalence of space and mass in terms of dimensional and energy value meaning. So in essence, absence of the infinite time of the metaphysical spacetime in the physical formula of Einstein is the gap preventing generalization beyond relativity, a gap that could be filled only by theorizing an energy-time principle with the symbolic CT, to be included in the ultimate proposed formula. In that way, the single symbol of C as the speed of light without any representation for time would be coupled with a symbol for time substantiating the possibility of energy-time to represent infinity and reaching beyond physical relativity of the Einstein's formula. This inclusion is now possible through the expression of gnosis as the product of energy and time shown in the formula above by using the energy component from Einstein's formula for the unit of gnosis that would give

2. GNOSIS = MC^2T

The foundation formula 1, expanded to represent Holognosis will give

3. HOLOGNOSIS = GNOSIS × TIME

The final formula for the holognostic unified creation therefore will be

4. HOLOGNOSTIC UNIFIED CREATION = MC^2T^2
5. $H = M(CT)^2$

In this presentation, the concept of ***the unified theory of the creation*** is assembling the metaphysical and the physical worlds in one holognostic primordium. The energy-time conceptual principle, as an isolated concept in $(CT)^2$, can be applicable to the infinite energy and infinite time of the hypothetical precreation spacetime and is projecting beyond possible relativity. In fact, the distinctive significance of the precreation energy-time principle abbreviated in $(CT)^2$ is marking the abstract beyond all relativity states for possible energy-time expansions. Creation of matter engendering physical forces naturally follows the norms of special and general relativity laws in energy-time and energy-matter expansions. In the Einstein's formula, there is already the tacit abstract time limit in MC^2, which basically confers relativity to the physical world in relation to mass. Energy-time, on the contrary, only indicates the precreation state, which, when including M as in the unified creation formula, extends the abstract meaning of infinity to space S, making the spacetime of the precreation infinite energy-time-space that could be therefore represented as:

6. $E = S\,(CT)^2$

In this understanding, S should be regarded the infinite space to include from infinitesimally small subatomic to extremely large cosmological dimensions as is in fact presently observed.

Another crucial significance of this energy-time principle is the suggestion that precreational state could be regarded only as a field of energy in potential value, without any polarity other than a uniform potential polarity of expansion or dissipation as for any energy. If the potential state is made to change into any kinetic reality, then a polarization may be conceived to occur initiated by a subatomic focus of material formation. Such nucleus could then create polarization in the field changing the uniformity and adding modalities of physical materials and energies keeping with the conservation of energy and the renovation principle.

This space contingency of CT principle is only clear, tangible in fact, with the created dimensionality by M. Before such detectable dimensionality is shown, what is thinkable to exist metaphysically is the root for it, which is the energy-time principle, a principle that can be seen in light assumed as without mass and its ultimate speed that surpasses all physical time dimensions metaphorically meaning an immeasurable timelessness. In this line of thinking, we can presume that the relativity in relation to the speed of light as thesis of known dimensionality matches the antithesis of infinite energy and infinite time as the hypothetical unknown dimensionality prior to the creation. Thus, the only separating boundary of finity versus infinity seems to be the condensation of time of "immaterial?" radiating energy prototype, representing connectivity principle in the material physics. In material biology, the analogue would be the condensed time in the ontogenesis repeating phylogenesis and for genetic transfer of the immortality principle as ground for Gnostic perpetuation through memory forming the "immaterial biology." This analogical comparison makes light and abstraction physical and metaphysical counterparts. It invokes the pristine autonomy and its vehicle, the time titan, as analogues of the pristine infinite energy and its carrier, infinite time, respectively, making the energy-time principle of the unified creation.

The holognostic formula must be regarded as an abstract logical concept, which holds true for the global holistic indivisible creation and can be used to interpret both the big bang and the subatomic before-big-bang creation. The formula suggests the conclusion that C in physical terms that imposes the relativity limit can be regarded in $(CT)^2$ as the energy-time analogue beyond the limits of relativity. It is evident that neither $C = 0$ nor $T = 0$ could be theorized as either C or T is the sine qua non for the other to be, making one unity as the energy-time principle, and both C and T represent the inseparable parts of the infinite energy—infinite time—infinite space of the precreation state.

As C is a physical reality that can theoretically only exist indefinitely through perpetual transfer from big bang to big bang, the light is the only physical analogue for the immortality principle of the metaphysical type, and by the same token, the abstract holognosis may be regarded as renovating with every big bang after the very initial holognosis of the hypothetical subatomic creation. Thus, the quantum photonic energy can be regarded to represent the energy unit of the holistic holognostic creation in physics, acting as the prototype force, and its analogue, the pristine autonomous energy in biology, the bioforce, can be interpreted as the immortality principle representing metaphysics from the same infinite energy of the creation. Thus, equivalence deduced to exist analogically between the light in physics and the immortality principle in life as prototypes of the material and nonmaterial worlds is remarkably imposing. Further to this, the concrete evidence of the light effect in photosynthesis confers credibility to this deduction and assures functional reality for the equivalent analogues, light and bioforce.

The entire issue can be further examined comparing Einstein's formula and the formula here presented for the unified creation. In the Einstein's formula, the speed of light is the only defined constant, but the mass is not constant, undefined, presenting any theoretical value, and there is no presentation for time in any defined form as there is also none for the space except in concrete mass. If theoretical assumptions for the sake of broader understanding of the physical spacetime are considered in this formula, the only other component, the mass, can be manipulated to vary the value of the energy. So in fact, E is dependent only on M, which is the essence of relativity binding energy to mass. If, in theory, M would be the hypothetical totality of the material world, the formula may be regarded as an all applicable global one for the hypothetical abstract totality of physics in metaphysics. In this viewing, the entire physical world could be regarded as a potential black hole of zero to infinite size between big bangs in an abstract way. This enormously vast variation of energy-mass exchange confers unique singularity to the accepted speed of light, as in fact, is shown with mini bang models evidencing stabilization with photonic formation.[3]

The same consideration seems warranted for the time as prototype metaphysical component of the immortality principle shown in CT in forming one unified entity. This immortality can be applicable to the nucleus of the early sensing, the presumed pristine instant Gnostic nucleus, with increasing memory time for any single concept initiation, recall, or intentional aiming to full abstractive Gnostic chain approaching ultimate holognosis. Furthermore, the time itself as inherent specificity of matter and energy interchange can be considered in the formula showing the subatomic indefiniteness, hypothetical, limitless, unconscious, or as definite for the memory time for the conscious mind. Thus, in this equation, the range of the variably measurable time must be regarded as an undefined abstract value that can vary from theoretically the reciprocal of the speed of light (C^{-1}) in indefinite time in the subatomic scale, to working memory time, to larger possible scales reaching attention-intention processes, to the chain of abstract reasoning in metaphysical concepts, and by extension to phasic big bang intervals. In this line of interpretation, the time is not limited to a single set of creation. This formula theoretically defines the hypothetical totality of the creation, matter, energy, and time for the assumed, unknown, and theoretically the all inclusive spacetime of physics and metaphysics covering material physics, material biology, and immaterial biology.

A theoretical zero point for time between two infinite scales of theoretical minus and plus ranges ($- \infty$ to $+ \infty$) can be taken to represent the essential time indefiniteness imaginably applicable at any range from the subatomic to the light-year scales and beyond. Thus, our time definite conscious abstraction can be seen as representing only the zero point of a unit of gnosis, the strict present time at each moment of attention starting to progress in the infinite variability of the energy-time principle. On the other hand, the theoretical question of variability in the speed of light, if raised, is incompatible with truth in appearance as evidenced in plasmonic experimentations allowing energy and wavelength changes at nanometer scales,[13] without impeding intervention to the light radiation. Impeding, interestingly can cause slowing or total transformation of the light energy to matter.[14] But more interestingly, if a retransformation from the material product is effected to take place, an exact duplicate of the original light form will be produced. This evidence further supports the analogy of light and consciousness and the time indefiniteness for scales between zero and infinity and for possible reversibility that is naturally innate in the energy-time principle.

This phenomenon confirms identical standing constancies for both C as prototype solid component of physical immortality principle and for T similarly as an identical component of immortality principle in the variable states from the pristine autonomy of the bioforce to the holognostic state. The quintessence of the nature of these constancies, however, remains the unsolved puzzle.

According to these theoretical considerations, the metaphysical gnosis originates in biology from functionally prepared grounds for the potential progressive directionality secured by the bioforce in pristine autonomy and reaches Gnostic abstraction seemingly as part of the infinite holognosis. Thus, the ***holognostic biothymic creation*** in these terms is regarded the whole and the physical world the part.

It is notable that the initial autonomy, establishing progressive evolving of the biognosis through the material biology and ultimately reaching the abstractive immaterial state and holognosis, seems pursuing an obligatory deterministic course through the evolution and not by the evolution that cannot be the cause of the autonomy. In fact, evolution reflects an extrinsic effect, but the innate autonomy is intrinsic to the live matter, is saved perpetually as the immortality principle and allows refinements necessitated by the environmental conditions. This trend can be interpreted as primarily deterministic and secondarily adaptable through evolution. It can imply the possibility of hard wiring in the brain for metaphysical beliefs[15] and can further accord respective credence to both Darwinism and Lamarckism.

In summary, the energy-time for the metaphysical, and the energy-matter for the physical world form two interrelated principles that present two different manifestations from one fundamental origin. Although that origin is undoubtedly unique, neither of the two manifestations alone can lead the observer to understand the uniqueness that, to be seen clear, the full account of the two principles of energy-time and energy-matter must be presented in one mathematical truth solidly formulated as shown in the equation for holognostic unified creation. This formula elaborated assuming metaphysics as primary and as the matrix, far from explaining the unique origin of the energy in energy-time principle and energy-matter format, nevertheless confirms the truth that the origin is indeed one and the same. The manifestations of metaphysical and physical nature can be satisfactorily understood through the logic analogically binding these two principles exposed with the hologonostic formula, $E = S (CT)^2$, and Einstein's formula, $E = mc^2$, regarding CT as the infinite beyond relativity precreation energy-time of which c would be a relativity bound creation-related to m as a relativity bound space occupying particle.

REFERENCES FOR CHAPTER THIRTEEN

1. Hawking SW. 2003. The Theory of Every thing. The Origin and Fate of the Universe. New Millennium Press. Beverly Hills, CA 90210.
2. Green B. 2005. The Fabric of the Cosmos. Vintage Books. New York.
3. McCutcheon M. 2004. The Final Theory. Rethinking our Scientific Legacy. Universal Publishers. Boca Raton.
4. Maldacena J. 2006. The illusion of Gravity. Scientific American. 293(5): 57-63.
5. Choi CQ. 2007. New Beginnings. Scientific American. 297(4): 26-29.
6. Riordan M, Zajc WA. 2006. The First Few Microseconds. Scientific American. 295(5): 34-41.
7. Nadis S. 2007. The Big Bang Plus One Second. Astronomy. 35(4): 38-43.
8. Weaver JH. 1987. Language and Reality in Modern Physics. Werner Heisenberg. In: The World of Physics. Vol. III. 851-864. Simon and Shuster, New York.
9. Weaver JH. 1987. The Development of Quantum Mechanics. Werner Heisenberg. In: The World of Physics. Vol. II. 353-355. Simon and Shuster, New York.
10. Quigg C. 2008. The Coming Evolution in Particle Physics. Scientific American. 298(2): 46-53.
11. Scannapieco E, Petitjean P, Broadhurst T. 2007. Scientific American Supplement. The Emptiest Places. Majestic Universe. Pp 32-37.
12. Bennett J, Donahue M, Schneider N, Voit M. 2005. The Essential Cosmic Perspective. Third Edition. Pearson, Addison Wesley. World Wide.
13. Atwater HA. 2007. The Promise of Plasmonics. Scientific American. 296(4): 56-61.
14. Minkel JR. 2007. Stop and Go Light. Scientific American. 296(4): 30.
15. Newberg A, D'Aquili E, Rause V. 2002. Why God Wont Go Away. Ballantine Books.

EPILOGUE

The story of creation is the tale of consciousness condensed in the infinity of time. You can read it through biology and physics as you probably realized reading it through this book. It is at once as inspiring and relaxing with elation, looking at it through your windows on the Pacific Ocean or to the night sky and the Milky Way with admiration and awe, or as harsh and hard to believe, seeing it at the death bed of a loved one with sorrow and pain.

The elation is sensed and the awe is sensed and the pain is sensed, consciously and inexorably without a word defining in your mind what is real in your reality sensing of the emotion. Yet you are sure of the feeling that takes you out of yourself for the time that is never definable and will be lost unknowingly as it appeared. This nucleus of consciousness is in our biology and in every live matter's biology that I have tried to materialize with the name of bioforce forming our pristine conscious state of autonomy that we only realize reaching and feeling it with psychognosis substantially and defining it with idiognosis with words, though incompletely.

This autonomy controls our material biology and expands our immaterial biology ultimately to full consciousness and the unlimited extent of metaphysical abstracts. Its expressivity is immense, it is all we are. It releases our inner energy that I have called Navand and have based my theory of language formation and processing of it on its basis. It expresses our inner problems in dream forms and forces our autobiographical memory to adjust to save our inner self. It gives us mutual telepathic impressions or visible scenes of things happening without seeing them in reality. It warns us of compatible or incompatible feelings of others toward us and reciprocally, without using physical means of communication but the sight. In short, our innate expressivity is part of our pristine autonomy developed proportional to our minds limits of consciousness.

Understanding scientific data through philosophical views exposed in this book leads us naturally to think of our origin and to look at the puzzle named creation. Helas, we only have the limited information that cannot lead us straightforward to find the right key and the right door to open the landscape to the hidden truth and solve the puzzle. Either the key(s) or the door(s) we imagine can be wrong, but our human reason usually accepts one right key for one right door. Now, what if there are two right keys for two right doors, and each must open the proper door to lead to the true holistic perspective showing the solution to this puzzle of creation that involves not only matter and energy and time, but beyond these materialities through time condensation as conscious processes that lead to gnosis. This was my approach crossmatching philosophy and science, each through its specific gate leading to the ultimate unified understanding of the holistic creation summarized in the ***holognostic unified theory*** for the creation of physics and metaphysics: ***the holognostic creation.*** Ironically, although the gates seemed different, the roads ultimately united to lead to one whole perspective of our origin.

The success depended solely on the philosophical truth that the hypothetical no mass energy-time unified the two roads, a unification that could not be reached with any of the other components of matter, energy, or time alone and needed the inseparability of energy and time forming the infinite space to be theoretically changed into infinite-reasoning matter with abstraction. This granted understanding is due to the omnipotent and omnipresent immortality principle that is in subatomic material level in the material biology and in the metaphysical expressive domain in immaterial biology. It seems that the more materially oriented we are, the more separate we will be from the truth, which is no mass energy: the imaginable energy-time principle in infinite undefined space. But, even in our material being, we must remember that not all energy is physical, and our innate bioforce is such an energy we can revive when we are meditating and reaching deep in our material being or perhaps even passing through and beyond our material limits with our natural curiosity for new horizons in search for a companion.

We can reach only one conclusion: that we are deterministically alive, living our consciousness in the matter with a connectivity principle, knowing that our material biology is but a relay for our consciousness energy to no mass energy of the creation. Therefore, living our conscious life and enjoying the natural benefit of our life commandment leaves us only one choice, which is ***"to live not as "I" please but to please as "WE" live,"*** as a one soul of the creation, sharing pleasure with other souls, humans or nonhumans.

We need no maxim in the name of religion to tell us this *truth* except perhaps the principle of **Asha** from the philosophy of Zoroaster. The ***TRUTH***, abbreviation of "**T**o **R**each the **U**ltimate **T**arget of **H**ope," in connection with the logic of immortality of the bioforce or as related to SR, is in fact our innate will to trigger our totipotent energy that can lead us to ultimate happiness if we use it wisely. We need simply realizing that we are living our consciousness as one parcel of the whole pure energy of ***the holognostic creation***.

We can fly high on the wings of our consciousness if we ever live to enjoy its weightless purity.

INDEX

A

C

D

E

F

G

H

I

J

K

L

M

N

O

P

Q

R

S